IET CONTROL, ROBOTICS AND SENSORS SERIES 108

Wearable Exoskeleton Systems

Other volumes in this series:

Volume 8 **A History of Control Engineering, 1800–1930** S. Bennett
Volume 18 **Applied Control Theory, 2nd Edition** J.R. Leigh
Volume 20 **Design of Modern Control Systems** D.J. Bell, P.A. Cook and N. Munro (Editors)
Volume 28 **Robots and Automated Manufacture J. Billingsley** (Editor)
Volume 33 **Temperature Measurement and Control** J.R. Leigh
Volume 34 **Singular Perturbation Methodology in Control Systems** D.S. Naidu
Volume 35 **Implementation of Self-Tuning Controllers** K. Warwick (Editor)
Volume 37 **Industrial Digital Control Systems, 2nd Edition** K. Warwick and D. Rees (Editors)
Volume 39 **Continuous Time Controller Design** R. Balasubramanian
Volume 40 **Deterministic Control of Uncertain Systems** A.S.I. Zinober (Editor)
Volume 41 **Computer Control of Real-Time Processes** S. Bennett and G.S. Virk (Editors)
Volume 42 **Digital Signal Processing: Principles, devices and applications** N.B. Jones and J.D.McK. Watson (Editors)
Volume 44 **Knowledge-based Systems for Industrial Control** J. McGhee, M.J. Grimble and A. Mowforth (Editors)
Volume 47 **A History of Control Engineering, 1930–1956** S. Bennett
Volume 49 **Polynomial Methods in Optimal Control and Filtering** K.J. Hunt (Editor)
Volume 50 **Programming Industrial Control Systems Using IEC 1131-3** R.W. Lewis
Volume 51 **Advanced Robotics and Intelligent Machines** J.O. Gray and D.G. Caldwell (Editors)
Volume 52 **Adaptive Prediction and Predictive Control** P.P. Kanjilal
Volume 53 **Neural Network Applications in Control** G.W. Irwin, K. Warwick and K.J. Hunt (Editors)
Volume 54 **Control Engineering Solutions: A practical approach** P. Albertos, R. Strietzel and N. Mort (Editors)
Volume 55 **Genetic Algorithms in Engineering Systems** A.M.S. Zalzala and P.J. Fleming (Editors)
Volume 56 **Symbolic Methods in Control System Analysis and Design** N. Munro (Editor)
Volume 57 **Flight Control Systems** R.W. Pratt (Editor)
Volume 58 **Power-Plant Control and Instrumentation: The control of boilers and HRSG systems** D. Lindsley
Volume 59 **Modelling Control Systems Using IEC 61499** R. Lewis
Volume 60 **People in Control: Human factors in control room design** J. Noyes and M. Bransby (Editors)
Volume 61 **Nonlinear Predictive Control: Theory and practice** B. Kouvaritakis and M. Cannon (Editors)
Volume 62 **Active Sound and Vibration Control** M.O. Tokhi and S.M. Veres
Volume 63 **Stepping Motors, 4th Edition** P.P. Acarnley
Volume 64 **Control Theory, 2nd Edition** J.R. Leigh
Volume 65 **Modelling and Parameter Estimation of Dynamic Systems** J.R. Raol, G. Girija and J. Singh
Volume 66 **Variable Structure Systems: From principles to implementation** A. Sabanovic, L. Fridman and S. Spurgeon (Editors)
Volume 67 **Motion Vision: Design of compact motion sensing solution for autonomous systems** J. Kolodko and L. Vlacic
Volume 68 **Flexible Robot Manipulators: Modelling, simulation and control** M.O. Tokhi and A.K.M. Azad (Editors)
Volume 69 **Advances in Unmanned Marine Vehicles** G. Roberts and R. Sutton (Editors)
Volume 70 **Intelligent Control Systems Using Computational Intelligence Techniques** A. Ruano (Editor)
Volume 71 **Advances in Cognitive Systems** S. Nefti and J. Gray (Editors)
Volume 72 **Control Theory: A guided tour, 3rd Edition** J.R. Leigh
Volume 73 **Adaptive Sampling with Mobile WSN** K. Sreenath, M.F. Mysorewala, D.O. Popa and F.L. Lewis
Volume 74 **Eigenstructure Control Algorithms: Applications to aircraft/rotorcraft handling qualities design** S. Srinathkumar
Volume 75 **Advanced Control for Constrained Processes and Systems** F. Garelli, R.J. Mantz and H. De Battista
Volume 76 **Developments in Control Theory towards Global Control** L. Qiu, J. Chen, T. Iwasaki and H. Fujioka (Editors)
Volume 77 **Further Advances in Unmanned Marine Vehicles** G.N. Roberts and R. Sutton (Editors)
Volume 78 **Frequency-Domain Control Design for High-Performance Systems** J. O'Brien
Volume 80 **Control-oriented Modelling and Identification: Theory and practice** M. Lovera (Editor)
Volume 81 **Optimal Adaptive Control and Differential Games by Reinforcement Learning Principles** D. Vrabie, K. Vamvoudakis and F. Lewis
Volume 83 **Robust and Adaptive Model Predictive Control of Nonlinear Systems** M. Guay, V. Adetola and D. DeHaan
Volume 84 **Nonlinear and Adaptive Control Systems** Z. Ding
Volume 86 **Modeling and Control of Flexible Robot Manipulators, 2nd Edition** M.O. Tokhi and A.K.M. Azad
Volume 88 **Distributed Control and Filtering for Industrial Systems** M. Mahmoud
Volume 89 **Control-based Operating System Design** A. Leva *et al.*
Volume 90 **Application of Dimensional Analysis in Systems Modelling and Control Design** P. Balaguer
Volume 91 **An Introduction to Fractional Control** D. Valério and J. Costa
Volume 92 **Handbook of Vehicle Suspension Control Systems** H. Liu, H. Gao and P. Li
Volume 93 **Design and Development of Multi-Lane Smart Electromechanical Actuators** F.Y. Annaz
Volume 94 **Analysis and Design of Reset Control Systems** Y. Guo, L. Xie and Y. Wang
Volume 95 **Modelling Control Systems Using IEC 61499, 2nd Edition** R. Lewis and A. Zoitl
Volume 96 **Cyber-Physical System Design with Sensor Networking Technologies** S. Zeadally and N. Jabeur (Editors)
Volume 99 **Practical Robotics and Mechatronics: Marine, space and medical applications** I. Yamamoto
Volume 100 **Organic Sensors: Materials and applications** E Garcia-Breijo and P Cosseddu (Editors)
Volume 102 **Recent Trends in Sliding Mode Control** L. Fridman, J.P. Barbot and F. Plestan (Editors)
Volume 104 **Control of Mechatronic Systems** L. Guvenc, B.A. Guvenc, B. Demirel, M.T. Emirler
Volume 105 **Mechatronic Hands: Prosthetic and robotic design** P.H. Chappell
Volume 107 **Solved Problems in Dynamical Systems and Control** D. Valério, J.T. Machado, A.M. Lopes and A.M. Galhano
Volume 111 **The Inverted Pendulum in Control Theory and Robotics: From theory to new innovations** O. Boubaker and R. Iriarte (Editors)
Volume 112 **RFID Protocol Design, Optimization, and Security for the Internet of Things** Alex X. Liu, Muhammad Shahzad, Xiulong Liu and Keqiu Li

Wearable Exoskeleton Systems

Design, control and applications

Edited by
Shaoping Bai, Gurvinder S. Virk, and Thomas G. Sugar

The Institution of Engineering and Technology

Published by The Institution of Engineering and Technology, London, United Kingdom

The Institution of Engineering and Technology is registered as a Charity in England & Wales (no. 211014) and Scotland (no. SC038698).

First published 2018

The Institution of Engineering and Technology
Michael Faraday House
Six Hills Way, Stevenage
Herts, SG1 2AY, United Kingdom

www.theiet.org

British Library Cataloguing in Publication Data
A catalogue record for this product is available from the British Library

ISBN 978-1-78561-302-9 (hardback)
ISBN 978-1-78561-303-6 (PDF)

Typeset in India by MPS Limited

Cover image supplied by ©2017 ATOUN Inc.

Contents

Preface

Wearable exoskeletons are electro-mechanical systems designed to assist, aid, strengthen, help people, and have a variety of human motion applications and scenarios to provide power supplementation and argumentation to the wearers. The applications cover a wide range of domains such as medical devices for compensating impairements or for rehabilitation training of patients recovering from trauma, movement aids for disabled persons, personal care robots for providing assistance for normal daily living to healthy persons, and reduction of physical burden in industrial and military applications.

The development of effective and affordable wearable exoskeletons poses challenges to researchers and manufacturers in the areas of mechanical design, ergonomic interfaces for comfort, and effectiveness, human-adaptive controllers for symbiotic support and human modelling and biomechanics. New technologies are required in novel materials and structures, adaptive motion controllers, human–robot interaction control, biological sensors, and new lightweight actuators.

In this book, we assembled a total of 14 chapters reporting the recent advances and technology breakthroughs in exoskeleton technology. The chapters are grouped into four sections, namely, Section I, review and overall requirements; Section II, design and control; Section III, devices; and Section IV, commercialization issues.

Section I includes four chapters. Chapter 1 reviews the state-of-the-art of lower limb exoskeletons; Chapter 2 overviews all European research projects on exoskeletons; Chapter 3 provides a review of soft wearable exoskeletons, discussing the challenges and emerging research problems. In Chapter 4, user requirements are analyzed for lower body exoskeletons. The requirements are described mainly for soft, lower body exoskeletons but the method is applicable for other types of exoskeletons as well.

Section II presents design and control methods for exoskeletons. Chapter 5 describes the design of a novel spherical shoulder exoskeleton. The robot design, development and control strategies are presented. Chapter 6 introduces a platform for calibration of wearable 3D motion sensors. In Chapter 7, the design and control of series elastic actuators (SEAs) are outlined. The application of SEAs in upper and lower body exoskeletons is described. In Chapter 8, a knee–ankle–foot exoskeleton for gait rehabilitation is introduced and gait synchronization and impedance control of the exoskeleton are described.

Section III introduces a variety of exoskeleton prototypes and systems. A lower limb exoskeleton robot is presented in Chapter 9. A real-time gait planning method is developed, in which stability issues are addressed. In Chapter 10, a soft wearable robot is described. In Chapter 11, an ankle exoskeleton is presented. User validation results for a rehabilitation exoskeleton are included and discussed.

Section IV includes three chapters to provide information on issues related to exoskeleton development and commercialization. Chapter 12 outlines legal and regulatory perspectives of the exoskeleton applications. Chapter 13 discusses suitable testing methods for exoskeletons. Chapter 14 shares the development story of the Ekso Bionics exoskeleton.

The exoskeleton research and development requires a wide range of knowledge, including mechanism design and control involving consideration of close human–robot interaction scenarios, advanced human motion intention detection and support, comfort and ergonomics and safety regulation in various wearable robot applications. With these topics covered in this book, we hope that the book will be of interest to engineers and researchers in academia as well as manufacturing companies and be helpful in creating and developing new markets for wearable exoskeleton robots.

We thank the authors who contributed very interesting and comprehensive chapters with a wide coverage on several areas of exoskeleton technology and commercial development and additionally for their cooperation in revising papers in due time in agreement with the editors comments.

We thank the publisher, the IET, and its Editorial staff for accepting and helping in the publication of this book. The project took nearly 2 years to finish, since its early steps in 2016.

We are grateful to our families, as without their patience and comprehension it would not have been possible for us to finish this project. Shaoping Bai and Gurvinder Virk acknowledge the support by the European AAL program, which created opportunities for the editors to explore new exoskeleton technologies for their application in the area of daily living assistance of the elderly.

Shaoping Bai
Dept. of Materials and Production, Aalborg University
Fibigerstraede 16
Aalborg 9220, Denmark
E-mail: shb@m-tech.aau.dk

Gurvinder S. Virk
Technical Director
Innovative Technology & Science Limited, North Wing,
The Old Livery, Hildersham Road, Cambridge, CB21 6DR, UK
E-mail: gurvinder.virk@innotecuk.com

Thomas G. Sugar
The Polytechnic School
Arizona State University, USA
6075 S. Innovation Way West
Mesa, AZ 85212
E-mail: thomas.sugar@asu.edu

Section 1

Review and overall requirements

Chapter 1: Lower-limb wearable robotics
Chapter 2: Review of exoskeletons for medical and service applications: ongoing Research in Europe on Wearable Robots, with focus on lower extremity exoskeletons
Chapter 3: Soft wearable robots
Chapter 4: Exploring user requirements for a lower body soft exoskeleton to assist mobility

The first section consisting of four chapters focuses on an overview of lower limb wearable robots, the development of soft-robotic systems, and their user requirements. Specifically, Chapter 1 gives an overview of lower limb wearable systems and explores the promises and challenges when developing these new types of inherently safe exoskeleton systems. Chapter 2 reviews the current medical- and non-medical (or service-based) exoskeletons being developed in R&D projects funded in Europe. Chapter 3 defines and explains the new field of soft, wearable robots including lower limb exosuits, covering systems to assist the elbow and hand, and devices that can be implanted into the body. Last, Chapter 4 focuses on the user requirements for developing lower limb exoskeletons with a case study focusing on XoSoft, a lower limb exoskeleton using soft robotic technology.

Defining wearable robotic systems is challenging because the design space is so large with passive, quasi-passive, quasi-active, and active/powered systems. Designs are being specialized for the upper extremity vs the lower extremity. Last, the systems can be categorized into medical-based devices (such as exoskeletons for spinal cord injured persons), rehabilitation devices (e.g., devices assisting mobility after a stroke), military devices (e.g., devices to carry large loads), industrial devices (e.g., devices to assist lifting or bending over for specific industrial tasks), and recreational devices (e.g., devices to assist leisure activities such as skiing).

Chapter 1 develops a definition of a wearable robotic system and focuses on lower limb systems. Promises and challenges are discussed next which include developing devices to assist human mobility for the ageing society problem. Challenges include portability, ease of use, cost, lightweight actuators, and controllers that are in synchrony with the user. Different systems are reviewed in key areas that include lower limb orthoses, lower limb prostheses, lower limb exoskeletons, and systems used in legged rehabilitation.

Chapter 2 reviews current European projects in the area of lower and upper limb exoskeletons. Systems include AXO-SUIT, BALANCE, BioMot, CORBYS,

CYBERLEGs, EXO-LEGS, Mindwalker, Spexor, Symbitron, and XoSoft. The AXO-SUIT project aims to design an assistive full-body exoskeleton for the elderly. The BALANCE project focuses on understanding human balance to develop an exoskeleton to augment balance when walking. The BioMot project is developing neuronal control and learning to provide new control methods that enable an exoskeleton to seamlessly interact with the human. The CORBYS project developed control architectures to allow a user to interact with the human. The CYBERLEGs project developed a pelvis orthoses, a knee-ankle foot orthosis, and a transfemoral prosthesis. A novel combined orthosis and prosthesis was demonstrated as well. The EXO-LEGS project developed a lower limb exoskeleton to assist the elderly and can be considered a personal care robot and aimed at complying with the recently published robot safety standard ISO 13482 because it does not make any medical claims. The Mindwalker project will develop an exoskeleton that uses brain control interfaces to control a lower limb exoskeleton. The Spexor project will develop a lower back exoskeleton to reduce back pain. The Symbitron project is developing an exoskeleton for people with a spinal cord injuries. Last, XoSoft is developing a lower limb exoskeleton based on new soft-robotic technologies.

In Chapter 3, an overview of soft, wearable robots is presented. Pneumatic and cable-driven exoskeletons assistting at the ankle and hip are described. Soft glove type exoskeletons are also presented that can assist stroke survivors in simple grasping tasks when performing activities of daily living. Last, soft fiber reinforced actuators are being inserted into the human body to assist the pumping action of the heart.

Chapter 4 focuses on user-requirements when developing lower body exoskeletons. A case study is developed based on the XoSoft project. Just as in Chapter 1, the authors describe challenges when designing wearable exoskeletons. A user defined design approach is used which studies the needs of primary, secondary, and tertiary users. Their user-centered approach incorporates the need to develop prototypes and develop validation studies along timeline when developing an exoskeleton product.

For their lower limb exoskeleton, the primary users include stroke survivors, people with incomplete spinal cord injury, and older adults with mild-moderate mobility impairments. Secondary users include health care professionals, formal and informal caregivers, spouses, family, and friends. Tertiary users include regulatory authorities, insurance companies, standardization bodies, advocate groups, resellers, hospitals, and rehabilitation centers.

Informative results included the need to assist housekeeping activities which involve standing while performing tasks with your hands. Current aids such as walkers, rollators, canes, walking sticks, etc. rely on the arms and hands to aid in balance. Thus, with the current mobility aids, users cannot use their hands to perform necessary tasks. Many primary users experienced limitations when walking outside in crowded environments, negotiating stairs, and performing activities that require upper arm function such as when picking items off of a shelf when shopping at a store.

The design requirements and must-have features for the primary and secondary users are defined. Important features in *both* lists, including easy to don and doff, improve safety and stability for users to prevent falls during gait and transfer movements (getting out of a chair or bed), lightweight, easy to use, low cost, and compatible with user footwear and clothing. The study also looks at the need to develop a system that does not restrict movement when doing other tasks such as sitting down or bending over.

These first four chapters define the advantages and promises of the new exoskeleton technology, define challenges that must be overcome, describe an overview of lower limb systems including the new area of soft-robotics, and lastly develop a method to create a set of user-requirements. The lower limb case-study defines a set of user-requirements that are needed by an academic or industrial creator of such devices.

Chapter 1

Lower-limb wearable robotics

Thomas G. Sugar[1,2], *David Armstrong*[3], *Bijan Najafi*[4], *Sangram Redkar*[2], *and Jeffrey A. Ward*[5]

Abstract

It is an exciting time in wearable robotics with new devices for spinal cord injury, gait assistance, rehabilitation, strength enhancement, manufacturing, construction, and recreation. This review will focus on lower limb wearable robotic systems that include orthoses, prostheses, and exoskeletons. The promises and challenges of these systems will be described. A review of some of the exciting systems will be presented.

Keywords: Orthoses, prostheses, lower-limb exoskeletons

1.1 Background

Aging in place, or the ability of older adults to continue to live safely and economically in their own residences for as long as possible, is one of the key issues of modern societies [1]. In 2029, more than 20% of the total US population will be over the age of 65 [2], and by 2050, this population will have increased to 83.7 million people, almost double the 2012 population estimate of 43.1 million [3]. Continued population growth of these older adults (age 65+) will lead to a greater need for preventive, acute, rehabilitative, and long-term health-care services, as well as a need for tools to enable them to function independently during everyday activities.

Advances in designing wearable robotics in recent years have opened new avenues to promote aging in place via assisting, enhancing, or rehabilitating gait. A hip exoskeleton has been able to rehabilitate gait of stroke survivors [4]. Some

[1]Department of Engineering, SpringActive, Inc., Wearable Robotics Association, Arizona State University, USA
[2]Department of Engineering, The Polytechnic School, Ira A. Fulton Schools of Engineering, Arizona State University, USA
[3]Department of Surgery, College of Medicine, University of Arizona, USA
[4]Interdisciplinary Consortium on Advanced Motion Performance (iCAMP), Michael E. DeBakey Department of Surgery, Baylor College of Medicine, USA
[5]SpringActive, Inc., USA

systems can assist mobility by using a combination of a motorized walker with robotic assistance at the joints [hips and knees (HKs)] [5]. There are however many challenges with the current exoskeleton systems and even more challenges to adapt such technologies for aging in place. For the elderly, these systems must be simple, easy to use, lightweight, economical, and autonomous. This review paper describes the current advances in the field, shortcomings of current technologies, and future directions.

1.2 Definition of wearable robotic system

Defining a wearable robotic system is challenging task because the system spans a large design space that includes exoskeletons, prosthetic systems, and non-anthropomorphic systems. They are "systems intended to exchange energy with the environment and human body in order to attain performance augmentation as well as for assistive, prosthetic, or rehabilitative purposes [6]." These systems are used to "extend, complement, substitute, or enhance human function and capability or empower or replace (a part of) the human limb where it is worn [6]."

Viteckova *et al.* grouped wearable exoskeletons into rehabilitative exoskeletons, assistive exoskeletons, and empowering exoskeletons [7]. One could imagine devices for rehabilitation after a stroke, devices to assist carrying heavy loads, or devices that enhance human function giving superior strength. Herr grouped the systems in the manner in which the actuators are attached to the body [8]. Some systems attach springs in series with the muscle-tendon structure while other systems attach structures in-parallel. Some systems transfer the load or weight to the ground while other systems rely on the human body to transfer the load while providing additional support to the joints.

There are many review articles on the topic of wearable robots [7–21], and this chapter will focus on lower limb systems only. This review will be organized into three areas: what are the positive aspects of lower limb systems, the challenges with lower limb systems, and what types of lower limb wearable robotic systems have been developed.

Many of the current systems have military or therapeutic purposes, but assistive type systems are now being developed to help people with their everyday activities of living [13]. Lines have been blurring between what is an assistive device and what is a therapeutic device [13]. For example, one could build a hip exoskeleton that flexes and extends the thigh to assist someone carrying a heavy load, although the same robot can be used to assist a person with a passive prosthetic ankle to achieve a therapeutic effect of making it easier to walk reducing metabolic cost and improving gait symmetry. In summary, it seems that wearable robots will be used in different applications such as: manufacturing, construction, military, therapeutic, assistive, and recreational applications. We will use these assistive devices when they are comfortable, robust, simple, and easy to use. For example, we use bicycles to increase mobility or we use glasses to read text, and soon, we will use wearable robots for mobility assistance as well.

1.3 Promise and potential of wearable robotic systems

We are presently victims of our own demographic success. In the developing world, in almost every nation, the population of older adults is the fastest growing segment of the population. As this group gets older, the number of people available to proportionally care for them is reducing. This raises both challenges and opportunities. It is expected that wearable robotic systems can and will assist and aid users as they get older. In 2009, for the first time in the history of humanity, more people died of "diseases of decay" such as diabetes, cancer, and heart disease than from all communicable diseases [22]. In the future, it will be possible to don a device to help with a sore knee or hip or wear a device to allow one to walk a longer distance than was usually possible. Powered orthoses can potentially assist walking with a total contact cast or other techniques used in diabetic care of foot ulcers [23,24]. Powered prostheses can enhance walking of a lower limb amputee and powered exoskeletons can allow someone to get up out of a wheelchair and walk for a couple of hours per day.

Other interesting applications of exoskeletons are envisioned. Dr. Ferris suggested that exoskeletons can be used to perturb the human system (while also protecting) so that clinicians and biomechanical experts can understand the human motor control system [11,12]. For example, a device can be used as a diagnostic tool to determine the amount of body sway or the amount of balance loss.

In the near-term, smart textiles and electronic systems can measure our movement and determine if we are about to fall, or just coax us into walking more daily.

In summary, in a paper on recent trends, it is argued that the goal of these devices is to be transformative [25]. The device is transformative if it "elevates mobility performance by people with a disability to that of their nondisabled peers [25]" or we hope that they can elevate mobility performance of the elderly back to a more youthful time.

1.4 Challenges

There is great fanfare with assistive systems that can make it easier to walk, run, jump, climb, or stand from a chair. However, there are major challenges with all of these systems such as

- Portability
- Ease of use
- Metabolic cost
- Lightweight actuators
- Actuators that can supply high torque at slow speeds
- Actuators that are soft and flexible
- Intuitive control systems
- Control systems that adapt to the changing dynamics of the human

Systems that are heavy cause too much energetic cost when wearing. Only a few systems have been able to overcome the metabolic burden of their weight and actually augment the user and make it easier to walk or run [26–30]. Most current systems actually increase the metabolic cost of use. Research efforts are needed to develop powerful, lightweight actuators that can be stiff and supply large torques when needed, or in an instant can become flexible and soft.

Other review articles agree that robotic exoskeletons must improve in multiple areas such as human–robot interaction, safe and robust actuators, portability and wearability, and user-centered design [19,25,31]. It was stated that exoskeletons must be lightweight, control algorithms must improve to handle varied terrain and speed changes, and user interfaces must be "intuitive, seamless, and nonobtrusive [25]."

Challenges include the need for better sensors and controllers that are tuned to the human [11,12]. Most controllers use standard technology with feedback loops driving an error to zero. Recently, the use of negative damping [32], reflex type controllers [33], and controllers based on phase signals [34] are being studied. The goal is to develop controllers that react quickly and allow the joints/actuators to move in sync with the user. In a different method, EMG (electromyogram) signals allow the user to control the robot [12,35–37].

A last challenge is how to link a controller with the human person. Improved human motor-neural models are needed. While walking is not something the average person thinks about, it is a highly complex function involving the coordination of multiple biological systems. Achieving a better understanding of the underlying biological and neurological control principles of human gait will allow roboticists to improve upon existing lower limb human–machine interfaces [11,12]. For example, should a robot drive in a particular pattern forcing the human to follow? Authors have been showing that an assist-as-needed strategy is a better learning regimen in stroke therapy [38]. Still questions remain. How to determine if a control algorithm is optimal? How best to train the human motor-neural system? How to determine the human's desired motion or intention? Ferris describes at a minimum to use elastic members, transfer energy using biarticular actuators, allow the leg to swing in a pendulum motion, powering push-off to reduce the collision at heel-strike, and allowing the user to control the robot using EMG-based control [12].

We believe that the best robotic systems follow the human and just apply small needed forces or torques at the correct time. If the user says that they cannot feel the robot working, then it is doing its job. If the user feels like a mannequin pulled by strings, the robot becomes uncomfortable and metabolic cost increases dramatically.

The objective of all of these systems is to interact with the dynamics of the human body. However, interacting with a changing set of dynamics especially as we age will be quite difficult. This poses a very challenging task where the dynamics of the human is changing, and the robot must somehow adapt.

As designers and roboticists in this field, we predict that new actuation schemes based on energy storage and smaller batteries will reduce device weight

Table 1.1 Lower limb wearable systems

	Population served	Objectives	Actuation
Orthoses	Stroke survivors, drop foot	Assist the ankle to improve gait and reduce metabolic cost	Motors and springs or hydraulics
Prostheses	Lower limb amputee	Assist the ankle to improve gait and reduce metabolic cost	Motors and springs or hydraulics
Exoskeletons for spinal cord injury	Spinal cord injury	Allow users to walk for a couple of hours per day	Motors at the hip, knee, and ankle
Assistive hip exoskeletons	Stroke, elderly, frail individuals	Small assistance at the hip	Motors and springs
Full exoskele-tons	Manufacturing, construction, military, elderly, stroke, spinal cord injury	Enhance, augment, or assist gait	Motors, hydraulics, and springs
Rehabilitation	Treadmill systems vs over ground systems	Repetitive task therapy	Motors, hydraulics, and springs

and increase wearability. Controller designs are becoming more robust allowing people to seamlessly walk and run [26]. Last, roboticists must work in collaboration with users, medical professionals, therapists, and industrial designers to truly develop systems that are user-friendly and user-acceptable.

1.5 Lower limb wearable systems

There are different types of lower limb systems: powered orthoses, prostheses, and rehabilitation, see Table 1.1.

This discussion will not be an exhaustive list of systems that have been developed, but will list relevant devices in three areas of lower limb assistance: powered orthoses, prostheses, and exoskeletons. Exhaustive reviews can be found in the literature [10,13,16,21,39].

1.6 Lower limb orthoses

In powered orthoses, researchers have been developing devices for drop foot, stroke therapy, and rehabilitation, see Figure 1.1 [40–45]. Devices include designs for multi-degree of freedom movement as well, see Figure 1.2 [46–49].

1.7 Lower limb prostheses

Powered prostheses have been making great strides allowing users to walk, run, and jump. Powered systems have been developed by Goldfarb, Herr, Sugar, and others

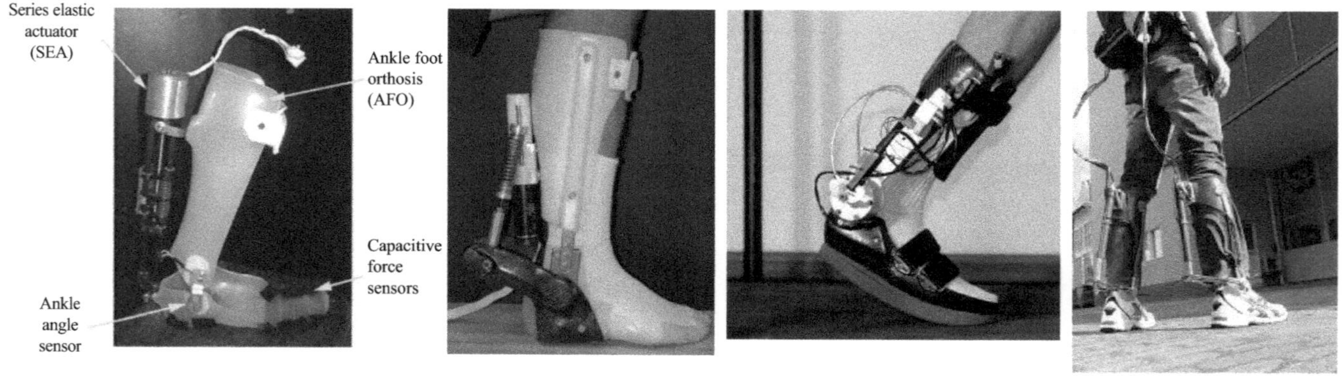

Figure 1.1 *A powered orthosis to assist drop foot [41]. A motor-spring orthosis to assist in gait for stroke rehabilitation [42]. A portable, powered orthosis using hydraulic actuation [40]. A leaf-spring-based ankle orthosis [45]*

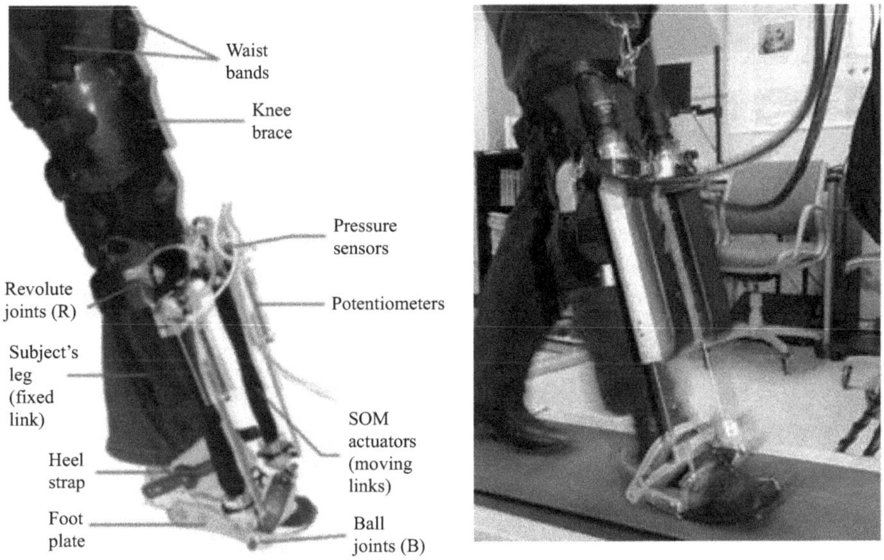

Figure 1.2 Powered ankle orthoses have been developed to assist plantarflexion, dorsiflexion, inversion and eversion. Designs have been developed at ASU and MIT [47,48]

Figure 1.3 A powered knee-ankle prosthesis [52]. Powered ankles developed at MIT and ASU [50,51]

[50–58]. Goldfarb has been developing a sophisticated, combined knee and ankle (Figure 1.3) [52]. The Cyberlegs project has been developing wearable hip exoskeletons to assist in push-off of a lower limb amputee [59]. Researchers have also been developing multi-degrees-of-freedom systems [60,61] (Figure 1.4).

1.8 Lower limb exoskeletons

Powered exoskeletons can be divided up into many different types of groups. One grouping could be systems that assist at the hip, knee, or ankle, or a

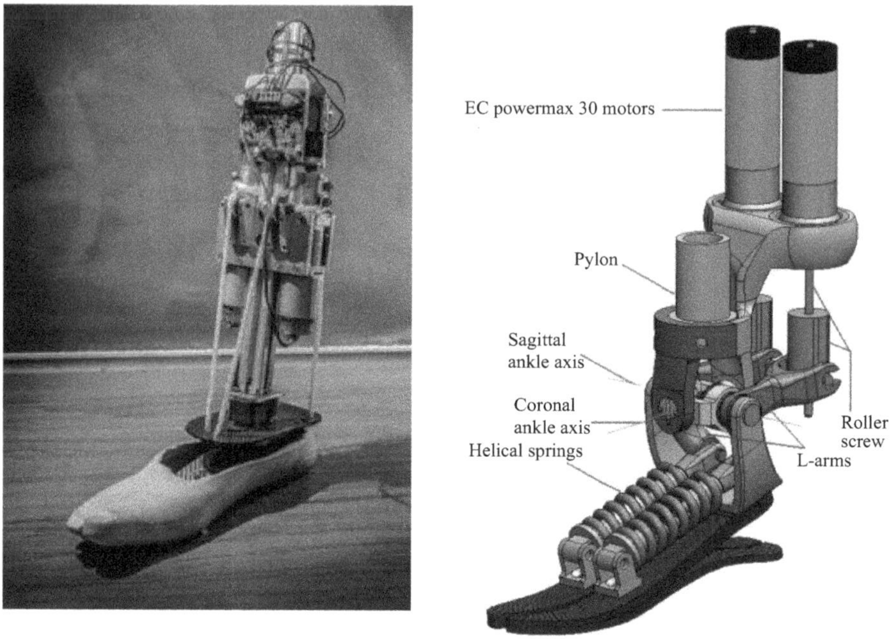

Figure 1.4 Multi-degrees-of-freedom powered prosthetic ankles [60,61]

combination of joints. Another grouping could be systems that have a medical purpose such as: assisting spinal cord injury, assisting someone with an amputation, or assisting a certain type of rehabilitation scheme. Systems could be grouped into their uses such as medical, manufacturing, construction, or recreation and fitness. Last, systems could be grouped if they transfer the load to the ground or if they use soft actuation that does not transfer the load to the ground. A nonexhaustive list of exoskeletons will be reviewed.

One grouping of exoskeletons is by systems that assist in spinal cord injury, see Figure 1.5. These systems allow one to walk for a couple hours a day and are well received by the users. Systems are being designed to allow people with spinal cord injuries to walk by Ekso Bionics, Parker Hannifin, ReWalk, RexBionics, and others [15,17,62–70].

Powering the hip allows one to assist gait, stair climbing, walking on inclines, and lifting objects. Devices have moved from academic research toward commercial devices, see Figure 1.6. The Honda hip exoskeleton has shown great benefit in assisting stroke survivors. The device improved gait allowing users to take more steps per day [4]. The hip exoskeleton from Cyberdyne is being used in airports [71]. Parker Hannifin is designing exoskeletons to power gait [63,72]. Research as part of the Evryon project designed systems where the actuators are not mounted at the joints, a non-anthropomorphic design [73]. As part of the DARPA Warrior Web Program, Walsh has been designing soft actuators that do not rely on

Figure 1.5 Exoskeleton systems allow users to walk. Pictures from Ekso Bionics, Indego, ReWalk, and Rex Bionics. Source: Company websites

rigid structures [74,75]. These systems hold promise in improving wearability and reducing metabolic cost.

Full system exoskeletons can power the HK or hip, knee, and ankle (HKA). These systems hold promise in aiding gait in rehabilitation, spinal cord injury, military, industrial, and recreational applications. A figure of 12 different systems is shown in Figure 1.7 showing large strength enhancing systems to smaller, lightweight systems.

The Hybrid Assistive Leg (HAL) system by Cyberdyne uses EMG signals to determine human intention and then amplifies these signals to apply a torque to the HK [37,76]. The BLEEX system was one of the first hydraulic systems to power the HKA [77]. Lockheed Martin and Ekso Bionics redesigned the system to create the HULC (Human Universal Load Carrier) exoskeleton [66]. The SARCOS X02 system is able to allow the user to lift very heavy objects [78]. The Hercule robot from RB3D assists in gait and is being used in construction [79]. The Power Assist Suit can be used to transfer people from the bed in medical applications [80]. The Power Loader by ActiveLink, a subsidiary of Panasonic will be used in manufacturing [81]. The Power Assist Suit by Kawasaki will also be used in manufacturing and agriculture to lift objects [82]. The PERCRO lab developed the Body Extender, a full upper and lower limb exoskeleton [83]. Boston Dynamics developed a soft actuation system to extract energy from the knee when walking down hills [84]. At the Arizona State University, researchers are developing systems to enhance running [26]. Walsh at Harvard has been developing soft, lower limb, wearable robots [74,75].

1.9 Legged rehabilitation

Robots are very repeatable devices which can be quite useful when performing repetitive type therapy [13]. Repetitive stroke therapy has been used in gait training on a treadmill to train a user's affected side and improve motor strength and coordination [38]. In fact, researchers discussed the need to train both the affected

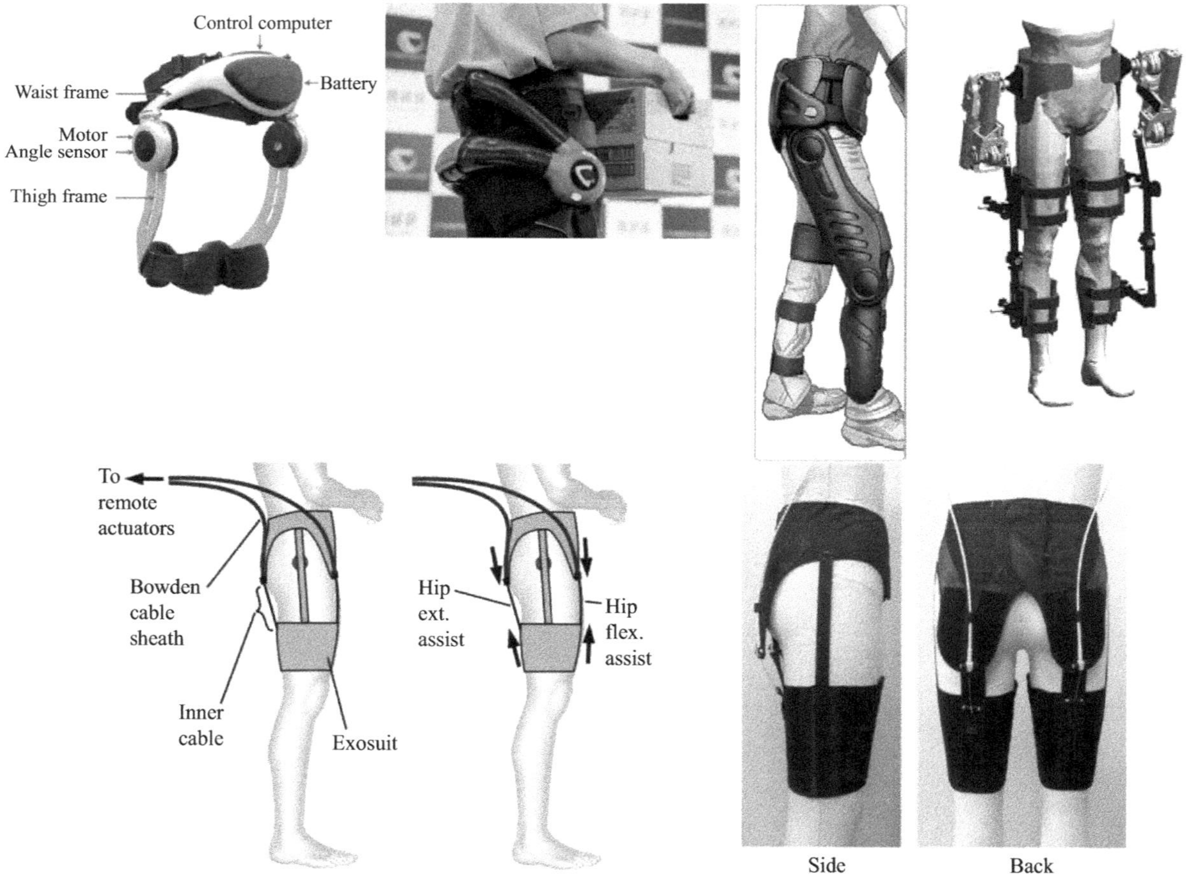

Figure 1.6 *Hip Exoskeleton Systems: The Honda Stride Assist System is being used in stroke rehabilitation [4]. The Cyberdyne hip exoskeleton is being used in the Tokyo airport [71]. Parker Hannifin is developing an exoskeleton [63,72]. Soft exoskeleton designs hold promise for improving comfort and wearability [74,75]*

Figure 1.7 Exoskeleton systems (top to bottom, left to right) HAL [37,76], BLEEX [77], HULC [66], SARCOS X02 [78], Hercule [79], Power Assist Suit [80], Power Loader [81], Power Assist Suit [82], Body Extender [83], Boston Dynamics knee [84], AirLegs [26], Soft, lower limb, wearable robots [74,75]

and unaffected sides to improve gait measures. These devices can also log relevant data needed for the outcomes-based research protocols [13].

Most current devices are stationary type devices such as the Lokomat, Auto-Ambulator, Gait Trainer, Haptic Walker, GaitMaster, G-EO System, LokoHelp, LOPES (Lower Extremity Powered Exoskeleton), ARTHur (Ambulation-Assisting Robotic Tool for Human Rehabilitation), POGO (Pneumatically Operated Gait Orthosis), PAM (Pelvic Assist Manipulator), ALEX (Active Leg Exoskeleton), and ALTACRO (Actuated Compliant Robotic Orthosis) [9,38]. Other devices do allow some over ground gait training such as the KineAssist, WalkTrainer, and exoskeletons such as HAL [9,38]. The goal of these devices is to aid in motor learning and rehabilitation and reduce the burden of repetitive training.

A systematic review article on the effectiveness of robotic therapy for ankle rehabilitation describes some success of wearable systems to improve gait measures by training the ankle in a seated position or when walking [85]. Different types of assistive strategies for lower limb, wearable robots are reviewed in detail [20].

Current therapy to improve gait, partial weight bearing training, is very labor intensive requiring more than three physical therapists. It was noted as the world population ages, new less labor intensive therapies are needed because the number of patients will overtake the capabilities of the limited number of therapists. However, it is controversial on how much improvement is gained in robotic therapy. One review article did note that robotic therapy is "as good as" manual therapy [38]. Lower limb therapy is still in its infancy and will need further research [86–88]. A combination of robotic therapy and conventional therapy will start to address this growing need for repetitive therapy in the future.

1.10 Future vision

Advances in miniaturized sensors, telecommunication, battery design, and wearable devices make wearable robots feasible, practical, and cost effective. These technologies could have a significant impact on the ability of older adults to age in place and maintain their independent and active lifestyle. We see the advent of new time where donning an assistive, wearable system will enhance mobility. These systems will assist in rehabilitation after surgery, enhance neural retraining after a stroke, or allow for more functional gait after an amputation. These systems will aid in tough, manual labor in construction and manufacturing tasks. Last, in recreation, a system can help when hiking a mountain or walking around a new town on that favorite vacation.

As the price of sensors, microprocessors, and batteries decrease substantially, wearable systems will become more affordable. Custom designs can be printed using new 3D printers to aid in wearability. New 3D printers can print compliant, hard, and a mixture of plastics with different material properties.

Intuitive control and sensing is still a challenging task especially when understanding the human dynamics of a person as they age or just get tired and fatigued. However, most systems still make it harder to walk and do not pay for their weight burden. Ending positively, we believe in the near future that users will wear assistive devices just like jumping on a bike to travel a longer distance.

References

[1] J. L. Wiles, A. Leibing, N. Guberman, J. Reeve, and R. E. Allen, "The meaning of 'aging in place' to older people," *Gerontologist*, vol. 52, 3, pp. 357–366, Jun 2012.

[2] J. M. Ortman, V. A. Velkoff, and H. Hogan, "An aging nation: The older population in the United States," *Current Population Reports*, pp. P25–1140, 2014.
[3] S. L. Colby and J. M. Ortman, "The baby boom cohort in the United States: 2012 to 2060," *Current Population Reports*, pp. P25–1141, 2014.
[4] C. Buesing, G. Fisch, M. O'Donnell, *et al.*, "Effects of a wearable exoskeleton stride management assist system (SMA®) on spatiotemporal gait characteristics in individuals after stroke: A randomized controlled trial," *Journal of Neuroengineering and Rehabilitation*, vol. 12, 2015.
[5] K. Kong and D. Jeon, "Design and control of an exoskeleton for the elderly and patients," *IEEE/ASME Transactions on Mechatronics*, vol. 11.4, pp. 428–432, 2006.
[6] G. Carpino, D. Accoto, N. L. Tagliamonte, G. Ghilardi, and E. Guglielmelli, "Lower limb wearable robots for physiological gait restoration: State of the art and motivations," *MEDIC*, vol. 21.2, pp. 72–80, 2013.
[7] S. Viteckova, P. Kutilek, and M. Jirina, "Wearable lower limb robotics: A review," *Biocybernetics and Biomedical Engineering*, vol. 33.2, pp. 96–105, 2013.
[8] H. Herr, "Exoskeletons and orthoses: Classification, design challenges and future directions," *Journal of NeuroEngineering and Rehabilitation*, vol. 6.1, p. 21, 2009.
[9] I. Diaz, J. J. Gil, and E. Sanchez, "Lower-limb robotic rehabilitation: Literature review and challenges," *Journal of Robotics*, vol. 2011, 2011.
[10] A. M. Dollar and H. Herr, "Lower extremity exoskeletons and active orthoses: Challenges and state-of-the-art," *IEEE Transactions on Robotics*, vol. 24.1, pp. 144–158, 2008.
[11] D. P. Ferris, "The exoskeletons are here," *Journal of NeuroEngineering and Rehabilitation*, vol. 6.1, p. 17, 2009.
[12] D. P. Ferris, G. S. Sawicki, and M. A. Daley, "A physiologist's perspective on robotic exoskeletons for human locomotion," *International Journal of Humanoid Robotics*, vol. 4.03, pp. 507–528, 2007.
[13] W. Huo, S. Mohammed, J. C. Moreno, and Y. Amirat, "Lower limb wearable robots for assistance and rehabilitation: A state of the art," *IEEE Systems Journal*, vol. 10.3, pp. 1068–1081, 2014.
[14] R. Jimenez-Fabian and O. Verlinden, "Review of control algorithms for robotic ankle systems in lower-limb orthoses, prostheses, and exoskeletons," *Medical Engineering & Physics*, vol. 34, pp. 397–408, 2012.
[15] K. H. Low, "Robot-assisted gait rehabilitation: From exoskeletons to gait systems," in *Defense Science Research Conference and Expo (DSR)*, pp. 1–10, 2011.
[16] S. Mohammed and Y. Amirat, "Towards intelligent lower limb wearable robots: Challenges and perspectives—State of the art," in *IEEE International Conference on Robotics and Biomimetics*, 2008, pp. 312–317.
[17] J. C. Moreno, G. Asin, J. L. Pons, *et al.*, "Symbiotic wearable robotic exoskeletons: The concept of the BioMot project," in *International Workshop on Symbiotic Interaction*, ed: Springer, Cham, 2014, pp. 72–83.

[18] J. L. Pons, *Wearable robots: Biomechatronic Exoskeletons*: John Wiley & Sons, 2008.

[19] J. L. Pons, "Rehabilitation exoskeletal robotics: The promise of an emerging field," *IEEE Engineering in Medicine and Biology Magazine*, vol. 29, 3, pp. 57–63, 2010.

[20] T. Yan, M. Cempini, M. O. Cologero, and N. Vitiello, "Review of Assistive Strategies in Powered Lower-Limb Orthoses and Exoskeletons," *Robotics and Autonomous Systems*, vol. 64, pp. 120–136, 2015.

[21] C.-J. Yang, J.-F. Zhang, Y. Chen, Y.-M. Dong, and Y. Zhang, "A review of exoskeleton-type systems and their key technologies," *Proceedings of the Institution of Mechanical Engineers, Part C: Journal of Mechanical Engineering Science*, vol. 222, 8, pp. 1599–1612, 2008.

[22] D. G. Armstrong, "Repair, regeneration and replacement revisited," in *Living Beyond 100 Lecture Series*, ed, 2012. *YouTube at* <https://www.youtube.com/watch?v=–iUH29YTs_g>.

[23] D. G. Armstrong, L. A. Lavery, B. P. Nixon, and A. J. Boulton, "It's not what you put on, but what you take off: Techniques for debriding and off-loading the diabetic foot wound," *Clinical Infectious Disease*, vol. 39, pp. 92–99, 2004.

[24] D. G. Armstrong and L. A. Lavery, "Evidence-based options for off-loading diabetic wounds," *Clinics in Podiatric Medicine and Surgery*, vol. 15, 1, pp. 95–104, 1998.

[25] R. E. Cowan, B. J. Fregly, M. Boninger, L. Chan, M. M. Rodgers, and D. J. Reinkensmeyer, "Recent trends in assistive technology for mobility," *Journal of NeuroEngineering and Rehabilitation*, vol. 9, 1, p. 20, 2012.

[26] J. Kerestes, T. G. Sugar, and M. Holgate, "Adding and subtracting energy to body motion – Phase oscillator," in *ASME Paper No. DETC2014-34405*, 2014.

[27] L. M. Mooney, E. J. Rouse, and H. Herr, "Autonomous exoskeleton reduces metabolic cost of human walking," *Journal of NeuroEngineering and Rehabilitation*, vol. 11, 1, pp. 151, 2014.

[28] P. Malcolm, W. Derave, S. Galle, and D. De Clercq, "A simple exoskeleton that assists plantarflexion can reduce the metabolic cost of human walking," *PLoS ONE*, vol. 8, 2, e56137, 2013.

[29] S. H. Collins, M. B. Wiggin, and G. S. Sawicki, "Reducing the energy cost of human walking using an unpowered exoskeleton," *Nature*, vol. 522, no. 7555, pp. 212–215, 2015.

[30] J. A. Norris, K. P. Granata, M. R. Mitros, E. M. Byrne, and A. P. Marsh, "Effect of augmented plantarflexion power on preferred walking speed and economy in young and older adults," *Gait & Posture*, vol. 25, 4, pp. 620–627, 2007.

[31] M. Cenciarini and A. M. Dollar, "Biomechanical considerations in the design of lower limb exoskeletons," in *IEEE International Conference on Rehabilitation Robotics*, pp. 1–6, 2011.

[32] G. Aguirre-Ollinger, J. E. Colgate, M. A. Peshkin, and A. Goswami, "A 1-DOF assistive exoskeleton with virtual negative damping: Effects on the kinematic response of the lower limbs," presented at the *IEEE/RSJ International Conference on Intelligent Robots and Systems*, San Diego, CA, USA, pp. 1938–1944, 2007.

[33] H. Geyer and H. Herr, "A muscle-reflex model that encodes principles of legged mechanics produces human walking dynamics and muscle activities," *IEEE Transactions on Neural Systems and Rehabilitation Engineering*, vol. 18, 3, pp. 263–273, Jun 2010.

[34] T. G. Sugar, A. Bates, M. Holgate, *et al.*, "Limit cycles to enhance human performance based on phase oscillators," *Journal of Mechanisms and Robotics-Transactions of the ASME*, vol. 7, 1, pp. 011001, Feb 2015.

[35] R. M. Singh and S. Chatterji, "Trends and challenges in EMG based control scheme of exokeleton robots – A review," *International Journal of Scientific & Engineering Research*, vol. 3, 9, pp. 933–940, 2012.

[36] G. Sawicki, K. Gordon, and D. P. Ferris, "Powered lower limb orthoses: applications in motor adaptation and rehabilitation," in *IEEE 9th International Conference on Rehabilitation Robotics*, 2005, pp. 206–211.

[37] K. Suzuki, G. Mito, H. Kawamoto, Y. Hasegawa, and Y. Sankai, "Intention-based walking support for paraplegia patients with robot suit HAL," *Advanced Robotics*, vol. 21, 12, pp. 1441–1469, 2007.

[38] A. Pennycott, D. Wyss, H. Vallery, V. Klamroth-Marganska, and R. Riener, "Towards more effective robotic gait training for stroke rehabilitation: A review," *Journal of NeuroEngineering and Rehabilitation*, vol. 9, 1, p. 65, 2012.

[39] R. Bogue, "Exoskeletons and robotic prosthetics: A review of recent developments," *Industrial Robot: An International Journal*, vol. 36, 5, pp. 421–427, 2009.

[40] K. A. Shorter, G. F. Kogler, E. Loth, W. K. Durfee, and E. T. Hsiao-Wecksler, "A portable powered ankle-foot orthosis for rehabilitation," *Journal of Rehabilitation Research and Development*, vol. 48, 4, pp. 459–472, 2011.

[41] J. A. Blaya and H. Herr, "Adaptive control of a variable-impedance ankle-foot orthosis to assist drop-foot gait," *IEEE Transactions on Neural Systems and Rehabilitation Engineering*, vol. 12, 1, pp. 24–31, 2004.

[42] K. W. Hollander, R. Ilg, T. G. Sugar, and D. E. Herring, "An efficient robotic tendon for gait assistance," *ASME Journal of Biomechanical Engineering*, vol. 128, 5, pp. 788–791, pp. 788–791, 2006.

[43] Q. Wang, J. Zhu, Y. Huang, K. Yuan, and L. Wang, *Segmented foot with Compliant Actuators and Its Applications to Lower-Limb Prostheses and Exoskeletons* In *Smart actuation and sensing systems-Recent advances and future challenges*. InTech 2012.

[44] D. P. Ferris, K. E. Gordon, G. S. Sawicki, and A. Peethambaran, "An improved powered ankle–foot orthosis using proportional myoelectric control," *Gait & Posture*, vol. 23, 4, pp. 425–428, 2006.

[45] C. Meijneke, W. van Dijk, and H. van der Kooij, "Achilles: An autonomous lightweight ankle exoskeleton to provide push-off power," presented at the *5th IEEE RAS & EMBS International Conference on Biomedical Robotics and Biomechatronics*, pp. 918–923, 2014.

[46] K. Bharadwaj, T. G. Sugar, J. B. Koeneman, and E. J. Koeneman, "Design of a robotic gait trainer using spring over muscle actuators for ankle stroke rehabilitation," *ASME Journal of Biomechanical Engineering*, vol. 127, 6, pp. 1009–1013, 2005.

[47] J. W. Wheeler, H. I. Krebs, and N. Hogan, "An ankle robot for a modular gait rehabilitation system," presented at the *Proceedings. 2004 IEEE/RSJ International Conference on Intelligent Robots and Systems (IROS 2004)*, vol. 2, pp. 1680–1684, 2004.

[48] K. Bharadwaj, K. W. Hollander, C. A. Mathis, and T. G. Sugar, "Spring over muscle (SOM) actuator for rehabilitation devices," presented at the *26th Annual International Conference of the IEEE Engineering in Medicine and Biology Society*, 2004. IEMBS '04, vol. 1, 2726–2729, 2004.

[49] A. Roy, H. I. Krebs, D. J. Williams, *et al.*, "Robot-aided neurorehabilitation: A novel robot for ankle rehabilitation," *IEEE Transactions on Robotics*, vol. 25, 3, pp. 569–582, 2009.

[50] J. K. Hitt, M. Holgate, R. Bellman, T. G. Sugar, and K. W. Hollander, "Robotic transtibial prosthesis with biomechanical energy regeneration," *Industrial Robot: An International Journal*, vol. 36, 5, pp. 441–447, 2009.

[51] S. Au, M. Berniker, and H. Herr, "Powered ankle-foot prosthesis to assist level-ground and stair-descent gaits," *Neural Networks*, vol. 21, 4, pp. 654–666, 2008.

[52] F. Sup, H. A. Varol, and M. Goldfarb, "Upslope walking with a powered knee and ankle prosthesis: Initial results with an amputee subject," *IEEE Transactions on Neural Systems and Rehabilitation Engineering*, vol. 19, 1, pp. 71–78, 2011.

[53] L. Flynn, J. Geeroms, R. Jimenez-Fabian, B. Vanderborght, N. Vitiello, and D. Lefeber, "Ankle-knee prosthesis with active ankle and energy transfer: Development of the CYBERLEGs alpha-prosthesis," *Robotics and Autonomous Systems*, vol. 73, pp. 4–15.

[54] J. Zhu, Q. Wang, and L. Wang, "PANTOE 1: Biomechanical design of powered ankle-foot prosthesis with compliant joints and segmented foot," presented at the *2010 IEEE/ASME International Conference on Advanced Intelligent Mechatronics (AIM)*, pp. 31–36, 2010.

[55] R. Versluys, P. Beyl, M. Van Damme, A. Desomer, R. Van Ham, and D. Lefeber, "Prosthetic feet: State-of-the-art review and the importance of mimicking human ankle–foot biomechanics," *Disability and Rehabilitation: Assistive Technology*, vol. 4, 2, pp. 65–75, 2009.

[56] P. Cherelle, A. Matthys, V. Grosu, B. Vanderborght, and D. Lefeber, "The amp-foot 2.0: Mimicking intact ankle behavior with a powered transtibial prosthesis," presented at the *4th IEEE RAS & EMBS International*

Conference on Biomedical Robotics and Biomechatronics (BioRob), pp. 544–549, 2012.
[57] B. J. Bergelin and P. A. Voglewede, "Design of an active ankle-foot prosthesis utilizing a four-bar mechanism," *ASME Journal of Mechanical Design*, vol. 134, 6, p. 061004, 2012.
[58] M. Grimmer and A. Seyfarth, "Stiffness adjustment of a Series Elastic Actuator in an ankle-foot prosthesis for walking and running: The trade-off between energy and peak power optimization," presented at the *IEEE International Conference on Robotics and Automation (ICRA)*, pp. 1439–1444, 2011.
[59] F. Giovacchini, F. Vannetti, M. Fantozzi, *et al.*, "A light-weight active orthosis for hip movement assistance," *Robotics and Autonomous Systems*, vol. 73, pp. 123–134, 2015.
[60] R. D. Bellman, M. A. Holgate, and T. G. Sugar, "SPARKy 3: Design of an active robotic ankle prosthesis with two actuated degrees of freedom using regenerative kinetics," presented at the *2nd IEEE RAS & EMBS International Conference on Biomedical Robotics and Biomechatronics*, pp. 511–516, 2008. BioRob 2008, 2008.
[61] E. M. Ficanha, M. Rastgaar, B. Moridian, and N. Mahmoudian, "Ankle angles during step turn and straight walk: Implications for the design of a steerable ankle-foot prosthetic robot," presented at the *ASME 2013 Dynamic Systems and Control Conference*, 2013.
[62] E. Strickland, "Good-bye, wheelchair," *IEEE Spectrum*, vol. 49, pp. 30–33, 2012.
[63] R. J. Farris, H. Quintero, and M. Goldfarb, "Preliminary evaluation of a powered lower limb orthosis to aid walking in paraplegic individuals," *IEEE Transactions on Neural Systems and Rehabilitation Engineering*, vol. 19, 6, pp. 652–659, 2011.
[64] A. J. del-Ama, A. D. Koutsou, J. C. Moreno, A. de-los-Reyes, A. Gil-Agudo, and J. L. Pons, "Review of hybrid exoskeletons to restore gait following spinal cord injury," *Journal of Rehabilitation Research & Development (JRRD)*, vol. 49, 4, pp. 497–514, 2012.
[65] A. Esquenazi, M. Talaty, A. Packel, and M. Saulino, "The ReWalk powered exoskeleton to restore ambulatory function to individuals with thoracic-level motor-complete spinal cord injury," *American Journal of Physical Medicine & Rehabilitation*, vol. 91, 11, pp. 911–921, 2012.
[66] L. Mertz, "The next generation of exoskeletons: Lighter, cheaper devices are in the works," *IEEE Pulse*, vol. 3, 4, pp. 56–61, 2012.
[67] J. Gancet, M. Ilzkovitz, E. Motard, *et al.*, "MINDWALKER: Going one step further with assistive lower limbs exoskeleton for SCI condition subjects," in *IEEE RAS/EMBS International Conference on Biomedical Robotics and Biomechatronics*, 2012, pp. 1794–1800.
[68] P. D. Neuhaus, J. H. Noorden, T. J. Craig, T. Torres, J. Kirschbaum, and J. E. Pratt, "Design and evaluation of Mina: A robotic orthosis for paraplegics,"

presented at the *IEEE International Conference on Rehabilitation Robotics (ICORR)*, pp. 1–8, 2011.

[69] S. Tanabe, E. Saitoh, S. Hirano, *et al.*, "Design of the Wearable Power-Assist Locomotor (WPAL) for paraplegic gait reconstruction," *Disability and Rehabilitation: Assistive Technology*, vol. 8, 1, pp. 84–91, 2013.

[70] W. Tung, Yi-Wei, M. McKinley, M. V. Pillai, J. Reid, and H. Kazerooni., "Design of a minimally actuated medical exoskeleton with mechanical swing-phase gait generation and sit-stand assistance," presented at the *ASME Dynamic Systems and Control Conference*, pp. V002T28A004–V002T28A004, 2013.

[71] J. Falconer, "Cyberdyne wants to offer robot suit HAL in the U.S.," in *IEEE Spectrum*, ed, Dec 9, 2014. Retrieved from https://spectrum.ieee.org/automaton/robotics/medical-robots/cyberdyne-to-offer-robot-suit-hal-in-the-us.

[72] Vanderbilt University, *Engineers' Exoskeleton Promises a New Degree of Independence for People with Paraplegia*, Oct 30, 2012. Retrieved from https://engineering.vanderbilt.edu/news/2012/engineers-exoskeleton-promises-a-new-degree-of-independence-for-people-with-paraplegia/

[73] F. Sergi, D. Accoto, N. L. Tagliamonte, G. Carpino, and E. Guglielmelli, "A systematic graph-based method for the kinematic synthesis of non-anthropomorphic wearable robots for the lower limbs," *Frontiers of Mechanical Engineering*, vol. 6, 1, pp. 61–70, 2011.

[74] A. T. Asbeck, R. J. Dyer, A. F. Larusson, and C. J. Walsh, "Biologically-inspired soft exosuit," in *IEEE International Conference on Rehabilitation Robotics*, pp. 1–8, 2013.

[75] A. T. Asbeck, K. Schmidt, and C. J. Walsh, "Soft exosuit for hip assistance," *Robotics and Autonomous Systems*, 73, pp. 102–110, 2014.

[76] T. Hayashi, H. Kawamoto, and Y. Sankai, "Control method of robot suit HAL working as operator's muscle using biological and dynamical information," presented at the *IEEE/RSJ International Conference on Intelligent Robots and Systems (IROS 2005)*, pp. 3063–3068, 2005.

[77] A. B. Zoss, H. Kazerooni, and A. Chu., "Biomechanical design of the Berkeley lower extremity exoskeleton (BLEEX)," *IEEE/ASME Transactions on Mechatronics*, vol. 11, 2, pp. 128–138, 2006.

[78] G. Mono. "Building the real Iron Man," *Popular Mechanics*, Apr 9, 2008. Retrieved from https://www.popsci.com/scitech/article/2008-04/building-real-iron-man

[79] (2015). http://www.rb3d.com/en/.

[80] M. Ishii, K. Yamamoto, and K. Hyodo, "Stand-alone wearable power assist suit-development and availability," *Journal of Robotics and Mechatronics*, vol. 17, 5, pp. 575–583, 2005.

[81] T. Ishida, T. Kiyama, K. Osuka, G. Shirogauchi, R. Oya, and H. Fujimoto., "Movement analysis of power-assistive machinery with high strength-amplification," in *IEEE Proceedings of the SICE Annual Conference*, 2010, pp. 2022–2025.

[82] E. Yagi, D. Harada, and M. Kobayashi., "Upper-limb power-assist control for agriculture load lifting," *International Journal of Automation Technology*, vol. 3, 6, pp. 716–722, 2009.
[83] S. Marcheschi, F. Salsedo, M. Fontana, and M. Bergamasco, "Body extender: Whole body exoskeleton for human power augmentation," presented at the *IEEE International Conference on Robotics and Automation (ICRA)*, pp. 611–616, 2011.
[84] Warrior Web Closer to Making Its Performance-Improving Suit a Reality, 2015. Retrieved from http://www.darpa.mil/news-events/2013-08-22.
[85] M. Zhang, T. C. Davies, and S. Xie, "Effectiveness of robot-assisted therapy on ankle rehabilitation – A systematic review," *Journal of NeuroEngineering and Rehabilitation*, vol. 10, 1, p. 30, 2013.
[86] E. L. Miller, L. Murray, L. Richards, *et al.*, "Comprehensive overview of nursing and interdisciplinary rehabilitation care of the stroke patient a scientific statement from the American Heart Association," *Stroke*, vol. 41, 10, pp. 2402–2448, 2010.
[87] Management of Stroke Rehabilitation Working Group. "VA/DOD Clinical practice guideline for the management of stroke rehabilitation," *Journal of Rehabilitation Research and Development*, vol. 47, 9, 2010.
[88] P. W. Duncan, K. J. Sullivan, A. L. Behrman, *et al.*, "Body-weight–supported treadmill rehabilitation after stroke," *New England Journal of Medicine*, vol. 364, 21, pp. 2026–2036, 2011.

Chapter 2

Review of exoskeletons for medical and service applications: ongoing Research in Europe on Wearable Robots, with focus on lower extremity exoskeletons

Jan F. Veneman, Dirk Lefeber** and Nicola Vitiello****

Abstract

This chapter presents an overview of European research projects that are involved with research and development of wearable robots, with an emphasis on exoskeletons for the legs. The structure and goals of public European Research funding are explained, the general directions of the recent research in wearable robots are summarized, and a more detailed overview of a selection of the larger European R&D projects in this field is provided. It covers exoskeletons for assistive use as well as for training, for medical as well as for service applications.

Keywords: Exoskeletons; Wearable Robots; European Research Programs

2.1 Introduction

This chapter aims to present an overview of European research projects that are involved with wearable robots (WRs), especially lower limb exoskeletons. A short introduction will be given on the general scope of the selected projects, as well as on the general background of European research in this area.

2.1.1 What are wearable robots and lower extremity exoskeletons?

WRs can be defined as 'actuated devices that are worn on the body and that mechanically interact with the sensorimotor system of a human user for the purpose of augmentation, assistance, or substitution of human motor functions'. Opposed to more straightforward passive mechanical devices, WRs are programmable and

*Medical Robotics, Health Division, Tecnalia Research and Innovation, Spain
**Robotics & Multibody Mechanics Research Group, Vrije Universiteit Brussel, Belgium
***The BioRobotics Institute, Scuola Superiore Sant'Anna, Italy

possess a degree of autonomy, making them more flexible and more 'intelligent', i.e. they have sensors, actuators and a decision making capability. This definition primarily points to so-called exoskeletons and exosuits – i.e. external WRs for applications such as locomotion or manipulation tasks, and not directly related to body-integrated devices such as robotic prosthetics, although these are sometimes also included in the category of WRs.

Following Hugh Herr's review (2009) [1], robotic exoskeletons can be described as 'mechanical devices that are essentially anthropomorphic in nature, as they are "worn" by a human, fitting closely to the body, and working in concert with intended movements. [...]. "Exoskeleton" within the {robotic} research community is taken to include mechanical structures, as well as associated actuators, visco-elastic components, sensors and control elements'.

Following this basic structure of exoskeletons, Herr distinguishes four types as follows:

1. devices that act in series with a human limb to increase limb length and displacement
2. devices that act in parallel with the human lower limb for load transfer to the ground
3. devices that act in parallel with human joint(s)
4. devices that act in parallel with a human limb for endurance augmentation.

The research projects presented here belong to these different categories, or even combine several.

2.1.2 European research funding structure

European-wide research on WRs and exoskeletons is for the largest part funded by the European Commission, through collaborative consortium projects of up to 5 years duration. The largest of these funding programmes are the EU Seventh Framework Programme for research, technological development and demonstration – FP7 for short which ended in 2013 – as well as the Horizon 2020 (or H2020) Framework Programme for Research and Innovation that followed in 2014. These are among the worlds' largest public funding instruments for research and innovation. Projects under FP7 or H2020 are typically shaped as collaborations among different academic, industrial or public partners from different countries in the European Union to meet specific calls for R&D that have been identified within the European Commission (EC). Next to this, in several European countries, projects on exoskeleton technology were funded through national funding schemes, both as collaborative and single-party projects. Currently the main funding source is H2020, although a number of FP7 projects are still active.

The novelty of Horizon2020 is characterised as: 'By coupling research and innovation, Horizon 2020 is helping to achieve {economic growth and creating jobs} with its emphasis on excellent science, industrial leadership and tackling societal challenges. The goal is to ensure Europe produces world-class science, removes barriers to innovation and makes it easier for the public and private sectors to work together in delivering innovation and making strong impact in society' [2].

For robotics-related research this is coordinated by the SPARC: the Public–Private Partnership for robotics in Europe. 'SPARC aims to make available European robots in factories, in the air, on land, under water, for agriculture, health, rescue services, and in many other applications in Europe which have an economic and societal impact' [3]. Next to the European Public partner, namely the European Commission, SPARC consists of the European Robotics Community, organised as 'EuRobotics AISBL', which represents the private industry and research community. EuRobotics produced and updates a research roadmap that directs the European funding via the so-called Multi-Annual Roadmap (MAR), which is the core document for prioritising the robotic research that should receive funding; WRs and exoskeleton-related applications and technologies are an integral part of it due to societal concerns about the ageing society.[1]

2.2 Review of recent wearable robot and exoskeleton-related research projects inside Europe

2.2.1 General directions

The general aims of the research can be found in the MAR in Section 3.1.4 Wearable Robotics. The development of a new generation of WRs is currently expected through the integration of diverse fundamental technical expertise and by developing cross domain co-operation. These directions of development are currently being explored by a European research coordination effort called COST Action 16116 "Wearable Robots for Augmentation, Assistance or Substitution of Human Motor Functions".[2] Such a new generation of WRs will be characterised by better integration and adjustment to human users, as well as to specific application domains which will allow WRs to become mainstream and thus largely expand Wearable Robotics' societal benefits. The cross-disciplinary knowledge required to develop WRs is not generally integrated nor is there currently sufficient inter-disciplinary communication on their impact on Wearable Robotics. A non-exhaustive list of the main topics is: human biomechanics, science of human sensorimotor control (understanding and modelling), human–robot interaction and interfacing, robot-ethics and philosophy of technology, application-related fields such as clinical medicine or industrial production, modelling of (interaction with) human tissues, mechatronics, control theory, wearable sensor technology, system energetics, system architecture and system integration, soft robotics design, ergonomics and human-centred design, materials and component technology.

A critical challenge in Wearable Robotics is to integrate these topics and domains to provide best practice for developing technical synergies that will be able impact on the market.

[1]More information on SPARC, EURobotics, and the MAR can be found on: https://eu-robotics.net/sparc/.

[2]More information on the COST Action CA16116 can be found on: https://wearablerobots.eu/.

An overview of active or recently finished WR projects covering these general areas are listed alphabetically in table format below. Websites, which are the main channel of dissemination of project results, are also shown and can be consulted for additional information.

Short name	Full name	Application domain	Website
AXO-SUIT	Assistive exoskeleton suitable for elderly persons	Assistive (elderly)	www.axo-suit.eu
BALANCE	Balance augmentation in locomotion, through anticipative, natural and co-operative control of exoskeletons	Cross-domain	www.balance-fp7.eu
BioMot	Smart wearable robots with bioinspired sensory-motor skills	Cross-domain	www.biomotproject.eu
CORBYS	Cognitive control framework for robotic systems	Neuro-rehabilitation	www.corbys.eu
CYBERLEGs	The CYBERnetic lower-limb cognitive ortho-prosthesis	Assistive (amputees)	www.cyberlegs.eu
EXO-LEGS	Exoskeleton legs for elderly persons	Assistive (elderly)	www.exo-legs.org
Mindwalker	Mind controlled orthosis and VR training environment for walk empowering	Assistive [spinal cord injury (SCI) patients]	mindwalker-project.eu
Robo-Mate	Modular Exoskeleton for Industry	Worker support in Industrial environments	www.robo-mate.eu
Spexor	Spinal exoskeletal robot for low back pain prevention and vocational reintegration	Preventive (healthy, industry) Rehabilitation (back pain)	www.spexor.eu
Symbitron	Symbiotic man-machine interactions in wearable exoskeletons to enhance mobility for paraplegics	Assistive (SCI patients)	www.symbitron.eu
XO-soft	Soft modular biomimetic exoskeleton to assist people with mobility impairments	Assistive (people with walking impairments)	www.xosoft.eu

For each project, a short overview will follow.

Assistive exoskeleton suitable for elderly persons	**AXO-SUIT**
Duration	1 September 2014–1 September 2017
Budget	2.79 M€ (1.64 M€ EU contribution)

(Continues)

(*Continued*)

Assistive exoskeleton suitable for elderly persons				**AXO-SUIT**
Funding programme	Active and Assisted Living (AAL programme)			
Website	www.axo-suit.eu www.aal-europe.eu/projects/axo-suit/			
Coordinator	Shaoping Bai, Associate Professor Dept. of Mechanical and Manufacturing Engineering Aalborg University, Denmark E-mail: shb@m-tech.aau.dk			
Participating institutes	**Aalborg University**	R&D	Denmark	http://www.aau.dk
	University of Gävle	R&D	Sweden	http://www.hig.se/
	University of Limerick	R&D	Ireland	www.ul.ie/
	Welldana A/S	End user	Denmark	http://www.welldana-innocare.com/
	Bioservo Technologies AB[3]	Business	Sweden	http://www.bioservo.se
	MTD Precision Engineering Ltd	Business	Ireland	http://www.tooling.ie/
	Hjälpmedelsteknik Sverige	End user	Sweden	www.hjalpmedelsteknik.se
	COMmeto bvba	Business	Belgium	http://www.commeto.be/
Project goals	The AXO-SUIT is meant to comprehensively supplement the strength of elderly persons with feasible exoskeletons in undertaking volunteer work, which will be achieved through six work packages: WP1 (end users) to get close involvement of the end-users throughout the project to determine the requirements, and final testing to determine user satisfaction, WP2 (lower body exoskeleton) to maintain elders mobility, WP3 (upper body exoskeleton) to supplement their physical abilities of holding, grasping, pushing or pulling involved for performing light-duty jobs, WP4 (system integration) to integrate all systems and test them in labs, WP5 (commercialisation) to develop and test potential AXO-SUIT products in the targeted countries (Belgium, Denmark and Sweden), Europe and beyond, and to develop business plans and create opportunities for further products, WP6 (project management) for overall work plan management and administration, finance, reporting, quality assurance, etc.			
Results	AXO-SUIT integrates recent advances in assistive technology to study and design exoskeletons and to meet the challenges in helping elderly workers. The results will include novel exoskeletons consisting of modules to allow integration to realise prototype upper, lower and full-body assistive suits. They will comprehensively supplement the strength of elderly persons with effective and affordable exoskeletons and improve directly their quality of life. The exoskeletons could also be extended to more aged persons, weak or disabled adults, or elder employees as their needs are similar.			

(Continues)

[3]Involvement is not funded from AAL Programme.

(Continued)

Assistive exoskeleton suitable for elderly persons	**AXO-SUIT**
	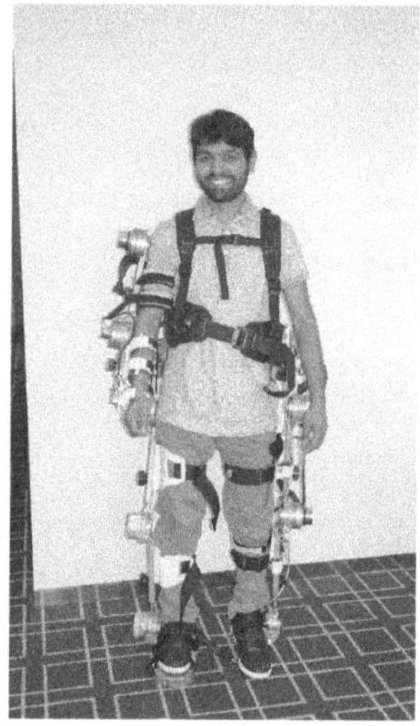
Publications	Christensen, S., & Bai, S. (2017). A Novel Shoulder Mechanism with a Double Parallelogram Linkage for Upper-Body Exoskeletons. In *Wearable Robotics: Challenges and Trends* (pp. 51–56). Springer International Publishing. O'Sullivan, L., Power, V., Virk, G., Masud, N., Haider, U., Christensen, S., & Vonck, K. (2015). End user needs elicitation for a full-body exoskeleton to assist the elderly. *Procedia Manufacturing*, *3*, 1403–1409.

Balance augmentation in locomotion, through anticipative, natural and co-operative control of exoskeletons	**BALANCE**
Duration	1 January 2013–31 July 2017
Budget	5.9 M€ (4.5 M€ EU contribution)
Funding programme	Seventh Framework Programme; ICT-2011.2.1 – Cognitive Systems and Robotics
Website	www.balance-fp7.eu cordis.europa.eu/project/rcn/106854_en
Coordinator	Jan F. Veneman Medical Robotics Area, Health Division Tecnalia Research and Innovation E-mail: jan.veneman@tecnalia.com or balance@tecnalia.com

(Continues)

(Continued)

Balance augmentation in locomotion, through anticipative, natural and co-operative control of exoskeletons				**BALANCE**
Participating institutes	**Tecnalia R&I**	RTO	Spain	www.tecnalia.com/en/health/
	Technische Universitat Darmstadt	Academic	Germany	lauflabor.ifs-tud.de/doku.php
	Eidgenoessische Technische Hochschule Zuerich	Academic	Switzerland	www.adrl.ethz.ch/doku.php
	Commissariat A L'Energie Atomique Et Aux Energies Alternatives	RTO	France	www-list.cea.fr/en/technological-research/research-programmes/advanced-manufacturing/collaborative-robotics
	Xsens Technologies B.V.	SME	Netherlands	www.xsens.com
	Universiteit Twente	Academic	Netherlands	www.utwente.nl/en/et/bw/
	Univerzitetni Rehabilitacijski Institut Republike Slovenije-Soca	Academic/end-user	Slovenia	www.ir-rs.si/en/
	Imperial College of Science Technology and Medicine	Academic	UK	www.imperial.ac.uk/human-robotics
Project goals	The goal of the BALANCE project is to realise an exoskeletal robot that improves the balance performance of humans, targeted at users facing balance-challenging conditions or suffering from a lack of ability to walk or maintain balance during walking. The proposed exoskeleton is human-co-operative in the sense that the control of the exoskeleton is complementary to the remaining human control. Depending on application it can either assist only in difficult conditions or in case of erroneous behaviour of the user, or can assist the user maximally. The ultimate goal is to have the exoskeleton seamlessly co-operate with the human. The exploitation of the results will focus on applications in neurorehabilitation and worker support.			
Results	In the fourth year of the project all major components of the project have been put in place. An exoskeleton, called EMY, has been engineered by the CEA-List, containing all specifications to be able to support postural balance, by passive and active degrees of freedom, and power of actuated degrees of freedom. An IMU-based motion capturing suit (XSENS) was adapted to move in highly magnetic environment, and algorithms where implemented to detect a loss of balance and to control supportive motion of the exoskeleton towards a stabilising foot placement. In parallel a large body of experimental work has been carried out studying stepping responses of healthy and impaired subjects following perturbations.			

(Continues)

(*Continued*)

Balance augmentation in locomotion, through anticipative, natural and co-operative control of exoskeletons	**BALANCE**

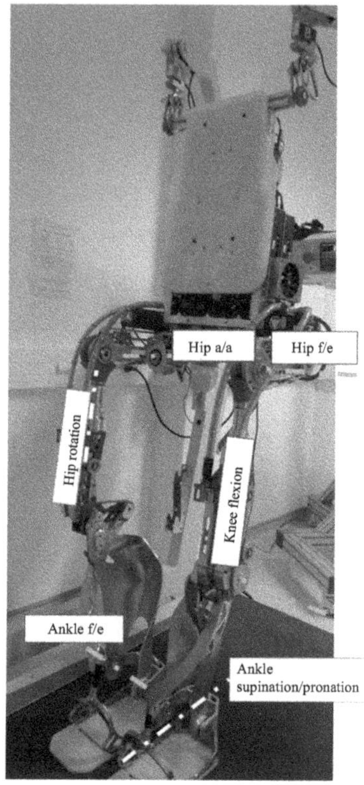

Publications

Olenšek, A., Zadravec, M., & Matjačić, Z. (2016). A novel robot for imposing perturbations during overground walking: Mechanism, control and normative stepping responses. *Journal of Neuroengineering and Rehabilitation*, *13*(1), 55.

Torricelli, D., Gonzalez-Vargas, J., Veneman, J. F., Mombaur, K., Tsagarakis, N., del-Ama, A. J., . . . , & Pons, J. L. (2015). Benchmarking bipedal locomotion: A unified scheme for humanoids, wearable robots, and humans. *IEEE Robotics & Automation Magazine*, *22*(3), 103–115.

Vlutters, M., Van Asseldonk, E. H. F., & Van der Kooij, H. (2016). Center of mass velocity-based predictions in balance recovery following pelvis perturbations during human walking. *Journal of Experimental Biology*, *219* (10), 1514–1523.

Smart wearable robots with bioinspired sensory-motor skills				BIOMOT
Duration	1 October 2013–30 September 2016			
Budget	3.87 M€ (2.68 M€ EU contribution)			
Funding programme	Seventh Framework Programme; ICT-2013.2.1 – Robotics, Cognitive Systems & Smart Spaces, Symbiotic Interaction			
Website	biomotproject.eu cordis.europa.eu/project/rcn/109702_en			
Coordinator	Dr. Juan Moreno Head of Human Locomotion Laboratory Spanish National Research Council Cajal Institute – Neural Rehabilitation Group E-mail: jc.moreno@csic.es			
Participating institutes	**Agencia del Consejo Superior de Investigaciones Científicas**	RTO	Spain	www.csic.es/web/guest/home
	Vrije Universiteit Brussel	Academic	Belgium	mech.vub.ac.be/multibody_mechanics.htm
	Università degli Studi di Padova	Academic	Italy	www.unipd.it/en/
	Össur hf.	Industry	Iceland	www.ossur.com/
	Fundación del Hospital Nacional de Parapléjicos	End-user	Spain	www.infomedula.org/index.php?
	RIKEN	Academic	Japan	www.riken.jp/en/
	Universidad Miguel Hernández de Elche	Academic	Spain	en.umh.es/
	Technaid S.L.	SME	Spain	www.technaid.com/
Project goals	The main objective of BioMot is to improve existing wearable robotic exoskeletons exploiting dynamic sensori-motor interactions and developing cognitive capabilities that can lead to symbiotic gait behaviour in the interaction of a human with a wearable robot. BioMot proposes a cognitive architecture for WRs exploiting neuronal control and learning mechanisms the main goal of which is to enable positive co-adaptation and seamless interaction with humans			
Results	A novel 6DoF bioinspired wearable exoskeleton for gait support and compensation has been produced. The exoskeleton includes a new generation of compliant actuators that can be interfaced with the human user through a computational neuromusculoskeletal model of human–robot interaction. A ROS-based framework has been designed and implemented in the BioMot exoskeleton to investigate the optimisation of human–robot interaction. This multilevel architecture allows monitoring human–robot status and performs estimations of human cognitive attention and biomechanical performance. On the one hand, a personalised biomechanical model is developed that provides online instantaneous joint stiffness estimation for prescribing robotic actuation and improving interaction. On the other hand, a tacit adaptability controller provides automatic adaptation of gait support as a function of human cognitive attention on the task. BioMot is tested with healthy and neurologically injured patients (incomplete spinal cord injured) demonstrating a hierarchical biomimetic controller that can provide flexibility to variations in given environmental constraints and also more seamless integration of volitional commands for self-adjustments of robotic support.			

(Continues)

(*Continued*)

Smart wearable robots with bioinspired sensory-motor skills	**BIOMOT**

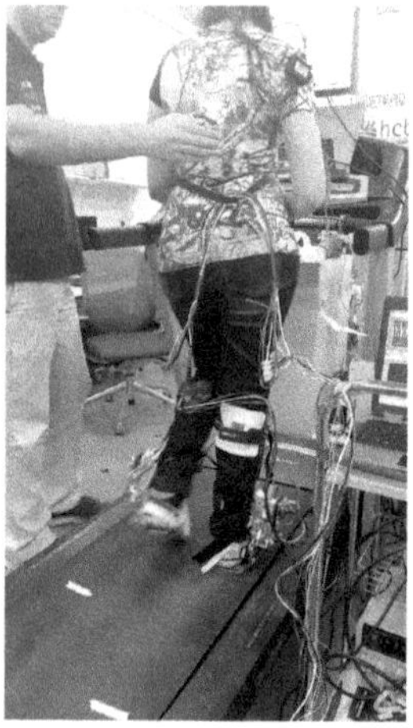

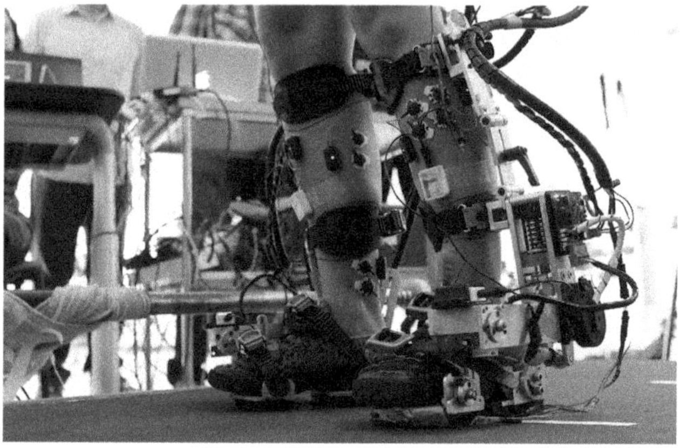

Publications

Bravo-Esteban, E., Taylor, J., Aleixandre, M., Simon-Martínez, C., Torricelli, D., Pons, J. L., & Gómez-Soriano, J. (2014). Tibialis anterior muscle coherence during controlled voluntary activation in patients with spinal cord injury: diagnostic potential for muscle strength, gait and spasticity. *Journal of Neuroengineering and Rehabilitation, 11*(1), 23.

Huo, W., Mohammed, S., Moreno, J. C., & Amirat, Y. (2016). Lower limb wearable robots for assistance and rehabilitation: A state of the art. *IEEE Systems Journal, 10*(3), 1068–1081.

Salazar-Varas, R., Costa, Á., Iáñez, E., Úbeda, A., Hortal, E., & Azorín, J. M. (2015). Analyzing EEG signals to detect unexpected obstacles during walking. *Journal of Neuroengineering and Rehabilitation, 12*(1), 101.

Torricelli, D., Gonzalez-Vargas, J., Veneman, J. F., Mombaur, K., Tsagarakis, N., del-Ama, A. J., . . . , & Pons, J. L. (2015). Benchmarking bipedal locomotion: A unified scheme for humanoids, wearable robots, and humans. *IEEE Robotics & Automation Magazine, 22*(3), 103–115.

Xu, R., Jiang, N., Mrachacz-Kersting, N., Lin, C., Prieto, G. A., Moreno, J. C., . . . , & Farina, D. (2014). A closed-loop brain–computer interface triggering an active ankle–foot orthosis for inducing cortical neural plasticity. *IEEE Transactions on Biomedical Engineering, 61*(7), 2092–2101.

Cognitive control framework for robotic systems				**CORBYS**
Duration	1 February 2011–31 January 2015			
Budget	8.76 M€ (6.10 M€ EU contribution)			
Funding programme	Seventh Framework Program, Area: Cognitive Systems and Robotics			
Website	www.corbys.eu cordis.europa.eu/project/rcn/97393_en			
Coordinator	Danijela Ristic-Durrant Institut für Automatisierungstechnik (IAT)/Institute of Automation (IAT) Universität Bremen/University of Bremen Bremen, Germany E-mail: ristic@iat.uni-bremen.de			
Participating institutes	**University of Bremen**	Academic	Germany	www.iat.uni-bremen.de
	University of Reading	Academic	United Kingdom	www.imss.reading.ac.uk
	University Rehabilitation Institute	Academic/ End-user	Slovenia	www.ir-rs.si
	University of Hertfordshire	Academic	United Kingdom	adapsys.cs.herts.ac.uk
	Vrije University Brussels	Academic	Belgium	mech.vub.ac.be/multibody_mechanics.htm
	SINTEF	SME	Norway	www.sintef.no/home
	Otto Bock Mobility Solutions	Industry	Germany	www.ottobock.de
	Neurological rehabilitation Centre Friedehorst	End-User	Germany	www.friedehorst.de/nrz
	BitBrain Technologies SL	SME	Spain	www.bitbrain.es
	SCHUNK	Industry	Germany	www.schunk.com
	Otto Bock Health-care	Industry	Germany	www.ottobock.de
Project goals	CORBYS aimed at design, development and validation of an integrated cognitive robot control architecture to support robot-human co-working with high-level cognitive capabilities such as situation-awareness, learning, reasoning, anticipation and decision-making. Five major objectives were targeted, namely: • Sensing systems for assessing dynamic robot environments including humans • Self-awareness as a basis for adaptation of robot behaviour • Anticipation in the context of human–robot synergy • Cognitive robot control • Cognitive mobile robot-assisted gait rehabilitation			
Results	The CORBYS architecture was developed as generic control architecture for a wide range of applications where robots work in synergy with humans. As a practical application of the control architecture, a novel robot-assisted gait rehabilitation system was developed during the project lifetime. It consists of an omnidirectional mobile platform, a powered robotic orthosis attached to the platform via a pelvis link, and a linear unit. During the project life time two operating modes were made functional, namely the Learning and Corrective modes. In both operating modes, the patient walks on a treadmill while wearing the powered orthosis. The CORBYS system is the first system which combines all the necessary mechanical degrees of			

(Continues)

(*Continued*)

Cognitive control framework for robotic systems	**CORBYS**
	freedom, needed to enable natural-like walking. Due to cognitive control, CORBYS demonstrator is a situation-aware system capable of learning and reasoning, that enables it to optimally match the requirements of the patient at different stages of rehabilitation in a wide range of gait disorders. The functionality of the integrated system was convincingly demonstrated at the University Rehabilitation Institute in Ljubljana, Slovenia.

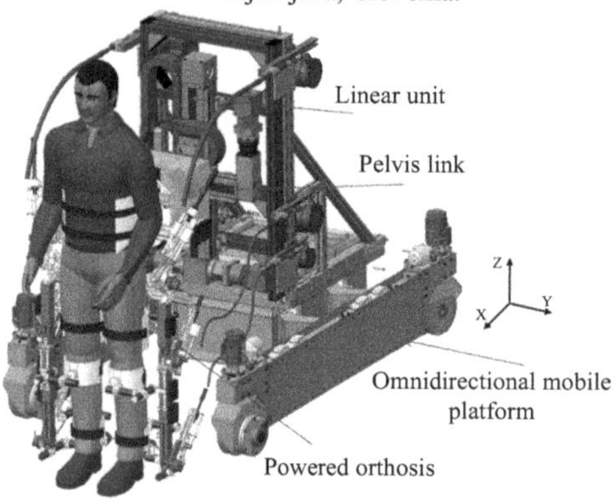

Publications	Grosu, S., Cherelle, P., Verheul, C., Vanderborght, B., & Lefeber, D. (2014). Case Study on Human Walking during Wearing a Powered Prosthetic Device: Effectiveness of the System "Human–Robot". *Advances in Mechanical Engineering*, 6, 365265. Pavčič, J., Matjačić, Z., & Olenšek, A. (2014). Kinematics of turning during walking over ground and on a rotating treadmill. *Journal of Neuroengineering and Rehabilitation*, *11*(1), 127. Sburlea, A. I., Montesano, L., de la Cuerda, R. C., Diego, I. M. A., Miangolarra-Page, J. C., & Minguez, J. (2015). Detecting intention to walk in stroke patients from pre-movement EEG correlates. *Journal of Neuroengineering and Rehabilitation, 12*(1), 113.

The CYBERnetic LowEr-Limb CoGnitive Ortho-prosthesis (and Cyberlegs plus plus)	**CYBERLEGS**
Duration	1 February 2012–29 January 2015 Cyberlegs ++: 1 January 2017–31 December 2012
Budget	3.5 M€ (2.5 M€ of EU contribution) 4.6 M€ (4.3 M€ of EU contribution) (plus plus)
Funding programme	Seventh Framework Programme; ICT-2011.2.1 – Cognitive Systems and Robotics H2020 ICT-25-2016-2017 – Advanced robot capabilities research and take-up
Website	www.cyberlegs.eu cordis.europa.eu/project/rcn/101213_en cordis.europa.eu/project/rcn/206351_en (plus plus)

(*Continues*)

(*Continued*)

The CYBERnetic LowEr-Limb CoGnitive Ortho-prosthesis (and Cyberlegs plus plus)				**CYBERLEGS**
Coordinator	Nicola Vitiello Head Wearable Robotics Laboratory The BioRobotics Institute, Scuola Superiore Sant'Anna Pisa, Italy E-mail: nicola.vitiello@santannapisa.it			
Participating institutes	**Scuola Superiore Sant'Anna**	Academia	Italy	sssa.biorobotics institute.it
	Université Catholique de Louvain	Academia	Belgium	www.uclouvain.be
	Vrije Universiteit Brussel	Academia	Belgium	mech.vub.ac.be/multibody_mechanics.htm
	Univerza v Ljubljani	Academia	Slovenia	www.robolab.si
	Fondazione Don Carlo Gnocchi	End User	Italy	www.dongnocchi.it
Project goals	Lower limb loss is a disabling condition affecting health and quality of life, particularly in older persons. There are multiple pathways leading to lower limb loss, including peripheral artery disease along with coronary artery and cerebrovascular diseases, both on a purely atherosclerotic basis and associated with diabetes, trauma, malignancy and congenital diseases (Ephraim et al., 2003) [4]. Noticeably, the incidence of vascular diseases increases with the expected life span of society: the proportion of individuals suffering from vascular diseases increases with advancing age. This aspect is a critical point if we consider that *the demographic shift of the European population is a major challenge*. Although all amputations lead to a disabling condition, thigh-level amputation (namely *transfemoral amputation*) is clearly the most challenging amputation level (about 20% of the total number of amputations). Persons living with transfemoral limb loss face distinct challenges: they need more physical and cognitive effort to perform any locomotion-related task and their locomotion is less stable. *Energetic, cognitive and stability challenges are not fully overcome by any artificial passive or active transfemoral prosthesis in the current state of the art.* The scientific-and-technological global goal of the CYBERLEGs project was therefore the development and early validation of *artificial cognitive technologies for dysvascular trans-femoral amputees' lower limb functional replacement and assistance in activities of daily living.* In particular the main objectives of the CYBERLEGs project, successfully addressed during the 3 years, were the development of a modular robotic system constituted of			

(*Continues*)

(Continued)

The CYBERnetic LowEr-Limb CoGnitive Ortho-prosthesis (and Cyberlegs plus plus)	**CYBERLEGS**
	an active cognitive artificial leg for the functional replacement of the amputated limb, and of wearable active orthoses for assisting the movement of both the hip joints and the knee and ankle of the contralateral sound limb. The CYBERLEGs project also addressed the development of a bi-directional human–robot interface and a cognitive control unit which could be capable of: (i) decoding the intended movement of the amputee from a multi-sensory fusion algorithm, and translating it into coherent motor commands for the robotic artefacts, (ii) mitigating the risk of fall and (iii) providing the amputee with an augmenting feedback, helping him or her to recover a more physiological and less cognitively demanding gait pattern
Results	The CYBERLEGs project started on 1 February 2012 and ended on 31 January 2015. In the 3-year framework, the CYBERLEGs project developed and successfully carried out a pre-clinical experimentation of a new set of wearable robotic technologies helping transfemoral amputees to regain a more efficient locomotion. In particular, the project invented a novel active pelvis orthosis and a novel knee–ankle–foot orthosis for movement assistance, and a novel powered transfemoral prosthesis controlled by means of a wearable sensory apparatus, the latter being used to decode the intended movement of the amputee. Finally, the project pioneered: (i) the combined use of a powered prosthesis and orthosis (named ortho-prosthesis) as a new means for helping the amputee during the post-amputation rehabilitation process, (ii) the use of vibrotactile feedback to help the amputee to recover a more physiological gait pattern and (iii) the use of wearable devices to mitigate the risk of fall.

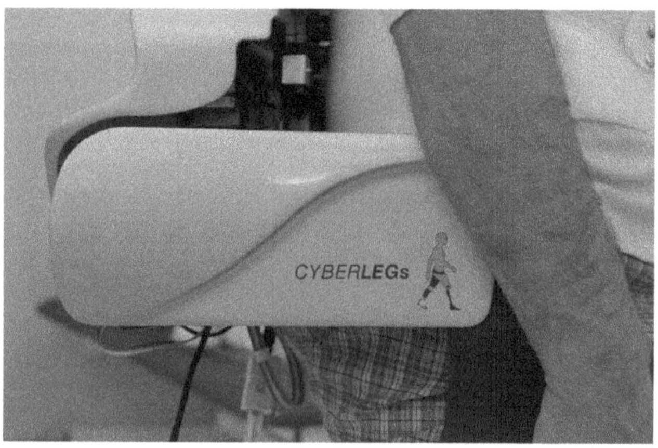

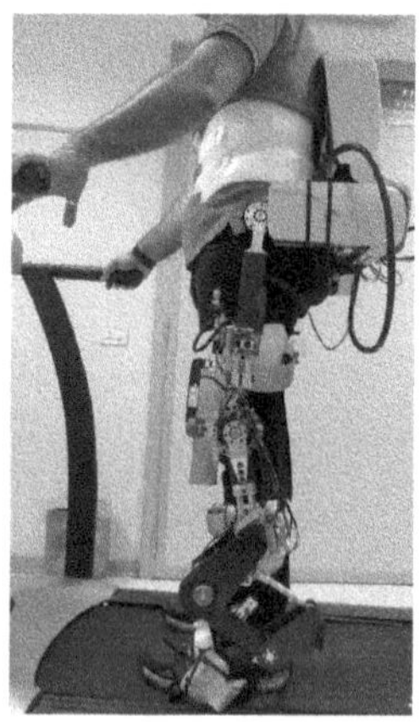

(Continues)

(*Continued*)

The CYBERnetic LowEr-Limb CoGnitive Ortho-prosthesis (and Cyberlegs plus plus)	**CYBERLEGS**
CYBERLEGs Plus Plus (CLs++)	On 1 January 2017 the new H2020 project CYBERLEGs Plus Plus started. In the next 4 years, this new project will validate the technical and economic viability of the powered robotic ortho-prosthesis modules developed within the framework of the FP7-ICT-CYBERLEGs project as a means to enhance/restore the mobility of transfemoral amputees and to enable them to perform locomotion tasks such as ground-level walking, walking up and down slopes, climbing/descending stairs, standing up, sitting down and turning in scenarios of real life. Restored mobility will allow amputees to perform physical activity thus counteracting physical decline and improving the overall health status and quality of life. By demonstrating in an operational environment (TRL = 7) – from both the technical and economic viability view point – a modular robotics technology for healthcare, with the ultimate goal of fostering its market exploitation, CYBERLEGs Plus Pus will have an impact on: • Society: CLs++ technology will contribute to increase the mobility of dysvascular amputees, and, more generally, of disabled persons with mild lower limb impairments; • Science and technology: CLs++ will further advance the hardware and software modules of the ortho-prosthesis developed within the FP7 CYBERLEGs project and validate its efficacy through a multi-centre clinical study; • Market: CLs++ will foster the market exploitation of high-tech robotic systems and thus will promote the growth of both a robotics SME and a large healthcare company. The CLs++ consortium is composed by all partners of the FP7 project and two additional industrial partners, namely the medical company Össur and the robotics spin-off company IUVO Srl
Publications	The CYBERLEGs project had a significant impact on the scientific community of 'wearable robotics'. Over the 3 years, the CYBERLEGs consortium published 29 ISI/Scopus journal papers and 28 papers in conference proceedings with peer-review, and filed five-patent applications. In addition, the consortium promoted the organisation of four international workshops, and three special issues on ISI journals (two special issues were on the wearable robotics, and one special issue was on the use of motor primitives for articulated robots, in the prestigious IEEE Robotics and Automation Magazine). More information about the dissemination of the CYBERLEGs results are available on the official project website: www.cyberlegs.eu

				EXO-LEGS
Duration	1 October 2012–30 March 2016			
Budget	4.56 M€ (2.78 M€ EU contribution)			
Funding programme	Active and Assisted Living (AAL programme) call 4			
Website	www.aal-europe.eu/projects/exo-legs			
Coordinator	Professor Gurvinder S. Virk Affiliation during project: Robotics and the Built Environment Faculty of Engineering and Sustainable Development University of Gävle, Sweden E-mail: gurvinder.virk@hig.se or now: gurvinder.virk@innotecuk.com			
Participating institutes	**University of Gävle**	University (End user)	Sweden	www.hig.se
	Karlsruhe Institute of Technology	University	Germany	rob.ipr.kit.edu
	Universidad Politécnica de Cartagena	University	Spain	www.upct.es
	Chas Blatchford & Sons Limited	Industry SME (End user)	UK	www.blatchford.co.uk
	Hocoma AG	Industry SME (End user)	Switzerland	www.hocoma.com
	Gigatronik Technologies GmbH	Industry SME	Germany	www.gigatronik.de
	MRK Systeme GmbH	Industry SME	Germany	www.MRK-Systeme.de
	Proyecto Control Montaje S.L.	Industry SME	Spain	www.pcmsl.com
	Gävle kommun	Municipality End-User	Sweden	www.gavle.se
Project goals	1. Design Basic, Standard and Deluxe mobility functionality exoskeleton prototypes 2. Develop, test and validate Basic exoskeleton prototype in research labs followed by validation via end users in Sweden, Germany, Spain, Switzerland and UK 3. Develop business models for potential market for commercialisation of the Basic exoskeleton in five target countries initially and then expanding to Europe and beyond. Basic exoskeleton will be an assistive device for supplementing up to 30% missing power to the elderly for daily motions such that it remains a low-risk non-medical device in compliance with ISO 13482			
Results	EXO-LEGS brought together end-users, companies and research organisations to identify key AAL mobility functionalities of normal daily living, and the help elderly persons need to enable active and independent living for as long as possible. The real-world mobility requirements were obtained via 118 members of an end-user group set up during the project. The top three AAL motions were studied and a modular framework			

(Continues)

(Continued)

EXO-LEGS

realised to design, build and test the basic exoskeleton prototype. The detailed work was organised in six work packages:

1. WP1 (end users and scenarios): Setting up and managing the stakeholder with ethical approval so the mobility requirements could be determined and the resulting exoskeletons tested
2. WP2 (exoskeleton components): Theoretical studies of the main motions, realisation of the main components comprising mechanical parts, sensors & controls, user interfaces, materials, etc.
3. WP3 (system integration, testing and validation): Configuration of the components developed to realise the Basic exoskeletons for end user testing as well as more advanced motions needed for the standard and deluxe exoskeletons
4. WP4 (pilot test beds): User testing carried out by end users in the five target countries
5. WP5 (commercialisation): Development of three business plans for medical (neuro & orthopaedic and rehabilitation) and non-medical (consumer) sectors, and dissemination of the project results and
6. WP6 (project management): Overall work plan management and administration, finance, reporting, quality assurance, etc.

The project concluded with two workshops held in Gävle on 17 March 2016, one in English and the other in Swedish; in total 180 persons participated and there was huge media attention.

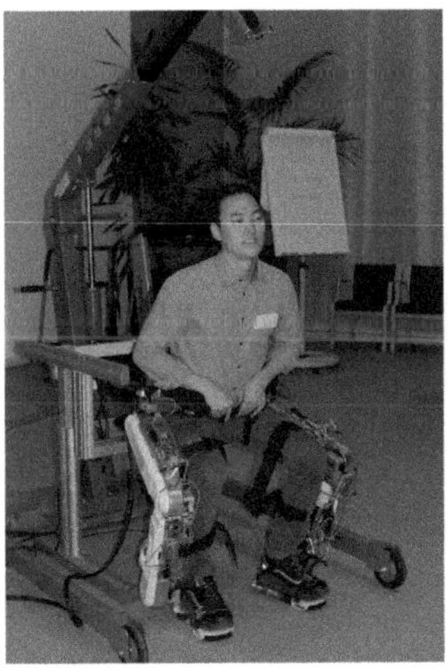
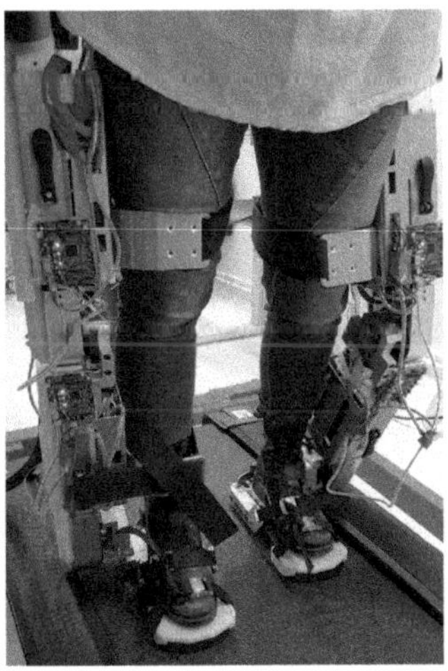

Publications

Haider, U., Nyoman, I. I., Coronado, J. L., Kim, C., & Virk, G. S. (2016). User-centric harmonized control for single joint assistive exoskeletons. *International Journal of Advanced Robotic Systems, 13*(3), 115.

Virk, G. S., Haider, U., & Nyoman, I. (2015). A modular universal joint and harmonised control method for an assistive exoskeleton, PCT/GB2015/050814, 19 March 2015.

(Continues)

(*Continued*)

	EXO-LEGS
	Virk, G. S., Haider, U., Nyoman, I., Krishnan, R., & Masud, N. (2014). EXO-LEGS for elderly persons. In *Proceeding International Conference CLAWAR'14*, pp. 85–92, Poznan, Poland, 21–23 July 2014. Virk, G. S., Haider, U., Nyoman, I., Masud, N., Mamaev, I., Hopfgarten, P., & Hein, B. (2015). Design of EXO-LEGS Exoskeletons. In *Proceedings CLAWAR'15*, Hangzhou, China, pp. 59–66, 6–9 Sept 2015. Winner of Application Innovation Award by Maxon Motors (Suzhou) Co. Ltd.

Mind controlled orthosis and VR training environment for walk empowering				**MINDWALKER**
Duration	1 September 2009–31 May 2013			
Budget	3.66 M€ (2.75 M€ EU contribution)			
Funding programme	EU Seventh Framework Programme; ICT-2009.7.2 – Accessible and Assistive ICT			
Website	mindwalker-project.eu cordis.europa.eu/project/rcn/93837_en.html			
Coordinator	Michel Ilzkovitz Space Applications Services Zaventem, Belgium E-mail: michel.ilzkovitz@spaceapplications.com			
Participating institutes	**Space Applications Services**	SME	Belgium	www.spaceapplications.com
	Université Libre De Bruxelles	University	Belgium	www.cheron.be
	Santa Lucia Foundation	Clinic End-user	Italy	www.hsantalucia.it
	University Of Twente	University	Netherlands	www.universiteittwente.nl
	Technical University Of Delft	University	Netherlands	www.tudelft.nl
	Eemagine Medical Imaging Solutions GmbH	SME	Germany	www.eemagine.com
	Össur	Industry	Iceland	www.ossur.com
Project goals	A lack of mobility often leads to limited participation in social life. The purpose of this project is to conceive a system empowering lower limbs disabled people with walking abilities that let them perform their usual daily activities in the most autonomous and natural manner.			

(Continues)

(Continued)

Mind controlled orthosis and VR training environment for walk empowering	**MINDWALKER**
	The project addresses three main different fields of expertise: • BCI technologies • Virtual reality • Exoskeleton mechatronics and control The project top level objective is to combine these expertises to develop an integrated MINDWALKER system. In addition, the system shall undergo a clinical evaluation process
Results	A torque controlled exoskeleton was developed with efficient series elastic actuators, which was operated with Ethercat. This hardware platform served as a research platform to implement and evaluate different human machine interfaces. Steps could be triggered by: a brain computer interface that used high-density EEG, muscle activity from the shoulder muscles, or the centre of mass location. These different interfaces were evaluated in group of healthy test subjects. From these experiments the interface using the centre of mass estimate was selected for an evaluation with spinal cord injured subjects.

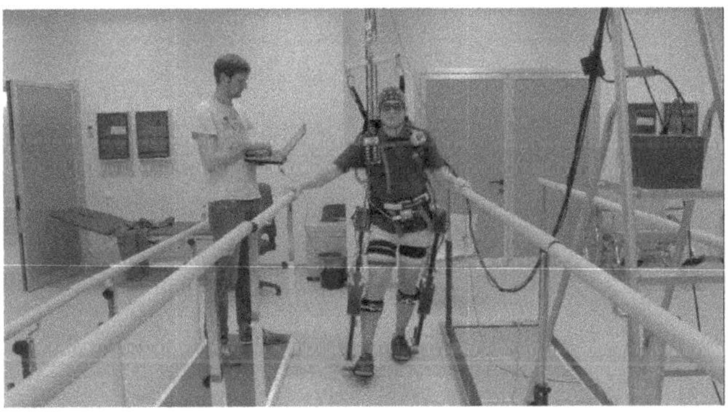

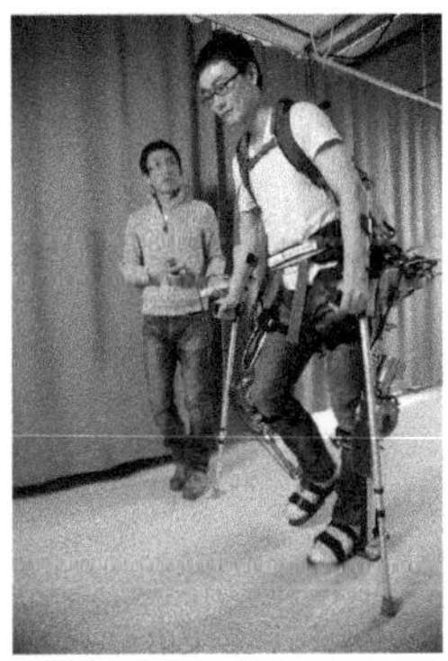

Publications	S. Wang, L. Wang, C. Meijneke, E. van Asseldonk, T. Hoellinger, G. Cheron, ... H. van der Kooij (2015). Design and control of the MINDWALKER exoskeleton, *IEEE Trans Neural Syst Rehabil Eng*, vol. 23, no. 2, pp. 277–286. F. Sylos Labini, V. La Scaleia, A. d'Avella, I. Pisotta, F. Tamburella, G. Scivoletto, ... Y. P. Ivanenko. (2014). EMG patterns during assisted walking in the exoskeleton. *Front Hum. Neurosci.*, vol. 8, pp. 1–12. S. Wang, C. Meijneke, & H. van der Kooij. (2013). Modeling, design, and optimization of Mindwalker series elastic joint. *IEEE Int. Conf. Rehabil. Robot*, vol. 2013, pp. 1–8.

Spinal exoskeletal robot for low-back pain prevention and vocational reintegration				**SPEXOR**
Duration	1 January 2016–31 December 2019			
Budget	4 M€			
Funding programme	EU Horizon2020; ICT-24-2015 – Robotics			
Website	www.spexor.eu			
Coordinator	Jan Babic Institut Jozef Stefan Ljubljana, Slovenia E-mail: jan.babic@ijs.si			
Participating institutes	**Jožef Stefan Institute**	RTO	Slovenia	www.ijs.si
	Heidelberg University	University	Germany	www.uni-heidelberg.de
	Vrije Universiteit Brussel	University	Belgium	http://mech.vub.ac.be/multibody_mechanics.htm
	VU University Amsterdam	University	Netherlands	www.vu.nl
	Otto Bock Healthcare GmbH	Industry	Germany	www.ottobock.com
	Otto Bock Healthcare Products GmbH	Industry	Austria	www.ottobock.com
	S2P Science To Practice d.o.o.	SME	Slovenia	www.s2p.si
	Heliomare	Clinic End-user	Netherlands	www.heliomare.nl
Project goals	Low-back pain is the leading cause of worker absenteeism after the common cold, accounting for 15% of sick leaves and hundreds of millions of lost work days annually [Walking 2.0, Nature, 2015]. Most of today's robotic assistive devices are in forms of exoskeletons that augment the motion of legs and arms and neglect the role of spinal column in transferring load from the upper body and arms to the legs. In SPEXOR, we will fill this gap and design a novel and revolutionary spinal exoskeleton to prevent low-back pain in able bodied workers and to support workers with low-back pain in vocational rehabilitation. The concept to realise the objectives of SPEXOR is driven by several interdisciplinary ideas that push current understanding of low-back pain intervention through several innovative research and technological stages. First, robot-centred requirements for low-back pain prevention are determined and a musculoskeletal stress monitoring system is developed to unobtrusively measure the associated key variables. Then, optimal design parameters and sensorimotor strategies are provided with respect to the robot-centred requirements and their associated key variables. Based on these aspects, a spinal exoskeleton mechanism, and its actuation is			

(Continues)

(*Continued*)

Spinal exoskeletal robot for low-back pain prevention and vocational reintegration	**SPEXOR**
	developed and adaptive control architecture is employed. Such research and development cycle is enclosed by multi-phase end-user evaluation, usability and satisfaction studies. The project builds upon the partner's extensive experience with work ergonomics, modelling and optimisation of human movement, design, control and evaluation of exoskeletons. Several beyond-the-state-of-art scientific approaches and technologies will be employed through a colourful mixture of research, industrial, SME and end-user partners of the consortium. Ultimately, the results of SPEXOR will have a significant impact well beyond the current scientific understanding and technological capabilities of assistive robots used in daily life and health care.
Results	Project recently started; no results to be reported.

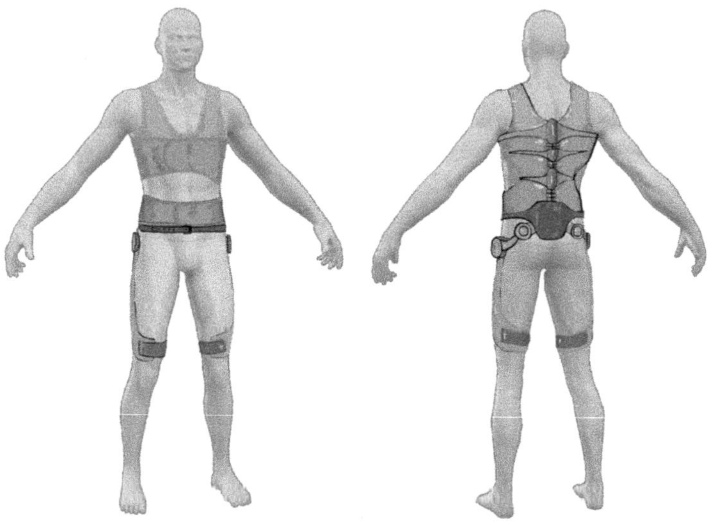

Publications	Babič, J., Oztop, E., & Kawato, M. (2016). Human motor adaptation in whole body motion. *Scientific Reports*, *6*: 32868. Peternel, L., Noda, T., Petrič, T., Ude, A., Morimoto, J., & Babič, J. (2016). Adaptive control of exoskeleton robots for periodic assistive behaviours based on EMG feedback minimisation. *PLoS ONE, 11*(2), e0148942.

Symbiotic man-machine interactions in wearable exoskeletons to enhance mobility for patients with spinal cord injury				**SYMBITRON**
Duration	1 October 2013–30 September 2017 (possible 6-month extension)			
Budget	4.4 M€ (3.1 M€ EU contribution)			
Funding programme	FP7-ICT project (contract PF7-ICT-2013-10, no. 661626) and part of the FET program of the EC			
Website	www.symbitron.eu and www.symbitron.eu/wiki cordis.europa.eu/project/rcn/110314_en			
Coordinator	Herman van der Kooij University of Twente Drienerlolaan 5 7522 NB Enschede, The Netherlands E-mail: H.vanderKooij@utwente.nl			
Participating institutes	**University of Twente**	Academic	Netherlands	www.utwente.nl/ctw/bw
	Ecole Polytechnique Fédérale de Lausanne (EPFL),	Academic	Switzerland	biorob.epfl.ch
	Technical University of Delft	Academic	Netherlands	www.3me.tudelft.nl/en/about-the-faculty/departments/biomechanical-engineering/research/dbl-delft-biorobotics-lab
	Imperial College of Science, Technology and Medicine	Academic	United Kingdom	www3.imperial.ac.uk/humanrobotics
	Fondazione Santa Lucia	End-user	Italy	www.hsantalucia.it/modules.php
	Össur	Industry	Iceland	www.ossur.com
Project goals	1. To develop an integrated neuromuscular model that describes the physiology of healthy versus impaired human gait 2. To design and manufacture personalised modular exoskeletons that compensate for spinal cord injury (SCI) impairments 3. To develop personalised human inspired neuro-muscular controllers for the wearable exoskeletons 4. To optimise the design & control, and bi-directional symbiotic man-machine interaction of wearable exoskeletons 5. To determine the safety and functionality of the personalised SYMBITRON wearable exoskeletons in a clinical study 6. To disseminate key findings to relevant stakeholders and to secure IP protection and exploitation of valuable innovations			
Results	10 Symbitron test pilots with varying levels of SCI were involved during the whole duration of the project in different measurement sessions and testing and training with the exoskeleton prototypes. Patient specific neuro-mechanical models have been constructed, which parameters have			

(Continues)

(*Continued*)

Symbiotic man-machine interactions in wearable exoskeletons to enhance mobility for patients with spinal cord injury — **SYMBITRON**

been derived from impedance and spasticity measurements. Neuromuscular controllers have been developed, which can be tuned for each subject, resulting in controllers that are intuitive, human-like and highly adaptive. Human-like balance control is achieved via momentum-based controllers and bidirectional human–machine interaction is ensured through vibrotactile sensory substitution. Besides subject specific controllers, the Symbitron exoskeleton itself can also easily be adjusted to the subjects' body characteristics. Furthermore, ankle, knee and hip modules can be added when needed by the patients. The resulting exoskeleton is lightweight (5 kg per module) and compact, but powerful (up to 100 N m). By the end of the project the test pilots will perform extensive training with the developed exoskeleton prototypes and the different controllers.

Publications

Dzeladini, F., van den Kieboom, J., & Ijspeert, A. (2014). The contribution of a central pattern generator in a reflex-based neuromuscular model. *Frontiers in Human Neuroscience, 8*, 371. doi: 10.3389/fnhum.2014.00371.

Dzeladini, F., Wu, A. R., Renjewski, D., Arami, A., Burdet, E., van Asseldonk, E., …, & Ijspeert, A. J. (2016, June). Effects of a Neuromuscular Controller on a Powered Ankle Exoskeleton During Human Walking. In *Biomedical Robotics and Biomechatronics (BioRob), 2016 6th IEEE International Conference on* (pp. 617–622). IEEE.

Meijneke, C., Wang, S., Sluiter, V., & van der Kooij, H. (2017). Introducing a Modular, Personalized Exoskeleton for Ankle and Knee Support of Individuals with a Spinal Cord Injury. In *Wearable Robotics: Challenges and Trends* (pp. 169–173). Springer International Publishing.

Soft modular biomimetic exoskeleton to assist people with mobility impairments				**XO-Soft**
Duration	1 February 2016–31 January 2019			
Budget	5.42 M€ (3.68 M€ EU contribution)			
Funding programme	Horizon2020; ICT-24-2015 – Robotics			
Website	www.xosoft.eu cordis.europa.eu/project/rcn/200108_en			
Coordinator	Jesus Ortiz Advanced Robotics Department, Istituto Italiano di Tecnologia, Genova, Italy E-mail: jesus.ortiz@iit.it			
Participating institutes	**Fondazione Instituto Italiano di Tecnologia**	RTO	Italy	www.iit.it
	Consejo Superior de Investigaciones Científicas	RTO	Spain	www.csic.es
	Saxion University of Applied Science,	Academic	Netherlands	www.saxion.nl
	University of Limerick	Academic	Ireland	www.ul.ie
	Zurich University of Applied Sciences	Academic	Switzerland	www.zhaw.ch/de/engineering/institute-zentren/ims www.zhaw.ch/en/health/institutes-centres/institute-of-physiotherapy
	Roessingh Research and Development BV	RTO/end-user	Netherlands	www.rrd.nl/en
	Accelopment AG	Management	Switzerland	accelopment.com
	Geriatrie-Zentrum Erlangen,	End-user	Germany	www.waldkrankenhaus.de/kliniken/fachbereiche/geriatrie-zentrum-erlangen/geriatrie-englisch.html
	Össur hf	Industry	Iceland	www.ossur.com
Project goals	Between 2000 and 2050, the older population (80 years+) is projected to almost qua-druple from approximately 100 million to 395 million people worldwide. Many of the elderly and patient groups such as patients with stroke or with incomplete spinal cord injuries experience varying degrees of mobility impairment. Assistive devices play a pivotal role in their lives and impact on their ability to live independently and perform basic tasks of daily living. There are currently 3.2 million wheelchair users in Europe and further 40 million people who cannot walk without an aid. Yet most assistive devices, such as powered wheel chairs, do not encourage or			

(Continues)

(*Continued*)

Soft modular biomimetic exoskeleton to assist people with mobility impairments	**XO-Soft**
	support the activation of legs. XoSoft, to be developed by a consortium of five research groups and three companies with EU project experience in exoskeleton/assistive orthotics development, is a class I medical device for these and other user groups to answer their need for low-to-moderate mobility assistance
Results	During the first months of the project, we analysed the XoSoft stakeholders and their needs, including demographics of the primary users, identification of the secondary users and the commercial analysis of the tertiary users. This information was of special importance for the definition of the system specifications and the conceptual design of the system. Following a user centred design approach, we developed the Alpha version of the system. This version was composed by different functional modules designed using existing technologies. These modules were tested in laboratory environment following the same testing protocol which will be used throughout the whole duration of the project. The information gathered during the testing of the alpha prototype modules is invaluable, and it is the basis of the design of the first soft prototype of the system.

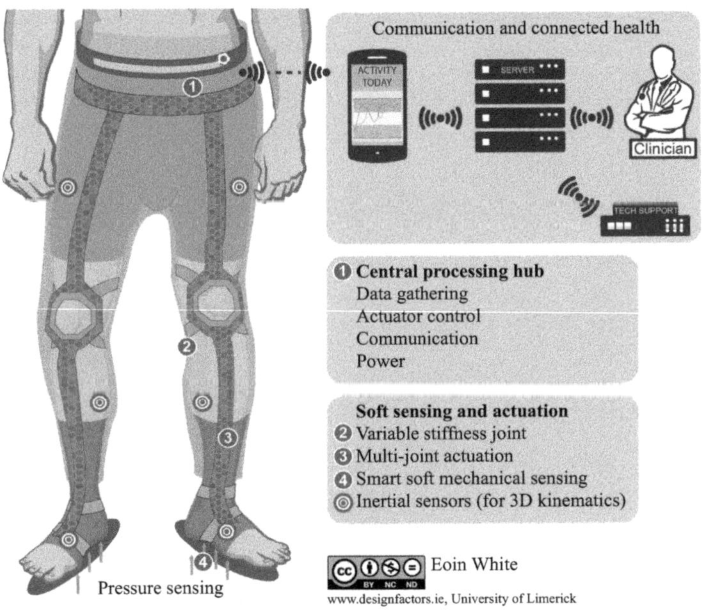

Publications	Ortiz, J., Rocon, E., Power, V., de Eyto, A., O'Sullivan, L., Wirz, M., . . . , & Teeuw, W. B. (2017). XoSoft-A Vision for a Soft Modular Lower Limb Exoskeleton. In *Wearable Robotics: Challenges and Trends* (pp. 83–88). Springer International Publishing.

Acknowledgement

This book chapter is based on work in the COST Action CA16116 "Wearable Robots for Augmentation, Assistance or Substitution of Human Motor Function", supported by COST (European Cooperation in Science and Technology). See also https://wearablerobots.eu.

References

[1] Herr, H. 'Exoskeletons and orthoses: classification, design challenges and future directions', *Journal of neuroengineering and rehabilitation*, 2009, 6(1) pp. 22.
[2] European Commission. What is horizon2020? [online] Available from: http://ec.europa.eu/programmes/horizon2020/en/what-horizon-2020 [Accessed 30/10/2017].
[3] EU-robotics. What is SPARC? [online] Available from: http://eu-robotics.net/sparc/about [Accessed 30/10/2017].
[4] Ephraim, P. L., Dillingham, T. R., Sector, M., Pezzin, L. E., MacKenzie, E. J., 'Epidemiology of limb loss and congenital limb deficiency: a review of the literature'. *Archives of physical medicine and rehabilitation*, 2003 84(5), pp. 747–761.

Chapter 3
Soft wearable robots

Conor J. Walsh[1,2]

Abstract

This chapter focuses on the recent and growing efforts in the field of soft wearable robotics and discusses how this technology can be used in a variety of contexts. This rapidly emerging field will not replace traditional exoskeletons but offers new possibilities to augment the performance of healthy individuals but also restore function for impaired individuals with residual capacity, i.e. where only small to moderate levels of assistance is needed to improve function ability (e.g. walking, grasping). The application requirements for soft wearable robots are fundamentally different than those for rigid exoskeletons, necessitating fundamental technological development in areas of actuation, human interfaces, sensing, control and system integration.

3.1 Introduction

Soft robotics is a rapidly growing research field that combines robotics and soft materials such as elastomers and textiles. These soft systems are engineered using low-cost fabrication techniques (e.g. moulding and sewing), provide adaptable morphology in response to environmental changes and are ideally suited for gripping and manipulating delicate objects as well as interacting directly with humans, for example in medical and wearable applications. Soft robots provide a particular advantage for assisting with human motion as their material properties minimize restrictions to the wearer and eliminate the need to carefully align a robot with biological joints. One approach of interfacing the human body is to use textiles and artificial muscles to create tensile forces over the body in parallel with biological muscles for a variety of applications such as assisting or restoring normal walking. Actuation can be achieved through a cable-based transmission driven by an electric motor and gearing system or other types of artificial muscles. Such an approach enables the limb-worn components of wearable robots to have very low

[1]Paulson School of Engineering and Applied Sciences, Harvard University, USA
[2]Wyss Institute for Biologically Inspired Engineering, Harvard University, USA

inertia and not restrict the wearer's natural kinematics. Due to the lack of a rigid exoskeleton, the wearer's bone structure would sustain all the compressive forces normally encountered by the body plus the forces generated by the wearable robot.

As an alternative to linear actuators, pneumatic and hydraulic powered elastomeric and fabric soft actuators are also promising candidates for wearable robotic applications. This approach again enables systems that are lightweight, have a high power-to-weight ratio, low material cost, easy to fabricate and can achieve complex output motions with single inputs. Specifically, when pressurized, embedded chambers in these soft actuators expand in the directions associated with low stiffness and give rise to bending, twisting and extending/contracting motions [1,2].

The field of soft wearable robotics is a relatively new area of research but is growing in interest in the robotics research community, as demonstrated by a research workshop on this topic at the IEEE International Conference on Biomedical Robotics and Biomechatronics in Singapore and a special session at the International Symposium on Wearable Robotics, both in 2016. Compared to traditional robotics, soft robotics offers advantages of inherent compliance and low weight; however, it also brings fundamental challenges, including sensing, calibration, actuation, efficiency and control. This chapter gives a brief description of soft wearable robots to assist the lower and upper body, organs inside the body and briefly mentions some emerging areas.

3.2 Soft wearable robots to assist locomotion

Over the last two decades, a number of exoskeletons have been developed for tasks ranging from heavy lifting [3] to helping the wearer to walk [4–10] and for providing robotic rehabilitation therapy in a hospital setting [11–13]. Recently, with improvements in actuator and sensor technology, we have seen these systems become portable, and begin to transition from academic to commercial applications. Several categories of exoskeletons exist, including those that provide the ability to replace human movements that have been completely lost, e.g. in the case of a patient paralysed below the waist. To achieve this, the device must provide sufficient control to ensure the user's full stability, making high speed and agility secondary concerns to balance and safety. In effect, these devices can be thought of as wheelchair replacements and offer an elegant and potentially life-changing tool for a specific group of users. By allowing the person to stand for a couple of hours a day, many co-morbidities are reduced such as pressure sores, bone loss, muscle atrophy, etc. [14].

Another type of exoskeleton is designed to assist able-bodied users perform tasks more easily or for longer duration. In particular, considerable work has been conducted in the area of active exoskeletons for augmenting load carriage capacity [5,10]. These traditional exoskeletons have transferred load to the ground via a parallel structure in parallel with the biological limb. For all these devices, a key challenge is minimizing the weight and power requirements and to this end some groups have proposed quasi-passive architectures in an effort to reduce the exoskeleton's energy consumption [10,15]. These previous exoskeletons all rely on rigid frameworks of linkages, coupled to the body at select locations via pads, straps or other interface techniques. As the wearer flexes or extends their biological joints, these rigid links add considerable inertia to movement which must be overcome by motors or by the user. More recently, a number of research groups have taken a different approach and focused on developing systems that assist a single joint. Through careful system level design, the last 3 years has seen exciting breakthroughs in assisted loaded and unloaded walking [16–18]. These systems function by supplementing normal muscle work at the hip or ankle during the stance phase and at the onset of the swing phase. With this approach, the systems are leveraging the natural dynamics associated with human walking that have been elegantly demonstrated with some walking robots.

As an alternative to rigid exoskeletons, a number of research groups are focusing on the development and evaluation of soft wearable robots to assist locomotion by targeting a limited number of joints [4,19–26]. A number of example devices are shown in Figure 3.1. Efforts have largely been focused on augmenting the normal muscle work of healthy individuals by applying assistive torques at the wearer's joints with the goal of reducing the metabolic cost of transport of the wearer [4,18–20,25]. These devices utilize flexible materials and actuators to specifically address human factors challenges and do not have a load bearing 'skeleton' but rather rely on the biological skeleton to assist with the application of forces and transfer of load. Exosuit is a term that has been used to define a class of these devices and most frequently, an electromechanical approach has been taken for actuation [19–21,26]. An exosuit typically consist of an integrated garment that includes attachment points to the body, a structured textile that transmits loads across the body and actuated segments that can reduce their relative length to provide controlled tensile forces in the suit. The suit creates moments around the joints as these forces are offset from the joint centres of rotation due to the tissue and bone structure surrounding the joints. A key feature of exosuits is that if the actuated segments are extended, the suit length can increase so that the entire suit is slack, at which point wearing an exosuit feels like wearing a pair of pants and does not restrict the wearer whatsoever.

Compared to traditional exoskeletons, soft wearable robots provide minimal additional mechanical impedance and kinematic restrictions. When considering how a soft wearable robot can augment healthy gait, there are several biomechanical studies that have been performed that provide insight on their potential. An initial study performed on a multi-joint soft exosuit used a reconfigurable multi-joint actuation platform that enabled the effects of assistance at different joints to be compared. The joint torques applied were relatively low, yet when comparing walking with the system powered to powered off, significant average metabolic reductions of 4.6% (hip extension only) and 14.6% (hip flexion/extension and ankle plantarflexion) were found [27]. This motivated the development of an autonomous version of the exosuit that could assist hip flexion/extension and ankle plantarflexion [21]. By exploiting that the muscle activity of each leg is out of phase and

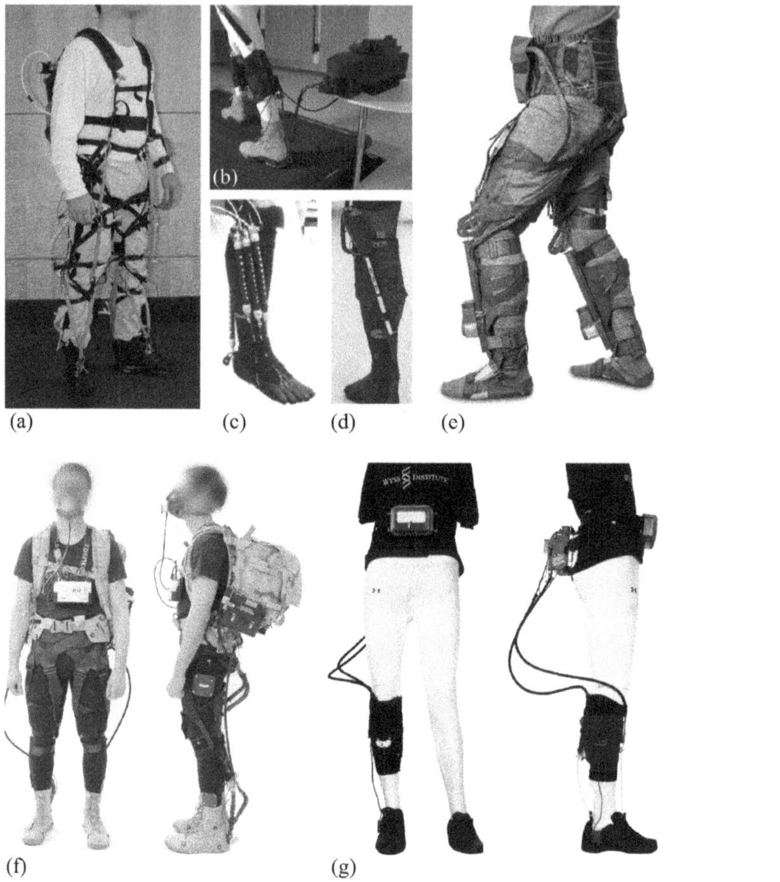

Figure 3.1 *Photos of various soft wearable robots for the lower extremity. (a) Suit is pneumatically powered and can assist each of the joints in the leg in both directions in the sagittal plane [4]. (b) A soft exosuit actuated with an off-board platform that demonstrated increasing metabolic reductions with increasing assistance [31]. (c) An active soft orthotic with a biologically inspired arrangement of pneumatic actuators [25]. (d) A soft exosuit for assisting the ankle joint that is actuated by a twisted string actuator [26]. (e) An autonomous soft exosuit design that provides joint assistance and support against collapsing [34]. (f) An autonomous multi-articular design aiding ankle plantarflexion and hip flexion and is actuated by geared motors driving Bowden cables with actuation units that attach to the side of a backpack [28]. (g) An autonomous soft exosuit that targets the paretic ankle for patients post-stroke [32]*

the synergy between hip flexion and ankle plantarflexion, the system used only two motors for actuation and had an overall weight of 6.5 kg including batteries. An evaluation of this system on seven subjects demonstrated a reduction in the metabolic cost of walking of 7.3% when comparing walking with the exosuit to walking with the exosuit unpowered with the equivalent mass removed [28].

More recent studies have utilized the lab-based hardware combined with optimized revisions of the functional apparel components of the exosuit to explore the effect of increased level of assistance at the hip and ankle during different walking conditions. For the hip joint [29], it was demonstrated that by increasing the assistance for hip extension by approximately a factor of two compared to [27], metabolic reductions of up to 8.5% could be achieved when comparing walking with the exosuit powered versus unpowered (almost a factor of two increase in metabolic reduction). Another aspect of the study in Ding [29] was to explore the effects of different timings for hip extension assistance and results showed benefit in beginning to apply forces just before heel strike. The same exosuit and off-board actuation platform was also evaluated in running and demonstrated a reduction in metabolic cost of 5.4% when compared to not wearing the exosuit [30]. For the ankle joint [31], a study was performed with seven subjects during treadmill walking where the peak moment applied at the ankle was varied from approximately 10% to 38% of the biological moment. Results showed that with increasing exosuit assistance, metabolic effort continually decreased within the tested range. Furthermore, when maximum assistance was applied, the metabolic rate of walking was reduced by 22.8% relative to the powered-off condition.

In addition to augmenting healthy gait, there is growing interest in how soft wearable robots can improve gait for patients with physical impairments, specifically with stroke [22,32,33] and spinal cord injury [23,34]. Often times in diseases such as stroke, a patient's gait is asymmetric, characteristically slow and metabolically expensive. Passive assistive devices such as ankle–foot orthoses are often prescribed to increase function and independence; however, walking remains highly impaired despite their use. In [32], a study was performed to evaluate the effect of the exosuit actively assisting the paretic limb of individuals in the chronic phase of stroke recovery during treadmill and overground walking. The level of assistance applied was relatively low (~12% of biological joint torques), yet the exosuit assistance was able to facilitate an immediate 5.33° increase in the paretic ankle's swing phase dorsiflexion and 11% increase in the paretic limb's generation of forward propulsion. These improvements in paretic limb function contributed to a 20% reduction in forward propulsion interlimb asymmetry and a 10% reduction in the energy cost of walking, compared to walking with the exosuit unpowered. Another study [33], demonstrated that a soft exosuit that targets the paretic ankle could reduce common poststroke gait compensations. Specifically, compared to walking with the exosuit unpowered, walking with the exosuit powered resulted in reductions in hip hiking (27%) and circumduction (20%).

Going forward, it will be important to perform studies with prototypes of future autonomous, body worn systems to understand the capabilities of soft exosuits in different populations. Future studies should involve comparing the walking with the exosuit powered to walking with the exosuit removed and there is merit to evaluating systems during overground walking. Many of the studies performed to date have been with tethered actuation platforms during treadmill walking and have compared wearing the exosuit powered to unpowered, with the assumption that there is a negligible or small effect to wearing the textile components of the exosuit. A key benefit to the functional apparel aspect of a soft exosuit is that it is lightweight and nonrestrictive but it seems likely that it has some effect on gait. While there have been preliminary studies evaluating effect of the textile components on biomechanics and energetics in healthy [29] and impaired [32] populations that did not find significant effects, further study is warranted. Another aspect of exosuit design that will be important to evaluate is the effectiveness of the apparel interface in transferring mechanical work to the wearer and early work is beginning to develop experimental methodologies that enable this [35].

3.3 Soft wearable robots to assist the upper extremity

Many activities of daily living (ADLs) involve precision grasping and manipulation, such as putting toothpaste on a toothbrush or feeding oneself. However, people afflicted by stroke, cerebral palsy, muscular dystrophy, amyotrophic lateral sclerosis or traumatic brain injury may lose the ability to actively and accurately control their hand and arm movement. Untreated, these deformities contribute to the loss of advanced grasps and the ability to perform many fundamental ADLs. Current therapy includes evaluating range of motion of the joints, muscle contractures and the ability to perform common manipulation tasks. Depending on the severity of the disability, treatment can range from passive stretching using corrective orthotics, to botulinum toxin (botox) injection to relax spastic muscles, to surgical procedures. Irrespective of the specific pathology and treatment, rehabilitation exercises are required to improve functionality. Patients with upper extremity impairments are served by occupational therapists in clinics where visits consist of a guided set of movements or practice at specific manipulation tasks that simulate ADLs.

A number of elegant robotic rehabilitation systems have been developed for the hand that consists of multi-degree-of-freedom exoskeletons, and a variety of designs have been proposed. Clinical studies have shown that stroke patients who have robotic assistance when performing intense repetitive movements demonstrate significant improvement in hand motor functions [36]. The rigid mechanical design associated with these systems provides robust and reliable devices capable of exerting high forces that allow more challenging rehabilitation scenarios to be executed, but often at the trade-off of these systems having limited portability.

Recently, a number of hand rehabilitation designs have followed an alternative approach to that of traditional exoskeletons. A soft wearable robotic device for ADLs and hand rehabilitation could lead to greater advances for assistive activity in the home. These designs combine soft gloves with cables connected to fingers that are driven by a number of motors located away from the hand [37–40] or soft pressurizable elastomeric actuators that support finger flexion or extension [2,41–45]. The cable-based designs follow a similar approach to that mentioned previously for the lower extremity with textile or plastic anchors attaching to the limbs and digits with cables anchored proximal and distal to a joint so that force can be generated when they contract. A number of example devices are shown in Figure 3.2.

In addition, a major focus in the area of soft wearable robots for the upper extremity has been the use of soft multi-material fluidic actuators due to their inherent compliance as well as their customizability and safety. The fiber-reinforced actuator is one type of actuator and is made from an elastomeric chamber with a flexible strain layer adhered to the bottom, which is then wrapped with a strain material to restrict expansion in certain directions and covered with a flexible outer protection layer. The design of many of these types of soft actuators

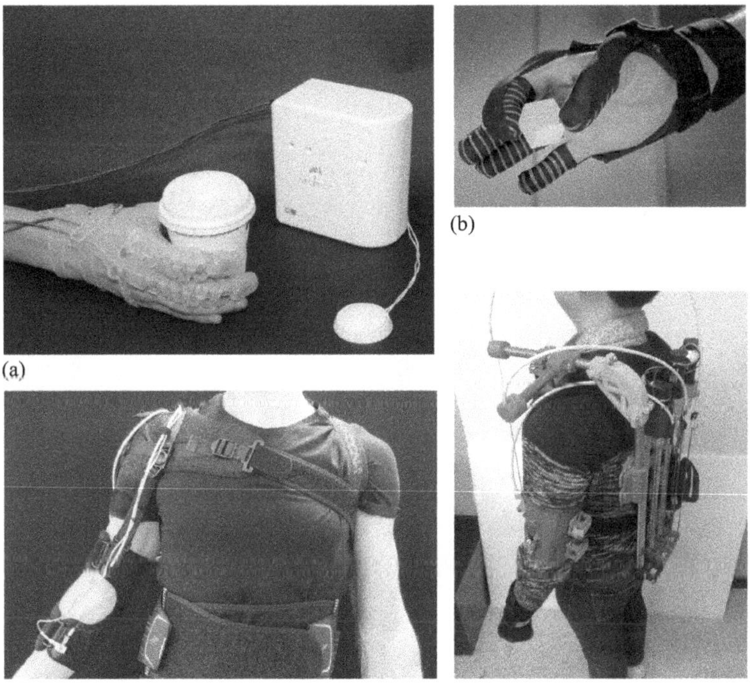

(a) (b) (c) (d)

Figure 3.2 Photos of various soft wearable robots for the upper extremity. (a) Cable-driven soft robotic glove with a polymer user attachment intended to assist grasping during activities of daily living [39]. (b) A soft robotic glove using fibre-reinforced elastomeric actuators attached to the hand through functional apparel [42]. (c) A cable-driven soft robot for assisting the elbow joint highlighting cable attachment points [46]. (d) A cable-driven soft robot for assisting the shoulder in multiple degrees of freedom actuated with twisted string actuators [47] has come via iterative experimentation in order to achieve embodiments capable of producing the desired bend, twist and extension motion paths with increasingly efforts being targeted at modelling [1].

When considering how these soft actuators could be integrated into a wearable device, it is clear that the kinematics of pure bending actuators would not match the kinematic movements of many limbs. To achieve more complex motion required for wearable applications, these types of actuators can be segmented and combined in series on a single actuator to produce an overall motion path to mimic that of the limb or digit being assisted. Using this approach, specific soft actuators have been designed and fabricated for each digit and demonstrated the ability to replicate the finger and thumb motions suitable for many typical motions [2]. These actuators were integrated into a robotic glove and gross and fine functional grasping abilities were evaluated through demonstrations in both free-space and by interaction with various objects that are encountered during ADLs.

Beyond the hand, the ability to move the upper limb and to interact with the environment is critical in sustaining a person's ability to perform ADLs. To this extent, the shoulder and elbow joints are particularly important because they are further up the limb's kinematic chain and their impairment drastically limits the function of the whole limb. Several soft robotic systems have also developed to address the size and mass requirements for a portable assistive device, which is particularly useful in cases where it is impossible to restore normal function through rehabilitation therapy. To date, the majority of these systems operate by having cables pull the arm upwards with reaction loads borne by the arm or torso [46–52]. As with the lower extremity and hand, the inherent advantages of these cable-driven systems are that they are lightweight at distal locations as they allow for remotely located actuation systems. Due to the anatomy of the arm and torso, they often require some form of structure protruding from the shoulder or upper limb to generate a sufficient lever arm to reduce the cable force required to achieve a given joint torque. More recently, there have been efforts exploring how to assist the shoulder with an inflatable actuator approach [53,54]. This leverages some of the concepts that have been applied to inflatable gloves but takes new approaches to actuator design and system integration, given the complex anatomy of the shoulder and the different force and sensing requirements.

3.4 Soft wearable robots for implantable applications

Ventricular assist devices (VADs) are implanted into the body and utilized as a life prolonging therapy, either as a bridge to transplant or in some cases, destination therapy in patients with cardiac disease. All current generation of VADs are based on pump and valve technology; the heart and great vessels are cannulated and blood is removed from the heart, and pumped through a one-way valve under pressure into the aorta. Because of the contact between blood and these artificial surfaces, anticoagulation is required. Despite best efforts at appropriate anti-coagulation, the risk of thromboembolic events including, stroke, may occur.

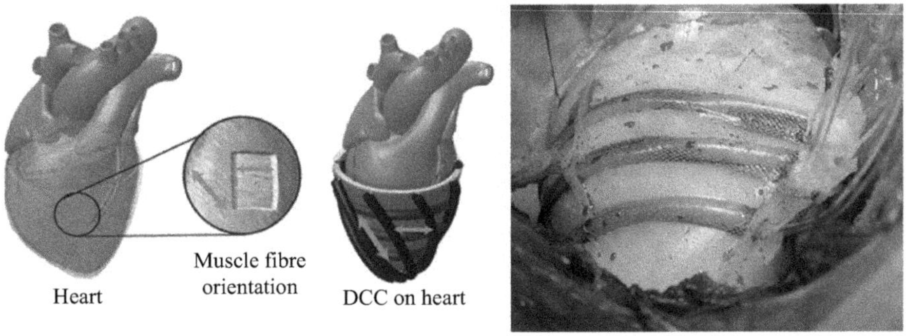

Figure 3.3 *Left, soft active material inspired by the muscle fibre orientation of the heart with an arrangement of linear contracting soft actuators arranged circumferentially and in a helix [56]. Right, soft actuators integrated into a prototype soft sleeve used in vivo to demonstration restoration of cardiac function [56]*

As an alternative to VAD technology, there have been efforts exploring how external assistance to the heart with an implantable wearable device. Assistance of the cardiac motion with traditional robotic systems can be complicated due to the delicate internal tissue and would present a challenge for implantation. Taking inspiration from how the complex motion in soft muscular structures is often achieved through the functional arrangement of many simple contractile elements, a soft robotic sleeve was developed that can mimic and assist the damaged myocardial architecture of the left ventricle as shown in Figure 3.3. It was demonstrated that by mimicking the orientation of the contractile elements in a soft elastomeric material in shape similar to the left ventricle, an accurate representation of apical twist could be achieved [55]. This work was then extended to create a device suitable for evaluation in vivo in an animal model where it was demonstrated that it can be surgically implanted around the heart, synchronized with the native heartbeat and used to restore cardiac output in a heart failure model [56]. A follow-up study with a variation on this device explored the effects of mechanical coupling and synchronization of assistance on the performance of an extracardiac soft robotic device for a failing heart. It found that adhesion of the actuators to the ventricles improves cardiac output and that contraction–relaxation ratio and rate of actuation have an impact on cardiac output [57].

3.5 Emerging directions in soft wearable robots

Given this field of soft wearable robots is relatively new, we do not yet have sufficient knowledge on how to most effectively tailor systems to match the need of different applications areas across populations. Specifically, there is benefit to understanding which joints/tissues (or combinations) can benefit from assistance and what levels of force and power are needed as this will also impact component and system level requirements (e.g. actuation modality). For example, soft wearable robots for some medical applications may not need to provide as much active assistance as applications designed to augment human performance during load carriage. To better understand the specific requirements for different applications, there is benefit in performing human subjects studies with versatile lab-based equipment where different levels of assistance can be delivered at different locations and data can be collected on benefits to patients.

As we learn more about different application requirements and the community grows [58], there will likely be an opportunity to explore new actuation approaches such as those made from electro-active materials that are suitable for generating low forces or modulating stiffness [59–61]. Such actuators are not yet commercially available so thus working with the material science community will be critical to achieve robust components suitable for integration into systems. As the field explores different applications for different populations, this will necessitate new control approaches that target the specific impairments. For example, some individuals may need more assistance with stability, while others may benefit more from assistance with forward propulsion. Furthermore, even within each population there will be variation at the level of impairment or disease progression and as their activities and environments vary. Manual tuning of parameters for all of these factors would be challenging. Thus, algorithms that automatically determine parameters for optimal assistance will be required.

Last, many of the examples of soft wearable robots described here are somewhat analogous to an exoskeleton in that they apply forces or torques to assist with the movement of the underlying biological muscle, but without the rigid structure. It is quite likely that we will see embodiments of soft wearable robots emerge that do not fit this category. Examples, could be devices that apply compressive or shear forces to the outside of the body to simulate the effects of massage in order to heal injured tissue [62] or supernumerary limbs or digits that enable completely new methods of human assistance and create exciting new opportunities for human–machine interaction [63,64]. This is a field that is just beginning and there is much opportunity to advance the underlying fundamental science of soft robotics and human–machine interaction, as well as identify new high impact applications.

References

[1] Connolly, F., Polygerinos, P., Walsh, C.J., Bertoldi, K. Mechanical Programming of Soft Actuators by Varying Fiber Angle. Soft Robotics, 2(1): 26–32, 2015.

[2] Polygerinos, P., Wang, Z., Galloway, K.C., Wood, R.J., Walsh, C.J. Soft Robotic Glove for Combined Assistance and at-Home Rehabilitation. Robotics and Autonomous Systems (RAS) Special Issue on Wearable Robotics, 73: 135–143, 2015.

[3] Wehner, M., Rempel, D., Kazerooni, H. Lower Extremity Exoskeleton Reduces Back Forces in Lifting. In ASME Dynamic Systems and Control Conference (pp. 49–56), October 12–14, 2009, Hollywood, California, USA.
[4] Wehner, M., Quinlivan, B., Aubin, P., *et al.* A Lightweight Soft Exosuit for Gait Assistance. In 2013 IEEE International Conference on Robotics and Automation (ICRA) (pp. 3347–3354), May 6–10, 2013, Karlsruhe, Germany.
[5] Chu, H.K., Zoss, A. On the Biomimetic Design of the Berkeley Lower Extremity Exoskeleton (BLEEX). In IEEE Int. Conf. Robotics and Automation (ICRA) (pp. 4356–4363). IEEE Press, Barcelona, Spain, 2006.
[6] Kawamoto, H., Lee, S., Kanbe, S., Sankai, Y. Power Assist Method for HAL-3 Using EMG-based Feedback Controller. In Systems, Man and Cybernetics, 2003. IEEE International Conference on (vol. 2, pp. 1648–1653). IEEE, 2003.
[7] Quintero, H., Farris, R., Goldfarb, M. Control and Implementation of a Powered Lower Limb Orthosis to Aid Walking in Paraplegic Individuals. In IEEE International Conference on Rehabilitation Robotics (pp. 1–6), July 2011.
[8] Sawicki, G.S., Daniel, P.F. Mechanics and Energetics of Level Walking with Powered Ankle Exoskeletons. Journal of Experimental Biology, 211(9): 1402–1413, 2008.
[9] Sugar, T.G., Fernandez, E., Kinney, D., Hollander, K. W., & Redkar, S. HeSA, Hip Exoskeleton for Superior Assistance. *Wearable Robotics: Challenges and Trends* (pp. 319–323). Springer International Publishing, Spain, 2017.
[10] Walsh, C., Endo, K., Herr, H.A. Quasi-Passive Leg Exoskeleton for Load Carrying Augmentation. International Journal of Humanoid Robotics, Special Issue: Active Exoskeletons, 4(3): 487–506, 2007.
[11] Banala, S.K., Agrawal, S.K., Scholz, J.P. Active Leg Exoskeleton (ALEX) for Gait Rehabilitation of Motor-Impaired Patients. In Proc. 2007 IEEE 10th Int. Conf. Rehabil. Robotics (pp. 401–407), 2007.
[12] Krebs, H.I., Hogan, N., Aisen, M.L., Volpe, B.T. Robot-Aided Neurorehabilitation. IEEE Transactions on Rehabilitation Engineering, 6(1): 75–87, 1998.
[13] Riener, R., Nef, T., Colombo, G. Robot-Aided Neurorehabilitation of the Upper Extremities. Medical and Biological Engineering and Computing, 43(1): 2–10, 2005.
[14] Raab, K., Krakow, K., Tripp, F., Jung, M., Effects of Training with the ReWalk Exoskeleton on Quality of Life in Incomplete Spinal Cord Injury: A Single Case Study, *Spinal Cord Series and Cases*, 2:15025, 2016.
[15] Van Dijk, W., van der Kooij, H., Hekman, E. A Passive Exoskeleton with Artificial Tendons: Design and Experimental Evaluation. In Rehabilitation Robotics (ICORR), 2011 IEEE International Conference on (1–6). IEEE, 2011.
[16] Collins, S.H., Wiggin, M.B., Sawicki, G.S. Reducing the Energy Cost of Human Walking using an Unpowered Exoskeleton. Nature, 522: 212–215, 2015.

[17] Mooney, L.M., Herr, H.M. Biomechanical Walking Mechanisms Underlying the Metabolic Reduction Caused by an Autonomous Exoskeleton. Journal of NeuroEngineering and Rehabilitation, 13: 4, 2016.
[18] Seo, K., Lee, J., Lee, Y., Ha, T., Shim, Y. Fully Autonomous Hip Exoskeleton Saves Metabolic Cost of Walking. In Proceedings of the 2016 IEEE International Conference on Robotics and Automation (pp. 4628–4635), 16–21 May 2016, Stockholm, Sweden.
[19] Asbeck, A.T., Dyer, R., Larusson, A., Walsh, C. Biologically-inspired Soft Exosuit. In 13th International Conference on Rehabilitation Robotics (ICORR), on the University of Washington Campus, June 24–26, 2013.
[20] Asbeck, A., De Rossi, S., Holt, K., Walsh, C.J. A Biologically Inspired Soft Exosuit for Walking Assistance. The International Journal of Robotics Research, 34(6): 744–762, May 2015.
[21] Asbeck, A., Schmidt, K., Galiana, I., Wagner, D., Walsh, C. Multi-joint Soft Exosuit for Gait Assistance. In Proc. Inter. Conf. on Robotics and Automation (pp. 6197–6204), 2015, Seattle, Washington, USA.
[22] Bae, J., De Rossi, S., O'Donnell, K., *et al.* A Soft Exosuit for Patients with Stroke: Feasibility Study with a Mobile Off-Board Actuation Unit. In 14th International Conference on Rehabilitation Robotics (ICORR), August 11–14, 2015, Singapore.
[23] Bartenbach, V., Schmidt, K., Naef, M., Wyss, D., Riener, R. Concept of a Soft Exosuit for the Support of Leg Function in Rehabilitation. In International Conference on Rehabilitation Robotics (ICORR), Singapore, 2015.
[24] Park, D., In, H., Lee, H., *et al.* Preliminary Study for a Soft Wearable Knee Extensor to Assist Physically Weak People. In Ubiquitous Robots and Ambient Intelligence (URAI), 2014 11th International Conference on (pp. 136–137). IEEE, 2014.
[25] Park, Y.-L., Chen, B.-R., Young, D., *et al.* Bio-Inspired Active Soft Orthotic Device for Ankle Foot Pathologies. In Intelligent Robots and Systems (IROS), 2011 IEEE/RSJ International Conference on (pp. 4488–4495). IEEE, 2011.
[26] Steven, M., Kernbaum, A. Twisted String Actuators for Exosuits. In Workshop at IROS 2016 on Twisted String Actuation: State of the Art, Challenges and New Applications, Oct, 2016.
[27] Ding, Y., Galiana, I., Asbeck, A.T., *et al.* Biomechanical and Physiological Evaluation of Multi-Joint Assistance with Soft Exosuits. IEEE Transactions on Neural Systems and Rehabilitation Engineering, 25(2): 119–130, 2017.
[28] Panizzolo, F.A., Galiana, I., Asbeck, A.T., *et al.* A Biologically-Inspired Multi-Joint Soft Exosuit that can Reduce the Energy Cost of Loaded Walking. Journal of NeuroEngineering and Rehabilitation, 13(1): 43, 2016.
[29] Ding, Y., Panizzolo, F., Siviy, C., *et al.* Effect of Timing of Hip Extension Assistance during Loaded Walking with a Soft Exosuit. Journal of NeuroEngineering and Rehabilitation, 2016(13): 87, 2016.

[30] Lee, G., Kim, J., Panizzolo, F.A., *et al.* Reducing the Metabolic Cost of Running with a Tethered Soft Exosuit. Science Robotics, 2(6): eaan6708, 2017.

[31] Quinlivan, B., Lee, S., Malcolm, P., *et al.* Assistance Magnitude Versus Metabolic Cost Reductions for a Tethered Multiarticular Soft Exosuit. Science Robotics, 2(2): eaah4416, 2017.

[32] Awad, L., Bae, J., O'Donnell, K., *et al.* A Soft Robotic Exosuit Improves Walking in Patients after Stroke. Science Translational Medicine, 9(400): p.eaai9084, 2017.

[33] Awad, L.N., Bae, J., Kudzia, P., *et al.* Reducing Circumduction and Hip Hiking During Hemiparetic Walking Through Targeted Assistance of the Paretic Limb Using a Soft Robotic Exosuit. American Journal of Physical Medicine and Rehabilitation, 96(10): S157–S164, 2017.

[34] Schmidt, K., Riener, R., MAXX: Mobility Assisting teXtile eXoskeleton that Exploits Neural Control Synergies. In Converging Clinical and Engineering Research on Neurorehabilitation II (pp. 539–543). Springer International Publishing, 2017.

[35] Yandell, M., Quinlivan, B., Popov, D., Walsh, C., Zelik, K. Physical Interface Dynamics Alter How Robotic Exosuits Augment Human Movement: Implications for Optimizing Wearable Assistive Devices. Journal of NeuroEngineering and Rehabilitation, 14(1): 40, 2017.

[36] Lo, A.C., Guarino, P.D., Richards, L.G., *et al.* Robot-Assisted Therapy for Long-Term Upper-Limb Impairment after Stroke. New England Journal of Medicine, 362(19): 1772–1783, 2010.

[37] Chiri, A., Vitiello, N., Giovacchini, F., Roccella, S., Vecchi, F., Carrozza, M.C. Mechatronic Design and Characterization of the Index Finger Module of a Hand Exoskeleton for Post-stroke Rehabilitation. IEEE/ASME Transactions on Mechatronics, 17: 884–894, 2012.

[38] Jeong, U., In, H.-K., Cho, K.-J. Implementation of Various Control Algorithms for Hand Rehabilitation Exercise using Wearable Robotic Hand. Intelligent Service Robotics, 6: 181–189, 2013.

[39] Kang, B.B., Lee, H., In, H., Jeong, U., Chung, J., Cho, K.-J. Development of a Polymer-based Tendon-Driven Wearable Robotic Hand. In Robotics and Automation (ICRA), 2016 IEEE International Conference on (pp. 3750–3755). IEEE, 2016.

[40] Xiloyannis, M., Cappello, L., Dinh, B.K., Masia, L. Towards the Design of an Underactuated Soft Exoskeleton for Grasp Assistance. In 2016 6th IEEE International Conference on Biomedical Robotics and Biomechatronics (BioRob), 26–26 June, Singapore.

[41] Connelly, L., Jia, Y., Toro, M.L., Stoykov, M.E., Kenyon, R.V., Kamper, D. G. A Pneumatic Glove and Immersive Virtual Reality Environment for Hand Rehabilitative Training after Stroke. IEEE Transactions on Neural Systems and Rehabilitation Engineering, 18: 551–559, 2010.

[42] Polygerinos, P., Galloway, K.C., Sanan, S., Herman, M., Walsh, C.J. EMG Controlled Soft Robotic Glove for Assistance during Activities of Daily

Living. In Rehabilitation Robotics (ICORR), 2015 IEEE International Conference on (pp. 55–60). IEEE, 2015.

[43] Vargas, P.A., Brasil, F.L., McConnell, A.C., *et al.* Combining Soft Robotics and Brain-Machine Interfaces for Stroke Rehabilitation. *Converging Clinical and Engineering Research on Neurorehabilitation II* (pp. 1257–1262). Springer International Publishing, Spain, 2017.

[44] Yap, H.K., Lim, J.H., Goh, J.C.H., Yeow, C.H. Design of a Soft Robotic Glove for Hand Rehabilitation of Stroke Patients with Clenched Fist Deformity using Inflatable Plastic Actuators. ASME Journal of Medical Devices, 10(4): 0445041, 2016.

[45] Zhao, H., Jalving, J., Huang, R., Knepper, R., Ruina, A., Shepherd, R. A Helping Hand: Soft Orthosis with Integrated Optical Strain Sensors and EMG Control. IEEE Robotics & Automation Magazine, 23(3): 55–64, 2016.

[46] Xiloyannis, M., Cappello, L., Binh, K.D., Antuvan, C.W., Masia, L. Preliminary Design and Control of a Soft Exosuit for Assisting Elbow Movements and Hand Grasping in Activities of Daily Living. Journal of Rehabilitation and Assistive Technologies Engineering, 4: 205566831668 0315, 2017.

[47] Gaponov, I., Popov, D., Lee, S.J., Ryu, J. Auxilio: A Portable Cable-driven Exosuit for Upper Extremity Assistance. International Journal of Control, Automation and Systems, 15(1): 73–84, 2016.

[48] Cappello, L., Binh, K.D., Masia, L. Design of SARCOMEX: A Soft ARm COMpliant Exoskeleton. In 2016 6th IEEE International Conference on Biomedical Robotics and Biomechatronics (BioRob), 26 June, Singapore.

[49] Galiana, I., Hammond, F.L., Howe, R.D., Popovic, M.B. Wearable Soft Robotic Device for Post-stroke Shoulder Rehabilitation: Identifying Misalignments. In Intelligent Robots and Systems (IROS), 2012 IEEE/RSJ International Conference on (pp. 317–322). IEEE, 2012.

[50] Kadivar, Z., Beck, C.E., Rovekamp, R.N., O'Malley, M.K., Joyce, C.A. On the Efficacy of Isolating Shoulder and Elbow Movements with a Soft, Portable, and Wearable Robotic Device. In Wearable Robotics: Challenges and Trends (pp. 89–93). Springer International Publishing, 2017.

[51] Park, D., Cho, K.-J. Development and Evaluation of a Soft Wearable Weight Support Device for Reducing Muscle Fatigue on Shoulder. PLoS ONE, 12(3): e0173730, 2017.

[52] Sugar, T.G., He, J., Koeneman, E.J., *et al.* Design and Control of RUPERT: A Device for Robotic Upper Extremity Repetitive Therapy. IEEE Transactions on Neural Systems and Rehabilitation Engineering, 15(3): 336–346, 2007.

[53] O'Neill, C., Phipps, N., Cappello, L., Paganoni, S., Walsh, C. Soft Robotic Shoulder Support: Design, Characterization, and Preliminary Testing. In 15th International Conference on Rehabilitation Robotics (ICORR), July, 2017, London.

[54] Simpson, C., Okamura, A., Hawkes, E. Exomuscle: An Inflatable Device for Shoulder Abduction Support. In IEEE International Conference on Robotics and Automation (ICRA), May 29–June 3, 2017, Singapore.

[55] Roche, E.T., Wohlfarth, R., Overvelde, J., *et al.* A Bioinspired Soft Actuated Material. Advanced Materials, 26(8): 1200–1206, February 2014.

[56] Roche, E.T., Hovarth, M.A., Wamala, I., *et al.* Soft Robotic Sleeve Supports Heart Function. Science Translational Medicine, 9(373): eaaf3925, 2017.

[57] Payne, C.J., Wamala, I., Abah, C., *et al.* An Implantable Extracardiac Soft Robotic Device for the Failing Heart: Mechanical Coupling and Synchronization. Soft Robotics. May 2017, ahead of print. https://doi.org/10.1089/soro.2016.0076.

[58] Holland, D., Park, E., Polygerinos, P., Bennett, G., Walsh, C. The Soft Robotic Toolkit: Shared Resources for Research and Design. Soft Robotics, 1(3): 224–230, September 2014.

[59] Rossiter, J., Knoop, E., Nakamura, Y. Affective Touch and Low Power Artificial Muscles for Rehabilitative and Assistive Wearable Soft Robotics. *Wearable Robotics: Challenges and Trends*. Springer International Publishing, Spain, 2017. 101–106.

[60] Ortiz, J., Rocon, E., Power, V., *et al.* XoSoft-A Vision for a Soft Modular Lower Limb Exoskeleton. *Wearable Robotics: Challenges and Trends*. Springer International Publishing, Spain, 2017. 83–88.

[61] Li, Y., Hashimoto, M. Design and Prototyping of a Novel Lightweight Walking Assist Wear Using PVC Gel Soft Actuators. Sensors and Actuators A: Physical, 239: 26–44, 2016.

[62] Cezar, C. A., Roche, E.T., Vandenburgh, H.H., Duda, G.N., Walsh, C.J., Mooney, D.J. Biologic-Free Mechanically Induced Muscle Regeneration. Proceedings of the National Academy of Sciences, 113(6), 1534–1539, 2016.

[63] Yap, H.K., Goh, J.C.H., Yeow, C.H. A Low-Profile Soft Robotic Sixth-Finger for Grasp Compensation in Hand-Impaired Patients. ASME Journal of Medical Devices, 10(3): 030914, 2016.

[64] Wu, F.Y., Harry Asada, H. Implicit and Intuitive Grasp Posture Control for Wearable Robotic Fingers: A Data-Driven Method Using Partial Least Squares. IEEE Transactions on Robotics, 32(1): 176–186, 2016.

Chapter 4

Exploring user requirements for a lower body soft exoskeleton to assist mobility

Valerie Power[1], Adam de Eyto[1], Christoph Bauer[2], Corien Nikamp[3], Samuel Schülein[4], Jeanette Müller[5], Jesús Ortiz[6], and Leonard O'Sullivan[1]

Abstract

Understanding the requirements of potential users is crucial to the successful design of wearable exoskeleton systems. Considering user requirements throughout the design process optimises the likelihood of user uptake and facilitates adherence to the use of wearable systems. This chapter describes the application of a user-centred design approach to the development of a soft lower body assistive exoskeleton for individuals with mild to moderate mobility impairment. Examples of the identification and characterisation of user groups, the use of qualitative and quantitative research methods to explore user requirements, and the implications of user requirements for soft exoskeleton technologies are presented.

Keywords: User requirements; user-centred design; soft robotics; assistive exoskeleton; gait; mobility

4.1 Introduction

Designing a successful wearable exoskeleton system requires extensive technical expertise and high levels of innovation. Yet, even the most advanced, innovative systems cannot be considered successful should they fail to be accepted by their intended users.

[1]School of Design, University of Limerick, Ireland
[2]Institut für Physiotherapie, ZHAW Zürcher Hochschule für Angewandte Wissenschaften, Switzerland
[3]Roessingh Research and Development, Netherlands
[4]Geriatrie-Zentrum Erlangen, Germany
[5]accelopment AG, Switzerland
[6]Department of Advanced Robotics, Istituto Italiano di Tecnologia, Italy

Technology acceptance has been studied across a variety of technologies, users and contexts of use, particularly over the past three decades, as technology has become increasingly pervasive in most facets of society. The Technology Acceptance Model – and its variants – is one of the most commonly-used frameworks for understanding user acceptance of technology [1–3]. Technology acceptance suggests that usage behaviour is strongly influenced by the user's perceived usefulness and perceived ease of use of a given technology, along with social factors. The challenges of designing highly complex wearable exoskeleton systems which users perceive as being both useful and easy to use are not insignificant and hence need to be carefully addressed.

The application of wearable exoskeleton systems as assistive technologies presents many opportunities, but also poses additional challenges. Assistive technology is considered from a 'person-first' approach, where the person, his/her occupation/activities and environment take precedence, with the technology adapting to individual needs e.g. the Human Activity Assistive Technology model [4]. To design wearable assistive exoskeletons that are fit for purpose, collaboration is required across multiple disciplines, including (but not limited to) technology, design, healthcare, psychology and industry.

This chapter will provide insights into the process of designing wearable assistive exoskeleton systems from a user-centred perspective. An example of the design of a soft lower body exoskeleton to assist mobility will be used throughout the chapter to present the principles, methods, opportunities and challenges associated with the user-centred design of assistive exoskeleton systems in a practical context. Since the focus of this chapter is on user-centred design, the user requirements discussed will be related to users and their needs, rather than technical specifications. Specifically, this chapter will present:

- Principles of user-centred design;
- An overview of the XoSoft project aimed at developing a lower body soft exoskeleton to assist human mobility;
- Identification of user groups for a lower body soft assistive exoskeleton;
- An example of mixed methods research to examine the requirements of potential users of a soft assistive exoskeleton for human mobility;
- The implications for soft exoskeleton technologies posed by user requirements.

4.2 User-centred design

When designing and developing products, particularly assistive devices such as XoSoft, it is recommended to involve both primary end users and secondary professional users in the design and development process. This helps to ensure that users' personal expectations are met, insofar as is possible, thereby maximising the possibilities for user uptake and acceptance [5]. To formulate a clear design brief, it is also necessary to explore and understand the demographic factors and experiences of users which contribute to the formation of their expectations and requirements. Observational design research methodologies, ethnographic research, clinical assessment and elements of co-design can be used or combined in mixed methods approaches, to enhance the early stage user involvement in the design process. These methods help to gather rigorous user feedback and rich contextual information on which to base the design specification [6].

In the early stages of development, a user-led approach ensures that the 'voice of the user' is both listened to and advocated for while developing the technological specification of the 'product' and the 'product service system' (PSS). The identification of real users' needs is an essential component to successfully adopting a user-centred approach – often small omissions or oversights in the development process can present the biggest challenges to primary and secondary users (SUs) when the product reaches the market. Using the time and resources on following a user-centred design approach at the beginning of a development process has been proven to be highly beneficial in avoiding costly mistakes and the need for re-designing in the later stages of development.

A user-centred approach to design is also recommended during the iterative phases of development and prototyping. This can be achieved by investigating if the subsequent prototype designs match the user requirements that were formulated at the beginning of the project, via a user requirements validation study. This approach allows developers to identify priorities for subsequent iteration cycles. Seeking user testing, feedback and response to early stage prototypes greatly enhances the chances of the product and PSS being relevant and successful within the marketplace but more importantly it verifies that the product is fit for purpose [7]. Highly regulated robotic assistive devices such as XoSoft require as part of regulatory approval to include user feedback and validation as formal elements of the design and development process.

SU feedback is also useful as a means of establishing what support and remote assistance clinicians, carers and personal assistants can give through the product's life cycle. In the case of assistive robotic devices, the technical nature of the product means that setup, customisation and maintenance of a product may be carried out by SUs. SU needs are often as critical as those of primary users (PUs), and the determination to invest in a technology can often weigh heavily on the ease of use from a SU's perspective.

4.3 XoSoft: a soft lower body exoskeleton to assist mobility

In Europe, it is estimated that over 51 million people experience a longstanding difficulty with walking [8]. People who experience mobility limitations can use a variety of assistive devices e.g. walkers, canes, manual and powered wheelchairs. However, currently available assistive devices have many limitations and frequently go unused, with up to 50% of users abandoning their devices soon after receiving them [9]. Reasons for this include excessive bulkiness of devices, usability issues with devices which occupy and/or load the upper limbs and the significant attentional demands which some devices place on the user. Such problems not only limit feasibility for everyday use, but can also lead to adverse effects, including falls, upper limb pain and potential injury [9–12]. Furthermore,

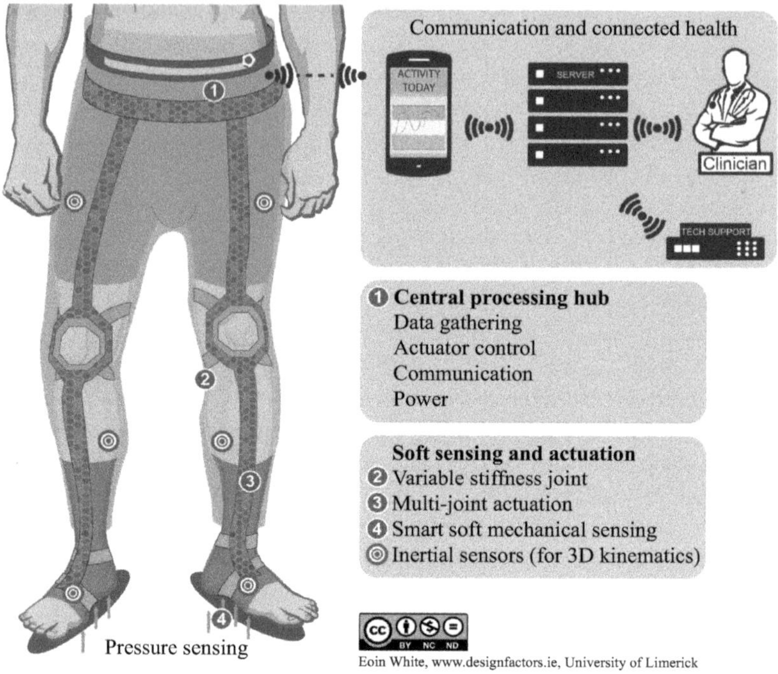

Figure 4.1 Image of the XoSoft concept and its components

many assistive devices are inappropriately prescribed and/or not prescribed by healthcare professionals, and users do not receive adequate training on safe and appropriate usage [9].

An international, multidisciplinary consortium of researchers set out to develop a soft, modular, lower limb exoskeleton (XoSoft) for older adults and people with disabilities. Ultimately, XoSoft's purpose is to help increase users' mobility, and thus improve their health and quality of life.

XoSoft was developed by a consortium of nine partners comprising clinical and research centres, universities, SMEs and an industrial partner with experience in developing exoskeletons/assistive devices. User-centred design was a core element of the XoSoft project. An iterative design process was undertaken, whereby user requirements drove technical innovations and user feedback informed cycles of design and prototype development.

Figure 4.1 displays an initial impression of the XoSoft concept and its key features. Advanced textiles and smart materials are employed to create sensing variable stiffness joints. Built-in sensors communicate the user's motion and intention to the controlling unit for analysis, to determine and provide the appropriate level of assistance, by way of actuators. Depending on the user's need at a given moment, the device will provide support, release or freedom of movement. Specific details of the sensing and actuation technologies used are not available, since XoSoft was in a conceptual stage of development at the time of writing.

4.4 Identifying users of a soft exoskeleton to assist mobility

In user-centred design of exoskeletons, it is crucial to clearly identify and understand the target users, their characteristics and their needs. The aim is to design an exoskeleton that is both functional and acceptable to users, and to ensure this, a variety of data should be gathered from multiple sources e.g. demographic data on the target user groups, user opinions on necessary and desired design features, data on the context in which the exoskeleton will be used, user anthropometrics and human motion data.

Users of a product may be categorised as primary, secondary and tertiary. The following sections describe these user groups, using the target users of XoSoft as examples of the populations who may be included each user group, and the reasoning used to identify and select them. This is not an exhaustive list of the potential users of a soft exoskeleton to assist mobility; the target users will vary depending on the functions of the exoskeleton (e.g. provision of assistance, monitoring and feedback on performance), its design (e.g. upper body, lower body, single or multi-joint) and its context of use (e.g. clinical, industrial, etc.).

4.4.1 Primary users

PUs are the people who actually use and directly benefit from the exoskeleton. XoSoft's PUs are adults who require low to moderate levels of assistance with mobility. Specifically, the XoSoft project focused on three distinct PU groups:

1. People with stroke;
2. People with incomplete spinal cord injury (SCI);
3. Older adults with mild–moderate mobility impairments.

These groups were selected using reasoning based on evidence of the opportunities, challenges and practicalities of creating an exoskeleton for these specific populations, as well as epidemiological data to identify potential global market sizes. Examples of this evidence-based reasoning are provided in the following sub-sections.

4.4.1.1 Stroke

In 2010, there were 33 million stroke cases worldwide, with 16.9 million people having a first stroke [13]. The nature and severity of post-stroke impairments vary between individuals, depending on the type and severity of stroke. Motor deficits are the most highly prevalent post-stroke impairment, commonly presenting as weakness affecting one side of the body (hemiparesis); approximately 72% of people experience lower limb motor deficits and 77% experience upper limb motor deficits [14].

Compared to healthy individuals, people who have experienced stroke generally exhibit slower, more asymmetrical gait [15]. A common pattern of motor deficits observed in post-stroke gait includes decreased ankle dorsiflexion [16–19], decreased knee flexion [18–20] and hip flexion [15] during the swing phases of gait, which combine to produce insufficient foot clearance and increased risk of trips and falls. Various compensatory movement strategies around the pelvis

[20–22] and hip [17,23] may be adopted to overcome the effects of these deficits, but such compensatory strategies can greatly increase the energy costs of walking [15]. Additionally, impairments related to sensory systems (e.g. vision, vestibular, somatosensory) and sensory organisation can further compromise gait in the stroke population [24,25].

Approximately one-third of community-dwelling people with stroke require supervision for community ambulation [26] indicating potential opportunities for assistive exoskeletons in this sector. Sensing and actuation technologies may be used to provide alternative sources of feedback to the user, as well as assistive lower limb joint stiffness/torques, which may offer valuable support during activities of daily living.

However, a number of challenges exist. The varying types and severities of clinical presentations following stroke present a design challenge, since design features required by some users may not be useful – or may present barriers to use – for others. Concomitant upper limb impairment may limit the ability to use devices to assist mobility, particularly exoskeletons which must be donned and doffed. Additionally, altered muscle tone and impaired voluntary movement following stroke can present challenges to the control of wearable devices like XoSoft. Furthermore, cognitive impairment has been observed in 35%–38% of individuals 3 months post-stroke and is reported to be even more prevalent in acute stages and among those with greater levels of disability [27,28]. Memory, orientation, language and attention are the cognitive domains most likely to be affected [28], which presents further challenges to usability and may increase the risk of inappropriate use of a wearable assistive exoskeleton.

4.4.1.2 Incomplete SCI

SCI may be categorised according to its aetiology: traumatic (T-SCI) or non-traumatic (NT-SCI). Motor and sensory impairments are experienced by both groups, thus wearable assistive devices offer great potential to assist all. However, the nature and extent of these impairments vary depending on the location and extent of the spinal cord lesion, therefore the potential requirements for assistance are diverse.

The age-adjusted incidence for T-SCI in Europe ranges from 8 to 131 (median 20.1) per million persons per year. The prevalence in Europe ranges from 227 to 526 (median 351) per million population [29]. These statistics concern both complete and incomplete SCI. Approximately 45% of individuals with T-SCI experience an incomplete tetraplegia, 21% an incomplete paraplegia; the remainder experience complete SCI [30]. The T-SCI population is generally young (average age: 42 years) and approximately 80% are male [30]. Cognitive impairment is not a feature of SCI, unless the trauma which led to the SCI also led to traumatic brain injury, and therefore does not generally need to be factored into the design requirements of this user group. In a European cohort, 8% of individuals of all lesion types regained walking function with one or two leg braces 6 months after injury [31]. Some rigid exoskeleton products are already available for individuals with SCI (e.g. ReWalk, Eksobionics, HAL). These products are primarily for use as rehabilitation tools at present, rather than as assistive devices, and although evidence of their effects on mobility is still limited, some promising research is beginning to emerge [32–34].

Reliable statistics concerning the incidence and prevalence of NT-SCI are less readily available. It is a growing group and outnumbers T-SCI. Patients with NT-SCI tend to be older and present with incomplete lesions [35,36]. Their changing needs over time present a challenge to the successful design of wearable assistive devices for this population, since NT-SCI or general health may deteriorate e.g. people with NT-SCI due to tumours. Age-related complications – as described in Section 4.4.1.3 – could also influence device usability in this group: for example, although NT-SCI has no direct influence on cognitive function, age-related cognitive impairment may be present in older adults with NT-SCI.

4.4.1.3 Older adults

The global population is ageing rapidly: the proportion of the world's population aged over 60 years is expected to rise from about 11% to 21% between 2000 and 2050, which represents an increase from 605 million to 2 billion older adults worldwide [37]. Unfortunately, increasing life expectancy does not necessarily equate to increasing healthy life expectancy. Multi-morbidity – living with two or more chronic health conditions – is reported to affect over 60% of adults aged 65 years and over [38]. Living with multiple chronic conditions reduces life expectancy, increases the risk of functional decline and reduces overall quality of life [39].

Mobility impairment is often a feature of age-related functional decline: data from the Health and Retirement Study in the United States shows that 8.5% of adults aged 65 years and over experience mobility impairment to the extent that they have difficulty walking across a room [40]. Disability is also becoming more prevalent among older adults in high-income countries, where obesity, diabetes and cardiovascular disease are widespread [41]. Given the prevalence of mobility impairments among older adults and the ageing global population, requirements for assistance to maintain mobility will continue to increase in the coming years, presenting considerable opportunities for assistive devices, including assistive exoskeletons.

Frailty among older adults may be considered as an opportunity and a challenge to the use of a soft assistive exoskeleton. Frailty is often defined as a syndrome, which presents as age-associated decline in functioning across multiple physiological systems [42]. It brings with it a high risk for falls, disability, hospitalisation and mortality. Frailty may be classified by the presence of three or more of the following criteria: slow gait speed, low physical activity, unintentional weight loss, exhaustion and muscle weakness. Older adults exhibiting one or two of these criteria may be considered as 'pre-frail' [43]. While the declining mobility seen with frailty suggests a potential need for assistive devices, exhaustion, general weakness and the multiple co-morbidities seen in frail older populations may act as barriers to uptake of, and adherence to, device usage.

An additional challenge to consider when designing for the older adult population is age-related cognitive decline. In Europe, the prevalence of dementia in adults aged 60–64 years is approximately 1%, but increases to almost 25% in those aged 85 years and over [44]. Mild cognitive impairment can occur in up to 20% of those aged 75 years and over [45]. This has clear implications for exoskeleton usability; designing a complex system such that older users can interact with it in a simple and intuitive manner is a significant design challenge.

4.4.2 Secondary users

SUs are those who have direct contact with PUs in professional or non-professional capacities e.g. healthcare professionals, formal/informal caregivers, spouses, family and friends. Professional SUs include healthcare professionals from multidisciplinary backgrounds, such as medical doctors, nurses, physiotherapists, occupational therapists, speech and language therapists, as well as professional care assistants, home help service providers and other support staff. Professional SUs may also be grouped according to the context in which they interact with PUs e.g. in hospitals, clinics, rehabilitation units, private practices, in the home environment, etc.

Depending on the context in which SUs encounter PUs, their expectations and requirements of a soft exoskeleton will differ. For example, professional SUs in acute healthcare settings will be focused on the short-term usability and efficacy of the device, since they often have very limited contact time with PUs. SUs in community settings may place greater emphasis on long-term factors such as the ability of the device to improve user safety, its robustness, and/or its adaptability to PUs during long-term use. Examples of some of the main expectations of professional SUs in relation to an assistive device are summarised in Table 4.1. These examples are based on previously published literature as well as expertise from professional SUs.

Table 4.1 Examples of professional secondary users' expectations in relation to a soft assistive exoskeleton and related indicators of success

Expectations	Indicators of success
Safety of device	Device is appropriate for its intended purpose, has a low incidence of adverse events, and evidence demonstrating safety and efficacy is available [46]
Efficacy of device	Improves clinical outcomes of PUs e.g. facilitates independent mobility, reduces disability, halts/delays functional decline, decreases the burden of care, maintains PUs independence and do so in a cost-effective manner [10,47,48]
Impact of device usage on PUs' quality of life	Positive impacts on factors that influence overall quality of life, aside from health status, such as social relations and participation in activities [49]
Accessibility of device to SUs and PUs	Available to PUs and SUs at a feasible cost or via appropriate reimbursement schemes
Alignment with professional aims and standards	Usage of the device is in keeping with the code of conduct of the relevant healthcare profession and its associated professional body/regulatory authority in a given jurisdiction

Non-professional SUs may include spouses, family members, friends, neighbours, community and/or voluntary organisations and other informal caregivers. When designing any assistive device for a person with impaired mobility, consideration must be given to the opinions and experiences of such informal caregivers, particularly spouses and immediate family members. Assistive devices – if adopted for long-term use by PUs – become entwined in the everyday lives of these SUs, become part of their shared home environments and can impart a sense of responsibility on SUs in relation to the devices and their usage [50]. Caregivers can also play a role in encouraging adoption of a device by PUs initially, can encourage, instruct and help with continued use of a device and ultimately promote user satisfaction with their device [51].

4.4.3 Tertiary users

Tertiary users (TUs) are defined as individuals and/or organisations that do not directly interact with the device, but have some influence on organising or enabling it e.g. involvement in regulatory frameworks and/or making financial provisions for devices. TUs are therefore often not considered as being part of the main focus when developing a new device, but nonetheless they need to be considered as stakeholders.

TUs' interests (Box 4.1) can be similar to those of SUs, but may be approached from a different perspective.

Box 4.1 Sample factors of interest to TUs in relation to a medical device, such as a soft exoskeleton

- **Safety of the device** – From a TU perspective, devices on the market must be safe to use, and the benefits of use must outweigh any potential risks. For medical devices, this includes obtaining the appropriate certification, registrations and conforming to all regulatory requirements.
- **Cost of the device** – Direct costs of the device may be considered, but TUs often focus on the costs relative to the benefits offered by the device.
- **Additional health/social care costs due to the device** – Devices that add costs must display clear benefits. TU support for a device that involves increased overall costs would be rare and would require a highly convincing case that the new device is replacing an older practice/technology with significant positive changes and fewer adverse effects.
- **Reduced health/social care costs due to the device** – TUs view reduced costs very favourably and often expect to see evidence of such effects. Cost reductions may be achieved in several ways:
  - By directly replacing an existing device but at a lower cost;
  - By hastening healing or rehabilitation, thus lowering the overall treatment cost;
  - By increasing user independence and reducing need for assistance, thus reducing costs of care.

- **Impact on functional ability of PU** – If an assistive device enables users to be active in society again, including working and paying taxes if possible, provision of the device is in the interest of TUs, as well as PUs and SUs. This is seen as a highly favourable outcome, as it lowers overall health/social care costs.
- **Impact on quality of life** – TUs may be primarily concerned with effects on quality of life of PUs at population levels, rather than for individual PUs. Overall, this would most likely not be a primary outcome measure from the TU point of view.
- **Willingness to pay for the device by PUs/general population** – The TUs to whom this relates will depend on the health/social systems in operation. In countries where there is high social involvement in providing assistive devices, political decisions are required, often informed by cost–benefit analyses. In countries with less social contribution, TUs in the insurance industry, in private healthcare services and others e.g. resellers, will play a greater role.

TUs may not be directly involved in the design and development of a device, however they may be consulted during design for advice on issues related to their interests e.g. when testing new devices. The interests of TUs must be considered to ensure the success of a proposed device at a later stage. TUs are a diverse group with varied interests, and Table 4.2 contains examples of some key types of TUs, their interests and some factors they may consider as indicators of success in the design of a soft assistive exoskeleton.

4.5 A mixed methods study to explore users' design requirements

The XoSoft project adopted a user-centred design focus: user needs formed the foundations of the concept and were the main drivers of the iterative design process. The needs of PUs and SUs are particularly crucial at this early stage, as discussed in Section 4.4, since these groups will be directly involved in using an assistive device such as XoSoft. Placing the design requirements of these users to the fore aims to ensure that the concept will not only meet users' basic requirements, but also their personal expectations, insofar as is possible, thereby maximising the possibilities for uptake and acceptance [5].

To formulate a clear design brief for the XoSoft concept, it was deemed necessary to conduct a study of user requirements. The aim of this study was to explore the functional needs and design requirements of a selection of individuals with mobility impairments (PUs), and those who have regular contact with them on professional and non-professional bases (SUs), in relation to a lower body soft exoskeleton to assist mobility.

Table 4.2 Examples of tertiary users (TUs) and their expectations and indicators of success with regard to a soft assistive exoskeleton

TU	Expectations	Indicators of success
Regulatory authorities	Ensure that devices are safe to use for intended purpose through legislation Prevent marketing of unsafe devices	Low incidence of adverse events
Authorities as health care providers	Provide best possible healthcare service on a fixed/limited budget	Keeping costs within budget while achieving acceptable levels of care Adequate cost–benefit ratio. Will need evidence of outcomes before accepting device for reimbursement
Insurance companies	Provide devices for their customers that offer benefits in relation to the device cost	Adequate cost–benefit ratio May define the level(s) of evidence that would be needed to accept a new device for reimbursement
Standardisation bodies	Create 'common rules' and provide framework for comparing through testing	Standards are widely used in the community
Influencers	Advocates for certain groups or products e.g. senior citizen groups or patient organisations	Their point of view is reflected in decisions from stakeholders that decide on use and reimbursement
Resellers (distributors)	Selling the device	Profitable distribution of device
Private hospitals and rehabilitation centres	Offering the device as a therapeutic tool Training PUs and SUs in the use of the device Highlighting the innovative character of their institution	Increase of overall profits and reputation of treatment centre

To achieve this aim, the following objectives were adopted:

- To conduct structured interviews with a range of potential PUs and SUs of a soft assistive exoskeleton;
- To collect information on users' requirements for and previous experiences of assistive devices for mobility;
- To obtain users' opinions on a soft assistive lower body exoskeleton, by presenting a visual impression of the XoSoft concept;
- To gather information on user characteristics which may contribute to their requirements and expectations of a device to assist mobility.

The following sections describe a study that was undertaken at the outset of the XoSoft project to achieve this aim and objectives. Details on the study methods are provided in Section 4.5.1, while the study results – presented in Section 4.5.2 – provide valuable insights into the opinions and requirements of a variety of potential users of a lower body soft assistive exoskeleton.

4.5.1 Methods

A cross-sectional mixed methods study was carried out across four European countries: Germany, Ireland, the Netherlands and Switzerland. A one-to-one structured interview was used to ascertain each user's opinions on and requirements for assistive devices for mobility, and to gather relevant background information to characterise users, and to provide context to their opinions and requirements. All research was conducted in accordance with local research ethics regulations, including obtaining written informed consent from individuals prior to participation, as appropriate.

4.5.1.1 Participants

Individuals who represented potential PUs and SUs of XoSoft were invited to participate in four European countries. The countries were selected because they were represented within the XoSoft consortium. Participants were recruited using a purposive sampling method initially, with additional snowball sampling if appropriate. Figure 4.2 illustrates the flow of participants through the study.

The PU participants included were:

- Older adults with moderate to high levels of mobility impairment recruited from eligible inpatients at a geriatric centre (termed 'frail older adults' hereafter);
- Older adults with mild to moderate mobility impairments recruited via older adults' community groups (termed 'pre-frail older adults' hereafter);
- Participants with mobility impairments due to stroke and incomplete SCI recruited from a database of previous stroke research participants and rehabilitation centre patients who had opted to receive invitations to take part in future research.

The presence of moderate to severe cognitive impairment – defined as a Mini-Mental State Examination score of less than 18 – was an exclusion criterion for PU participants, as this was considered to be a potential limiting factor to the individual's capacity to provide informed consent to participate and/or to complete the interview successfully. Individuals with scores of 18–24 were not excluded, since mild cognitive impairment is reasonably prevalent among the selected PU groups [45,52].

SU participants were sourced via personal and professional networks of the XoSoft consortium and included:

- Healthcare professionals from multidisciplinary backgrounds who specialise in working with frail older adults at a geriatric centre;
- Healthcare professionals and human movement scientists experienced in working with people with stroke and SCI in both research and rehabilitation settings;

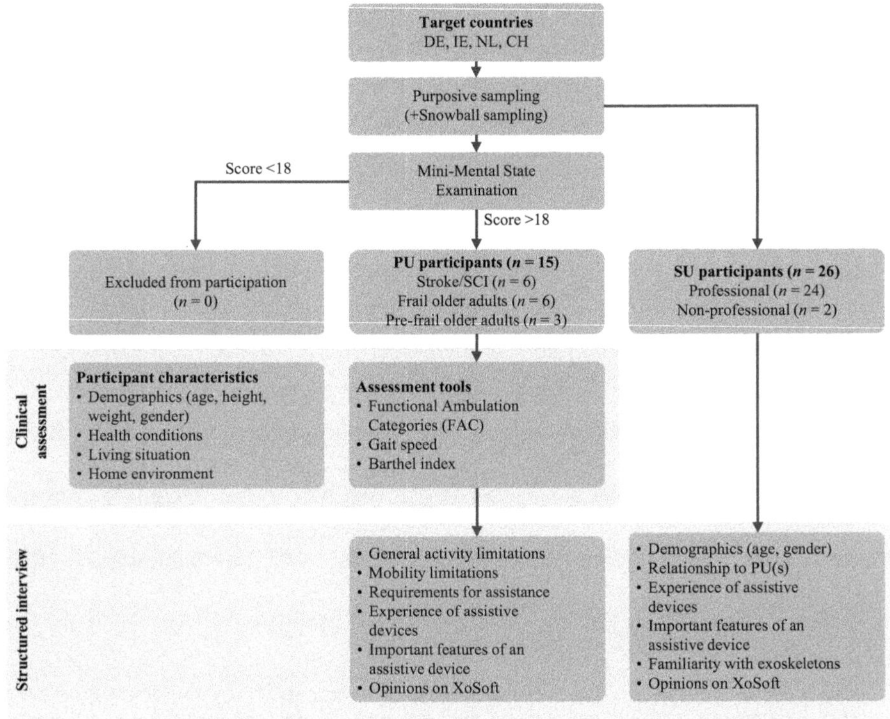

Figure 4.2 Flow chart illustrating participants and components of the mixed methods study

- Allied health professionals with experience across multiple fields – neurology, musculoskeletal, geriatrics – in a range of settings, from acute to long-term care and in both public and private healthcare practice;
- Non-professional SUs, including spouses, family members and friends of PUs.

4.5.1.2 Primary user assessment and interview

As shown in Figure 4.2, PU participants completed a brief clinical assessment and a structured interview. The clinical assessment gathered information to characterise PUs and their levels of mobility and independence. The following standardised clinical assessment tools were included:

- Mini-Mental State Examination: A measure of cognitive function rated by the researcher and used to screen participants for potential cognitive impairment. Maximum score is 30. A score of less than 18 was considered to indicate moderate to severe cognitive impairment, and therefore ineligibility to participate further in the study [53].
- Functional Ambulation Categories (FAC): A researcher-rated tick-box scale used to categorise participants' independence with regard to walking ability. Comprised of six broad categories, ranging from zero (participant cannot

walk, or needs help from two or more people) to five (participant can walk independently in any environment) [54]. Walking aid usage was recorded in addition to FAC score, since the FAC does not take this into account.

- Gait speed: Participants' usual gait speed recorded using either a 5 or 10 m walk test, depending on the setting and space available. It is a robust continuous measure of gait performance, and normative data is available for a variety of populations [55].
- Barthel Index: A questionnaire measuring independence in activities of daily living. Ten items are rated, and scores range from 0 to 20, with higher scores indicating greater levels of independence [56].

The structured interview schedule – implemented across all data collection sites – included questions on the following topics:

1. General activity limitations experienced by participants at home and in the community;
2. Activity limitations specifically related to mobility and participants' perceptions of these;
3. Participants' requirements for assistance with mobility, including assistive devices, if used;
4. Important features of an assistive device for mobility, as perceived by the participant;
5. Participants' opinions on a visual impression of the XoSoft concept (Figure 4.1) and their perspectives on perceived benefits and challenges to its usage.

Cues were included in the schedule for each topic so that all interviewers could provide standard clarification of topics and/or prompt discussion among participants, as required. Interview data were recorded in the form of detailed field notes, which summarised all points made by participants in relation to each question on the interview schedule.

4.5.1.3 Secondary user interview

SUs completed a structured interview only (Figure 4.2). The SU structured interview schedule differed slightly from that of PUs and included questions on the following:

1. The nature of participants' relationship(s) to potential PU(s) of XoSoft;
2. Participants' experiences of assistive devices for mobility;
3. Challenges encountered in relation to assistive devices for mobility;
4. Important features of an assistive device for mobility, as perceived by the participant;
5. Participants' familiarity with assistive exoskeletons and perceptions of usefulness;
6. Opinions on a visual impression of the XoSoft concept, as shown to PUs (Figure 4.1).

SUs, who were healthcare professionals and were familiar with exoskeletons, were also asked an additional question regarding the device(s) with which they were familiar, if they had used any such devices in their treatment of patients, and their experiences of doing so. Cues were also used to clarify topics and/or prompt discussion, as required, similar to the PU interview.

4.5.1.4 Data analysis

For PU clinical assessments, assessment tools were scored as per guidelines and summarised using descriptive statistics. Summaries of each participant's data for each question in the structured interview were created by the interviewers. These data were then compiled for each PU and SU group by a single researcher, and analysed using thematic analysis to identify the main themes in each user group for each question. Themes across user groups were then broadly categorised as

- Needs for assistance and experiences of assistive devices;
- Design requirements for wearable assistive devices for mobility;
- Perspectives on a soft assistive exoskeleton concept.

4.5.2 Results

A total of 15 PUs were interviewed: 6 in Germany, 6 in the Netherlands, and 3 in Ireland. An overview of the characteristics of the PUs interviewed is displayed in Table 4.3, split according to the distinct PU groups of interest. Frail and pre-frail older PUs are presented separately, since frail PUs were considerably older than pre-frail PUs and exhibited poorer mobility and greater levels of functional disability, which would be expected to influence interview findings.

In total, 26 SUs participated (18 female, 8 male): 2 in Ireland, 6 in the Netherlands, 12 in Germany and 6 in Switzerland. There were 24 professional and 2 non-professional SUs. Professional SUs included 12 physiotherapists, 5 physicians, 3 occupational therapists, 2 nurses and 2 other support staff. Twelve listed geriatrics as their main area of work, four primarily worked in stroke rehabilitation, two in SCI and eight worked with the general population. Three of the professional SUs also noted that they currently work in clinical health research involving older adults and neurological populations. Professional SUs were based in a variety of

Table 4.3 Overview of PU participant characteristics

	Stroke/SCI ($n=6$)	**Frail older adults ($n=6$)**	**Pre-frail older adults ($n=3$)**
Age (years)	55 (42–64)	87.5 (82–96)	74 (71–75)
Gender (m/f)	4/2	2/4	0/3
Height (m)	1.79 (1.70–1.96)	1.57 (1.50–1.66)	1.65 (1.60–1.68)
Weight (kg)	76.0 (63.0–100.0)	71.6 (48.2–78.2)	76.2 (76.0–88.9)
Gait Speed (m/s)	0.70 (0.34)*	0.61 (0.34–0.79)	1.40 (1.26–1.53)
FAC (score/5)	5 (5–5)	4 (3–4)	4 (3–5)
Barthel index (score/20)	19 (17–20)	12.5 (12–17)	19 (18–20)

*Gait speed of PUs with stroke is presented as mean (standard deviation).

Data are presented as median (minimum–maximum). SCI, spinal cord injury; FAC, functional ambulation categories.

settings: 12 in a specialist geriatric centre, 5 in rehabilitation centres, 4 in private/general practice, 2 in a hospital and 1 in a nursing home. Professional SUs had a mean age of 37.1 ± 9.0 and 12.0 ± 8.3 years of experience. Non-professional SUs included one spouse and one close friend of a PU.

4.5.2.1 Existing needs for and experiences of assistive devices

Although PUs with stroke and SCI were all independently mobile (FAC scores = 5, Table 4.3), the impairments of body structures and functions which they experienced were not uniform. Therefore, they used a wide variety of assistive devices to achieve independent mobility, including ankle–foot orthoses, walking sticks, four-wheeled walkers and sometimes wheelchairs or mobility scooters. Device usage among these groups also varied depending on environmental context e.g. an individual may have used a walking stick indoors, a four-wheeled walker outdoors and a wheelchair for long distances. Additionally, all PUs with stroke had upper limb impairments, which influenced the range of assistive devices which were feasible for them to use.

All frail PUs used a four-wheeled walker, although some preferred to use smaller devices like a walking stick around the home for convenience. Only one pre-frail PU regularly used devices to assist with mobility, however, those who did not use devices sought assistance from other persons for specific mobility tasks e.g. getting on and off buses. All PUs, except frail older adults, expressed concerns about their balance and risk of falling. While this finding may seem counter-intuitive, it can be attributed to a lack of risk exposure among frail PUs. Only one frail PU was independently mobile in the community – all others either required supervision or did not go out.

Inside the home, almost all PUs reported difficulties with housekeeping activities which involved standing while performing tasks with their hands e.g. food preparation and cooking. Many PUs reported sitting to complete these tasks, since their walking aids could not provide the necessary hands-free support. PUs across all groups reported receiving assistance from family or caregivers for housework such as cleaning, vacuuming and laundry. A number of PUs also reported having home adaptations to facilitate independent mobility e.g. ramps in/out of the home, stair lifts, grab rails and adapted toilet and shower seats. Gardening was also challenging for many PUs – some due to problems negotiating uneven terrain, and others due to issues with bending or crouching down, and rising from such positions.

In the community, all PUs experienced limitations in walking long distances, walking in crowded environments, negotiating stairs, and performing any activities that required hand/arm function while using a walking aid e.g. shopping. Most PUs relied on others for transport, as they did not or could not drive independently.

SUs had previous experience with a number of mobility aids, particular walking sticks, crutches, four-wheeled walkers and wheelchairs, as well as ankle–foot orthoses among SUs who worked with stroke populations. Most SUs reported positive experiences, although challenges reported included the limited contexts in which some devices can be used, lack of hands-free mobility aids, the bulkiness of some devices, difficulty in training PUs in the correct usage of a device, and limited access to devices due to costs or reimbursement options.

4.5.2.2 Design requirements for wearable assistive devices for mobility

Table 4.4 summarises the main design requirements noted by PUs and SUs for a wearable device to assist mobility. Design features are prioritised as 'must-have' and 'nice-to-have' based on the frequency and emphasis with which they were cited

Table 4.4 Prioritised design requirements for potential users of a soft lower body assistive exoskeleton

Primary users	Secondary users
Must-have design features	
Effective	Improves safety and stability for users, preventing falls during gait and transfers
Facilitates better quality, safer, faster walking, for longer and with less effort	Improves users' independence
Safe to use in the presence of other devices and health conditions	Usable in a range of different environments (home, clinic, community, etc.)
Hands-free usage	Adaptable to different users and their individual needs for assistance
Supports the appropriate joints/areas of the lower limb, as required by the user	Easy to use (particularly for PUs with cognitive impairment)
Help with foot drop and coordination during gait (PUs with stroke)	Easy to don and doff (even for users with upper limb impairments)
Easy to don and doff	Adjustable sizing
Worn under users' usual clothing	No extensive training and support required to use
Comfortable	User support available to PUs and SUs, if required
Lightweight	Lightweight
Easy to clean	Robust
Should not limit the user's own movements	Compatible with usual footwear and clothing
Easy to use	Easy to clean
Compatible with users' preferred clothing	Low cost/reimbursed through public and/ or private schemes
Not bulky to wear	
Low cost, or accessible at no cost to PU	
Nice-to-have design features	
Provides help and support with bending activities	Improves users' confidence in mobility and transfers
Robust materials	PU can don and doff independently
Not bulky to store	Worn under users' clothing
Possible to use independently	Not bulky, particularly if used in the home

by users of each type i.e. features noted by multiple users and/or highlighted as being vitally important were considered as 'must-haves'; features noted by very few users or considered as being not entirely necessary were considered to be 'nice-to-have'. Both PUs and SUs cited very similar requirements overall, with some notable discrepancies.

The effectiveness of the device in improving mobility and the safety of the device were high priorities for both user groups. Effectiveness was favoured by all PUs, whereas SUs felt that improving user safety – including prevention of falls – was the most important function of a device to assist mobility.

Both PUs and SUs felt strongly that being easy to use was an essential feature of any mobility-related assistive device. This includes being sufficiently simple and intuitive to use for PUs with cognitive impairments, and being easy to handle for users with upper limb impairments.

Of particular relevance to wearable devices was the requirement to be easy to don and doff, as noted by both PUs and SUs. Both groups felt that it should ideally be possible to don and doff a device independently, although PUs with stroke/SCI felt that requiring some assistance from another person would be acceptable, likely due to the higher prevalence and severity of upper limb impairments among these users.

Mixed opinions on aesthetics were noted; some PUs felt that appearance was very important, while others felt that a device does not have to look nice, as long as it is effective. SUs considered aesthetics to be more important if a device is to be used in community settings, rather than in hospital or rehabilitation settings, due to the potential for stigma to be felt by PUs in the community. Despite these mixed opinions, both user groups agreed that a device should be lightweight and not excessively bulky. Also, PUs with stroke/SCI noted that a device should be compatible with their usual clothing, as many had experienced problems using ankle–foot orthoses with their usual footwear and trousers.

Both user groups agreed that a device must be affordable, either by being available to purchase at low cost, or being available at no cost to PUs via public health services or reimbursement by health insurance companies. For expensive devices, SUs believed that reimbursement was critical to facilitate uptake, as they believed that users would very rarely have the means or willingness to pay for such devices.

4.5.2.3 User perspectives on a soft assistive exoskeleton concept

Most PUs had positive opinions on the XoSoft concept, and felt that it could be helpful for assisting walking and general mobility. The hands-free nature of the device was seen as a distinct benefit over current mobility aids. PUs with specific needs perceived other more specific benefits e.g. PUs with stroke felt it could help with foot drop and lower limb coordination during gait.

Frail older adult PUs were the least positive about the concept, with only one such individual expressing an interest in using such a device. They perceived few benefits to the concept over and above their current mobility aids. Most frail older PUs felt that XoSoft would be too complex for them to use, too difficult to don and doff, and considered learning to use such a device to be an unnecessary burden for them.

In general, SUs perceived a large range of positive functions and benefits to the XoSoft concept, particularly the flexibility and the adaptive nature, which could allow it to be used by many different patients with varying conditions in a range of settings.

SUs working in neurological rehabilitation had mixed opinions on the use of XoSoft, as some felt there was a sufficient array of exoskeleton products available in this field already. However, a device which could be used outside of a laboratory setting, takes little time to set up for use and can provide adjustable partial support to users was perceived as being beneficial, as it would offer functions beyond those of existing products.

The main concerns of SUs with XoSoft were ease of donning, doffing and general ease of use, particularly in frail users or those with cognitive impairments. SUs acknowledged the technical challenges of creating a sufficiently simple, yet reliable and effective interface for both PUs and SUs.

SUs also recognised the potential to use a soft lower body assistive exoskeleton with individuals outside the specific PU groups targeted in the XoSoft project. These broader potential applications could include use as a rehabilitation tool rather than as an assistive device for multiple patient populations in clinics and hospitals and for home and/or clinical use in musculoskeletal rehabilitation.

4.6 User needs: implications for soft exoskeleton design

The study described in Section 4.5 has revealed information of value to all design stakeholders on the requirements of potential PUs and SUs of a lower body soft exoskeleton to assist adults with mobility impairments. The study generated unique insights into users' desired functional and design requirements for such devices, from which many opportunities and challenges for soft exoskeleton technologies can be drawn.

4.6.1 Functional requirements

A clear understanding of users' design requirements (e.g. Table 4.4) is necessary to develop an effective device, since effectiveness is relative to the user's intended function or goal [57]. For PUs, an effective device would support and assist them in obtaining gross improvements in overall mobility and carrying out activities of daily living with greater ease and independence. SUs placed a greater emphasis on improving user safety and preventing falls. These findings echo those of a previous survey of wheelchair users' and healthcare professionals' perspectives on exoskeleton technology [58]. The emphasis on fall prevention and risk minimisation from healthcare professionals is understandable, given the proliferation of evidence, clinical guidelines and policies on this topic in recent years [59–61]. It is clear that providing adequate assistance to users while ensuring stability and safety are the key functional requirements.

It is also imperative that wearable devices that aim to assist mobility incorporate design features that support – and certainly do not impede – users' ability to undertake other activities. For example, aside from walking, support for bending and crouching activities were highlighted as priority functions by PUs in the present study and in a previous study of older adults' requirements of a rigid assistive exoskeleton [62]. There are considerable design and technical challenges associated with developing a user-friendly system that provides adequate assistance without restricting intended movements, while maintaining dynamic stability and ensuring user safety.

4.6.2 Design and aesthetics

A notable design feature sought after by the majority of users in the present study was the possibility to wear a device under their usual clothing. This is not a common feature of current exoskeletons, possibly because it would be unfeasible for bulky, rigid exoskeletons. However, when designing a device for use in home and community environments, aesthetics cannot be dismissed as a trivial aspect of device design. In home and community contexts, an assistive device may be a source of stigma for users [57], more so than in clinical contexts, a matter highlighted by SUs in the present study. This presents a distinct opportunity for lightweight, soft exoskeletons to be developed for use in the home and the community. It also emphasises the need to continually seek to reduce the size, weight and noise of rigid exoskeletons to maximise their viability in these contexts.

As seen previously in relation to rigid exoskeleton design, comfort and ease of putting on/taking off the device were of great concern to all users [58]. Ease of donning and doffing the device was greatly emphasised in the present study, most likely due to the high prevalence of upper limb impairments among the PUs interviewed and SUs' extensive experience working with such individuals. Soft exoskeleton technologies offer great potential to meet these users' needs, as they present opportunities to integrate sensing and actuation into comfortable lightweight garments that are fit for everyday life, can be easily donned and doffed, and/or worn for moderate periods of time. Further research to identify optimal soft exoskeleton garment designs to facilitate users with a variety of impairments is required.

4.6.3 Willingness to use the concept

The majority of PUs and SUs in the present study were positive about the concept of a soft assistive exoskeleton and indicated willingness to use such a device provided it meets their functional and design requirements. Frail older PUs were the least positive about the prospect of using such a device. These individuals were considerably older than other PU groups, had a greater number of co-morbidities, and experienced greater levels of disability and low levels of social participation. These factors can all contribute to these users' perceptions of the usability and useworthiness of an assistive device, particularly when considering the activities which the device aims to assist [57]. For example, if a frail PU with poor general health considers increasing their mobility to be neither possible to do, nor worth doing, then using a device to assist mobility will not be a priority.

Cost was identified as a concern for both PUs and SUs, in line with previous user studies on assistive exoskeletons [58] and can also affect willingness to use a concept. While soft exoskeletons are not yet at a sufficient technology readiness level to be placed on the market, potential avenues to create accessible products should be considered as part of overall PSS development. For example, offering leasing and or rental services to lower costs of ownership for PUs, while incorporating refurbishment and circular economy options for TUs.

4.6.4 Alternative assistive devices

Any proposed exoskeleton intended for use by stroke and SCI populations will face competition from existing exoskeleton products aimed at these user groups. Therefore, designers of new exoskeletons in this space must strive to create systems which offer new and/or improved features in comparison to existing products. The limited range of currently available devices are rigid exoskeletons for use in the rehabilitation of individuals with moderate–severe motor impairments and predominantly provide full assistance and support to users, while generally being only suitable for use in an intensive therapeutic context [63,64]. Opportunities therefore exist for assistive devices, devices for users with low–moderate levels of impairment, devices that offer adaptable support and devices which can be used in home or community settings. Also, considering the comfort and wearability requirements of users (Table 4.4), current hard exoskeletons are unlikely to be readily accepted by PUs with low–moderate levels of impairment for use in activities of daily living. A soft exoskeleton which provides adaptable support and can be used in the home and/or community would therefore provide a unique and innovative solution for users, with greater probability of user acceptance.

4.6.5 Current challenges for soft exoskeleton technologies

Soft robotics is an emerging area with a wide variety of potential applications across many fields, however, the technologies are very much still in development [65,66]. At present, it is difficult to foresee where the limits of soft robotic technologies lie, and thus what solutions may be realised or may not be possible, as the field is rapidly advancing. Considering user requirements together with technological limitations, some of the challenges that currently exist with respect to soft exoskeletons as assistive devices include:

- **Soft sensing (motion):** The use of soft sensors to measure the motion of the human body is relatively easier than their use in the measurement of soft robots: the human body can be considered as a relatively rigid structure in comparison to soft robots, which present highly complex motions with infinite degrees of freedom and non-linearities. Even where human joint motion is not ideal, it can often be approximated by a hinge or other simple kinematic chain. The challenge of measuring human joint motion using soft sensors lies in the fact that we need to measure it through indirect variables such as elongation and bending [67], since it is not always possible to place a sensor in the centre

of rotation of the human joint. For this reason, and in order to get representative information, a detailed calibration process is required to correlate the sensor reading with the joint angle. This calibration might also include the identification of hysteresis and other non-linear behaviours.
Most technologies are based on the electrical properties of certain materials which change under stress or strain conditions. These types of sensors can be classified as resistive and capacitive sensors. However, it is possible to use the measurement of other physical variables such as light transmission e.g. using optic fibres [68].

Another common challenge of measuring motion using soft sensors is the placement of the sensors, since it is not always possible to know the exact location of the sensor over time. For this reason, sensory fusion algorithms are required to combine the information coming from a redundant set of sensors to get a sufficiently reliable estimation of the motion.

- **Soft sensing (force/pressure):** Similar technologies as those used to measure motion can be used to measure force/pressure in a soft lower body exoskeleton, since the body areas where force/pressure will be measured will – in most cases – present very low levels of deformation [69]. Typically, the measurement of the external forces is only required on the contact between the foot and the floor. The level of information required depends on the design of the controller. A simple sensor that is able to detect the heel-strike and the toe-off moments is the minimum required for reliable gait detection. The robust measurement of the contact force at the feet is still a technical challenge, even more when considering the detection of the direction of the force.
- **Soft actuation:** Pneumatic systems have been widely used for soft actuation in wearable robots. Artificial muscles – comprised of pneumatic components combined with structured materials – have been successfully used for the actuation of assistive exoskeletons [70]. Their power–weight ratio is excellent (500–2,000 W/kg), and they can transmit high forces (100–500 N/cm^2) at a reasonable level of efficiency (32%–50%) [71]. The behaviour and interaction is quite natural; however, the control bandwidth is limited (typically around 5 Hz), due to limitations of the flow rate [72]. These systems also require of a source of pressure, commonly an air compressor or pressurized gas tank, which is not easy to integrate on a wearable device, and the level of noise is not suitable for some applications e.g. for home or community use.

  Cable-driven actuation is a common alternative which permits the use of classical actuation systems (electrical motors) in a soft wearable device. The actuators are placed in a rigid structure, usually carried on the back of the user, while the motion/force is transmitted to the target joints by means of cables [73]. These systems can provide a high level of assistance, only limited by the force exchange between the wearable device and the user. The weight and energy efficiency of cable-driven actuation depends mostly on the electrical motors and mechanical components used in the system, which can be in the order of 100–300 N m/kg for commercial motors.

Other alternatives for soft actuation systems in wearable devices exist, but this is still an open research topic with scarce applications in real scenarios. Technologies such as Shape Memory Alloys or Electroactive Polymers look promising, but it remains challenging to implement them on wearable devices [74]. However, when a low–moderate level of assistance is required, it is possible to use quasi-passive elements such as variable damping [75], variable stiffness, clutches or elastic materials, which can be implemented using non-classical actuation systems. In practice, the challenge of providing sufficient support for specific users and their varied individual needs over time will be highly complex (e.g. in stroke), since the requirements for support will differ between users depending on the nature and extent of their pathology.

- **Force transmission:** The assistive forces generated by the wearable device will be transmitted to the user through the body attachments. One of the major challenges is that the human-body surface is highly heterogeneous; each individual user is completely different, even more so when considering users with various pathologies. This makes the design of the body attachments highly important and challenging, since their design will limit the total amount of force the system can provide to the user. In most of the cases, the body attachments are designed taking advantage of designs present in other areas, such as orthosis, wheelchair postural supports and trekking backpacks.

  It is also important to consider that all the forces generated by the system will load the human rigid structure (bones and joints) because of the absence of a rigid structure in the wearable device. For this reason, the maximum expected assistive force cannot exceed the one generated by the human muscular system. That means, using a soft wearable assistive device, in the best case scenario, can restore the mobility of a disabled limb, but we cannot augment the capabilities of the human body.
- **Control:** The same kind of algorithms used in hard exoskeletons can be applied to soft wearable devices, but with some added challenges. As indicated previously, the estimation of the motion and interaction forces might not be fully reliable. Also, due to the soft nature of the system, the actuation may have a longer reaction time. However, the control system should maintain a high level of stability to avoid any discomfort for the user.
- **Human-robot interaction:** The use of soft robotics in assistive exoskeletons may offer improved safety over rigid exoskeletons when interacting with humans, but the aforementioned issues with sensing, actuation and control will also create more complexity when considering the safety of human-robot interaction.

Current technologies allow soft exoskeletons to provide low–moderate assistance, which would be appropriate for assisting users with low–moderate levels of impairment in most common tasks (e.g. standing, walking, etc.). These levels of assistance can represent a partial restoration of the gait pattern, implying an improvement of the stability and reduction of the energy expenditure, which is, in most of the cases, a major improvement in the quality of life of the patient.

The short-term future of soft assistive exoskeletons must involve compromise between the high-level demands and expectations of users and the technical feasibility of realising these demands [76]. The long-term future looks promising, considering the user requirements stated above. With further advances in soft assistive technologies more assistance could be provided for more complex tasks (e.g. sit-to-stand, stair climbing, etc.), with the ultimate aim of creating safe to use and user-friendly product-service systems.

4.7 Chapter summary

Using the framework of a project aiming to develop a soft lower body exoskeleton to assist mobility, this chapter has presented:

- An overview of the main principles of user-centred design;
- An example of the use of a mixed methods study design to engage users in soft exoskeleton design by exploring their needs and design requirements;
- Insights into the perspectives of users with mild–moderate mobility impairments, and individuals close to these users, on their requirements of an assistive device for mobility, and the implications of these for exoskeleton technologies;
- The main opportunities and challenges currently facing soft exoskeletons as assistive devices.

References

[1] Davis F.D., Bagozzi R.P., Warshaw P.R., 'User acceptance of computer technology: a comparison of two theoretical models'. *Management Science*, 1989. **35**(8): pp. 982–1003.

[2] Davis Jr F.D., *A technology acceptance model for empirically testing new end-user information systems: Theory and results*. 1986, Massachusetts Institute of Technology, Cambridge, MA.

[3] Venkatesh V., Davis F.D., 'A theoretical extension of the technology acceptance model: Four longitudinal field studies'. *Management Science*, 2000. **46**(2): pp. 186–204.

[4] Cook A.M., Polgar J.M., *Assistive technologies – E-Book: Principles and practice*. 4th ed., 2014, Elsevier Health Sciences.

[5] Shah S.G.S., Robinson I., AlShawi S., 'Developing medical device technologies from users' perspectives: a theoretical framework for involving users in the development process'. *International Journal of Technology Assessment in Health Care*, 2009. **25**(04): pp. 514–521.

[6] de Eyto A., Ryan A., McMahon M., Hassett G., Flynn M., 'Health futures lab: transdisciplinary development of T shaped professionals through "wicked problem" challenges'. *8th Engineering Education for Sustainable Development (EESD)*, UBC, Vancouver, Canada, 2015.

[7] Abras C., Maloney-Krichmar D., Preece J., 'User-centered design'. *Bainbridge, W. Encyclopedia of Human-Computer Interaction. Thousand Oaks: Sage Publications*, 2004. **37**(4): pp. 445–456.

[8] Eurostat, *European Health and Social Integration Survey*. 2012.

[9] Bateni H., Maki B.E., 'Assistive devices for balance and mobility: benefits, demands, and adverse consequences'. *Archives of Physical Medicine and Rehabilitation*, 2005. **86**(1): pp. 134–145.

[10] 1Bradley S.M., Hernandez C.R., 'Geriatric assistive devices'. *American Family Physician*, 2011. **84**(4): pp. 405–411.

[11] Hoenig H., Morgan M., Montgomery C., Landerman L.R., Caves K., 'One size does not fit all-mobility device type affects speed, collisions, fatigue, and pain'. *Archives of Physical Medicine and Rehabilitation*, 2015. **96**(3): pp. 489–497.

[12] Stevens J.A., Thomas K., Teh L., Greenspan A.I., 'Unintentional fall injuries associated with walkers and canes in older adults treated in US emergency departments'. *Journal of the American Geriatrics Society*, 2009. **57**(8): pp. 1464–1469.

[13] Feigin V.L., Forouzanfar M.H., Krishnamurthi R., *et al.*, 'Global and regional burden of stroke during 1990–2010: findings from the Global Burden of Disease Study 2010'. *The Lancet*, 2014. **383**(9913): pp. 245–255.

[14] Lawrence E.S., Coshall C., Dundas R., *et al.*, 'Estimates of the prevalence of acute stroke impairments and disability in a multiethnic population'. *Stroke*, 2001. **32**(6): pp. 1279–1284.

[15] Woolley S.M., 'Characteristics of gait in hemiplegia'. *Topics in Stroke Rehabilitation*, 2001. **7**(4): pp. 1–18.

[16] Lamontagne A., Malouin F., Richards C.L., Dumas F., 'Mechanisms of disturbed motor control in ankle weakness during gait after stroke'. *Gait & Posture*, 2002. **15**(3): pp. 244–255.

[17] Roche N., Bonnyaud C., Geiger M., Bussel B., Bensmail D., 'Relationship between hip flexion and ankle dorsiflexion during swing phase in chronic stroke patients'. *Clinical Biomechanics (Bristol, Avon)*, 2015. **30**(3): pp. 219–225.

[18] Burdett R.G., Borello-France D., Blatchly C., Potter C., 'Gait comparison of subjects with hemiplegia walking unbraced, with ankle–foot orthosis, and with Air-Stirrup brace'. *Physical Therapy*, 1988. **68**(8): pp. 1197–1203.

[19] Yamamoto S., Ibayashi S., Fuchi M., Yasui T., 'Immediate-term effects of use of an ankle–foot orthosis with an oil damper on the gait of stroke patients when walking without the device'. *Prosthetics and Orthotics International*, 2015. **39**(2): pp. 140–149.

[20] Chen G., Patten C., Kothari D.H., Zajac F.E., 'Gait differences between individuals with post-stroke hemiparesis and non-disabled controls at matched speeds'. *Gait Posture*, 2005. **22**(1): pp. 51–56.

[21] Stanhope V.A., Knarr B.A., Reisman D.S., Higginson J.S., 'Frontal plane compensatory strategies associated with self-selected walking speed in

individuals post-stroke'. *Clinical Biomechanics (Bristol, Avon)*, 2014. **29**(5): pp. 518–522.

[22] Cruz T.H., Dhaher Y.Y., 'Impact of ankle–foot-orthosis on frontal plane behaviors post-stroke'. *Gait & Posture*, 2009. **30**(3): pp. 312–316.

[23] Kim C.M., Eng J.J., 'Magnitude and pattern of 3D kinematic and kinetic gait profiles in persons with stroke: relationship to walking speed'. *Gait & Posture*, 2004. **20**(2): pp. 140–146.

[24] Connell L.A., Lincoln N., Radford K., 'Somatosensory impairment after stroke: frequency of different deficits and their recovery'. *Clinical Rehabilitation*, 2008. **22**(8): pp. 758–767.

[25] Rowe F., Brand D., Jackson C.A., *et al.*, 'Visual impairment following stroke: do stroke patients require vision assessment?'. *Age and Ageing*, 2009. **38**(2): pp. 188–193.

[26] Lord S.E., McPherson K., McNaughton H.K., Rochester L., Weatherall M., 'Community ambulation after stroke: how important and obtainable is it and what measures appear predictive?'. *Archives of Physical Medicine and Rehabilitation*, 2004. **85**(2): pp. 234–239.

[27] Patel M.D., Coshall C., Rudd A.G., Wolfe C.D., 'Cognitive impairment after stroke: clinical determinants and its associations with long-term stroke outcomes'. *Journal of the American Geriatrics Society*, 2002. **50**(4): pp. 700–706.

[28] Tatemichi T., Desmond D., Stern Y., Paik M., Sano M., Bagiella E., 'Cognitive impairment after stroke: frequency, patterns, and relationship to functional abilities'. *Journal of Neurology, Neurosurgery & Psychiatry*, 1994. **57**(2): pp. 202–207.

[29] Furlan J.C., Sakakibara B.M., Miller W.C., Krassioukov A.V., 'Global incidence and prevalence of traumatic spinal cord injury'. *Canadian Journal of Neurological Sciences*, 2013. **40**(4): pp. 456–464.

[30] National Spinal Cord Injury Statistical Center, *Spinal Cord Injury (SCI) Facts and Figures at a Glance*. 2016; Available from: https://www.nscisc.uab.edu/Public/Facts%202016.pdf.

[31] Van Hedel H., Wirz M., Dietz V., 'Standardized assessment of walking capacity after spinal cord injury: the European network approach'. *Neurological Research*, 2008. **30**(1): pp. 61–73.

[32] Louie D.R., Eng J.J., Lam T., Spinal Cord Injury Research Evidence Research Team, 'Gait speed using powered robotic exoskeletons after spinal cord injury: a systematic review and correlational study'. *Journal of NeuroEngineering and Rehabilitation*, 2015. **12**: pp. 82.

[33] Fisahn C., Aach M., Jansen O., *et al.*, 'The effectiveness and safety of exoskeletons as assistive and rehabilitation devices in the treatment of neurologic gait disorders in patients with spinal cord injury: a systematic review'. *Global Spine Journal*, 2016. **6**(8): pp. 822–841.

[34] Grasmucke D., Zieriacks A., Jansen O., *et al.*, 'Against the odds: what to expect in rehabilitation of chronic spinal cord injury with a neurologically controlled Hybrid Assistive Limb exoskeleton. A subgroup analysis of

55 patients according to age and lesion level'. *Neurosurgical focus*, 2017. **42**(5): pp. E15.

[35] New P., Simmonds F., Stevermuer T., 'A population-based study comparing traumatic spinal cord injury and non-traumatic spinal cord injury using a national rehabilitation database'. *Spinal Cord*, 2011. **49**(3): pp. 397–403.

[36] New P.W., Rawicki H.B., Bailey M.J., 'Nontraumatic spinal cord injury: demographic characteristics and complications'. *Archives of Physical Medicine and Rehabilitation*, 2002. **83**(7): pp. 996–1001.

[37] United Nations Department of Economic and Social Affairs Population Division, *World Population Ageing 2013*, New York: United Nations. 2013.

[38] Salive M.E., 'Multimorbidity in older adults'. *Epidemiologic Reviews*, 2013. **35**: pp. 75–83.

[39] Gijsen R., Hoeymans N., Schellevis F.G., Ruwaard D., Satariano W.A., van den Bos G.A., 'Causes and consequences of comorbidity: a review'. *Journal of Clinical Epidemiology*, 2001. **54**(7): pp. 661–674.

[40] Hung W.W., Ross J.S., Boockvar K.S., Siu A.L., 'Recent trends in chronic disease, impairment and disability among older adults in the United States'. *BMC Geriatrics*, 2011. **11**: pp. 47.

[41] Chatterji S., Byles J., Cutler D., Seeman T., Verdes E., 'Health, functioning, and disability in older adults – present status and future implications'. *The Lancet*, 2015. **385**(9967): pp. 563–575.

[42] Xue Q.L., 'The frailty syndrome: definition and natural history'. *Clinics in Geriatric Medicine*, 2011. **27**(1): pp. 1–15.

[43] Fried L.P., Tangen C.M., Walston J., *et al.*, Cardiovascular Health Study Collaborative Research Group, 'Frailty in older adults: evidence for a phenotype'. *The Journals of Gerontology. Series A, Biological Sciences and Medical Sciences*, 2001. **56**(3): pp. M146–M156.

[44] Ferri C.P., Prince M., Brayne C., *et al.*, 'Global prevalence of dementia: a Delphi consensus study'. *The Lancet*, 2005. **366**(9503): pp. 2112–2117.

[45] Busse A., Bischkopf J., Riedel-Heller S.G., Angermeyer M.C., 'Mild cognitive impairment: prevalence and incidence according to different diagnostic criteria. Results of the Leipzig Longitudinal Study of the Aged (LEILA75+)'. *British Journal of Psychiatry*, 2003. **182**: pp. 449–454.

[46] Storz-Pfennig P., Schmedders M., Dettloff M., 'Trials are needed before new devices are used in routine practice in Europe'. *BMJ: British Medical Journal*, 2013. **346**: f1646.

[47] Faruqui S.R., Jaeblon T., 'Ambulatory assistive devices in orthopaedics: uses and modifications'. *Journal of the American Academy of Orthopaedic Surgeons*, 2010. **18**(1): pp. 41–50.

[48] de Vries O.J., Peeters G.M., Elders P.J., *et al.*, 'Multifactorial intervention to reduce falls in older people at high risk of recurrent falls: a randomized controlled trial'. *Archives of Internal Medicine*, 2010. **170**(13): pp. 1110–1117.

[49] Wilhelmson K., Andersson C., Waern M., Allbeck P., 'Elderly people's perspectives on quality of life'. *Ageing & Society*, 2005. **25**(04): pp. 585–600.

[50] Pettersson I., Berndtsson I., Appelros P., Ahlström G., 'Lifeworld perspectives on assistive devices: Lived experiences of spouses of persons with stroke'. *Scandinavian Journal of Occupational Therapy*, 2005. **12**(4): pp. 159–169.

[51] Chen T.-Y.A., Mann W.C., Tomita M., Nochajski S., 'Caregiver involvement in the use of assistive devices by frail older persons'. *OTJR: Occupation, Participation and Health*, 2000. **20**(3): pp. 179–199.

[52] Sun J.-H., Tan L., Yu J.-T., 'Post-stroke cognitive impairment: epidemiology, mechanisms and management'. *Annals of Translational Medicine*, 2014. **2**(8): p. 80.

[53] Folstein M.F., Folstein S.E., McHugh P.R., '"Mini-mental state". A practical method for grading the cognitive state of patients for the clinician'. *Journal of Psychiatric Research*, 1975. **12**(3): pp. 189–198.

[54] Holden M.K., Gill K.M., Magliozzi M.R., 'Gait assessment for neurologically impaired patients. Standards for outcome assessment'. *Physical Therapy*, 1986. **66**(10): pp. 1530–1539.

[55] Middleton A., Fritz S.L., Lusardi M., 'Walking speed: the functional vital sign'. *Journal of Aging and Physical Activity*, 2015. **23**(2): pp. 314–322.

[56] Mahoney F.I., Barthel D., 'Functional evaluation: the Barthel Index'. *Maryland State Medical Journal*, 1965. **14**: pp. 56–61.

[57] Krantz O., 'Assistive devices utilisation in activities of everyday life – a proposed framework of understanding a user perspective'. *Disability and Rehabilitation: Assistive Technology*, 2012. **7**(3): pp. 189–198.

[58] Wolff J., Parker C., Borisoff J., Mortenson W., Mattie J., 'A survey of stakeholder perspectives on exoskeleton technology'. *Journal of NeuroEngineering and Rehabilitation*, 2014. **11**(1): pp. 169.

[59] American Geriatrics Society, British Geriatrics Society, 'Summary of the Updated American Geriatrics Society/British Geriatrics Society Clinical Practice Guideline for Prevention of Falls in Older Persons'. *Journal of the American Geriatrics Society*, 2011. **59**(1): pp. 148–157.

[60] Avin K.G., Hanke T.A., Kirk-Sanchez N., *et al.*, Academy of Geriatric Physical Therapy of the American Physical Therapy Association, 'Management of falls in community-dwelling older adults: clinical guidance statement from the Academy of Geriatric Physical Therapy of the American Physical Therapy Association'. *Physical Therapy*, 2015. **95**(6): pp. 815–834.

[61] NICE, *Falls: assessment and prevention of falls in older people.* NICE clinical guideline 161. 2013, National Institute for Health and Care Excellence, Manchester.

[62] O'Sullivan L., Power V., Virk G., *et al.*, 'End user needs elicitation for a full-body exoskeleton to assist the elderly'. *Procedia Manufacturing*, 2015. **3**: pp. 1403–1409.

[63] Sale P., Franceschini M., Waldner A., Hesse S., 'Use of the robot assisted gait therapy in rehabilitation of patients with stroke and spinal cord injury'. *European Journal of Physical and Rehabilitation Medicine*, 2012. **48**(1): pp. 111–121.

[64] Chen G., Chan C.K., Guo Z., Yu H., 'A review of lower extremity assistive robotic exoskeletons in rehabilitation therapy'. *Critical Reviews in Biomedical Engineering*, 2013. **41**(4–5): pp. 343–363.

[65] Trivedi D., Rahn C.D., Kier W.M., Walker I.D., 'Soft robotics: biological inspiration, state of the art, and future research'. *Applied Bionics and Biomechanics*, 2008. **5**(3): pp. 99–117.

[66] Majidi C., 'Soft robotics: a perspective—current trends and prospects for the future'. *Soft Robotics*, 2014. **1**(1): pp. 5–11.

[67] Lucarotti C., Totaro M., Sadeghi A., Mazzolai B., Beccai L., 'Revealing bending and force in a soft body through a plant root inspired approach'. *Scientific Reports*, 2015. **5**: pp. 8788.

[68] Salceda-Delgado G., Van Newkirk A., Antonio-Lopez J., Martinez-Rios A., Schülzgen A., Correa R.A., 'Compact fiber-optic curvature sensor based on super-mode interference in a seven-core fiber'. *Optics Letters*, 2015. **40**(7): pp. 1468–1471.

[69] Viry L., Levi A., Totaro M., *et al.*, 'Flexible three-axial force sensor for soft and highly sensitive artificial touch'. *Advanced Materials*, 2014. **26**(17): pp. 2659–2664.

[70] Tsagarakis N.G., Caldwell D.G., 'Development and control of a "soft-actuated" exoskeleton for use in physiotherapy and training'. *Autonomous Robots*, 2003. **15**(1): pp. 21–33.

[71] Tsagarakis N., Caldwell D.G. *Improved modelling and assessment of pneumatic muscle actuators. Robotics and Automation, 2000. Proceedings. ICRA'00. IEEE International Conference on*, 2000. IEEE.

[72] Davis S., Tsagarakis N., Canderle J., Caldwell D.G., 'Enhanced modelling and performance in braided pneumatic muscle actuators'. *The International Journal of Robotics Research*, 2003. **22**(3–4): pp. 213–227.

[73] Asbeck A.T., Dyer R.J., Larusson A.F., Walsh C.J., 'Biologically-inspired soft exosuit'. *2013 IEEE international conference on rehabilitation robotics (ICORR)*, 2013, IEEE. pp. 1–8.

[74] Cianchetti M., Mattoli V., Mazzolai B., Laschi C., Dario P., 'A new design methodology of electrostrictive actuators for bio-inspired robotics'. *Sensors and Actuators B: Chemical*, 2009. **142**(1): pp. 288–297.

[75] Hauser S., Robertson M.C., Ijspeert A., Paik J. 'JammJoint: A variable stiffness device based on granular jamming for wearable joint support'. *IEEE Robotics and Automation Letters*, 2017. DOI: 10.1109/LRA.2017.2655109.

[76] Viteckova S., Kutilek P., Jirina M., 'Wearable lower limb robotics: A review'. *Biocybernetics and Biomedical Engineering*, 2013. **33**(2): pp. 96–105.

Section 2
Design and control of exoskeletons

Creating wearable robotic systems requires unique robot designs and controllers. While industrial manipulators and mobile robots have been extensively studied for over a half century, exoskeleton robots are relative new, with a very short history of less than two decades and there are still many challenges in design and control to be solved. Some major challenges, as the editors foresee, details of include the mechanism design to facilitate and ease physical human–robot interaction, the development and selection of sensors for motion and human motion intention detection, novel actuators with reduced inertia and impedance to make the exoskeletons safe and sufficiently comfortable for short-term and long-term use, and controllers that facilitate the interaction between the wearer and the exoskeleton, among others. In this section, we included four chapters to address separately the above issues.

Chapter 5 presents a novel design of a spherical shoulder mechanism and its control requirements. In designing exoskeletons, a big challenge is to generate the desired motion while structurally complying with the human biomechanics and details fitting to individual humans. In the presented design, a hybrid mechanism that consists of two revolute joints connected together via a double parallelogram linkage is developed, and a spherical mechanism with three-degrees-of-rotations is built. The new design features a lightweight structure, large workspace free of singularities, and reasonable structural stiffness for force/load transfer.

Chapter 6 addresses the application of wearable motion sensors (WMSs). These types of sensors are commonly used in exoskeletons to detect the position and orientations for both the human and exoskeletons. The chapter presents an evaluation platform for assessing the accuracy of WMSs. The platform is an instrumented gimbal with three rotation axes. Each axis is equipped with a DC motor and an absolute encoder and it can be used for accurate motion analysis without any additional equipment.

Chapter 7 presents compliant actuators and exoskeleton control. Contrary to traditional industrial robots where stiff actuators are used, exoskeletons adopt mostly compliant actuators for better human–robot interaction. This chapter starts with an overview of compliant actuators. Series elastic actuators (SEAs) are described in connection with an exoskeleton, namely, NEUROEXO from Scuola Superiore Sant'Anna, Italy. Its mechanics and control are described as well as assessing the performance of the SEA in the NEUROEXO exoskeleton.

Chapter 8 describes the control method for a knee–ankle–foot exoskeleton robot. The control methods in exoskeletons can be classified into trajectory-based and interaction-based approaches. This chapter presents an interaction-based control method and outlines the mechanics of the exoskeleton, including the biomechanics, the robot design, including the compliant actuator design. The interaction control in this chapter aims to achieve human–robot gait synchronization. The research includes gait-event detection, development of than adaptive oscillator, and an assistive controller are introduced, with experimental results presented for different conditions.

The four chapters in this section give an introductory knowledge to the design and control of exoskeletons. In addition to these topics mentioned, there are many other issues highly relevant to exoskeleton design and control, for example, modelling and simulation [1], EEG/EMG based controllers [2–5], specialized soft exoskeleton control method [6], pneumatic muscle actuation strategies [6], and exoskeleton hand [7]. As it is only possible to include a limited selection of papers here, the readers are directed to more papers as given in the references here and others in the four chapters.

References

[1] Human Modeling for Bio-Inspired Robotics, Jun Ueda and Yuichi Kurita (eds), Academic Press, 2016.

[2] Jiménez-Fabián R, Verlinden O. Review of control algorithms for robotic ankle systems in lower-limb orthoses, prostheses, and exoskeletons. *Medical Engineering & Physics*. 2012; 34(4):397–408.

[3] Tucker MR, Olivier J, Pagel A, *et al.* Control strategies for active lower extremity prosthetics and orthotics: a review. *Journal of Neuro Engineering and Rehabilitation*. 2015;12(1):1. doi:10.1186/1743-0003-12-1.

[4] Kawase T, Sakurada T, Koike Y and Kansaku K, A hybrid BMI-based exoskeleton for paresis: EMG control for assisting arm movements. *Journal of Neural Engineering*. 2017;14:016015

[5] Lalitharatne T, Teramoto K, Hayashi Y, *et al.* Towards hybrid EEG-EMG-based control approaches to be used in bio-robotics applications: current status, challenges and future directions. *Paladyn, Journal of Behavioral Robotics*, 2013;4(2):147–154.

[6] Tsagarakis N, Caldwell DG. Development and control of a 'Soft-Actuated' exoskeleton for use in physiotherapy and training. *Autonomous Robots* 2003;15(1):21–33.

[7] Wege A, Kondak K, Hommel G. Mechanical design and motion control of a hand exoskeleton for rehabilitation. *IEEE International Conference Mechatronics and Automation*. 2005; pp. 155–159.

Chapter 5

Design and control of spherical shoulder exoskeletons for assistive applications

Shaoping Bai[1], Simon Christensen[1], and Muhammad Raza Ul Islam[1]

Abstract

The shoulder complex is the most complex joint in human's four limbs and this poses a big challenge in the exoskeleton design to achieve a mechanism able to generate the desired motion while structurally complying with human anatomy. This chapter presents a novel design of an exoskeleton shoulder mechanism and its control. The new design is a hybrid mechanism that consists of two revolute joints connected together via a double parallelogram linkage (DPL). By virtue of a DPL, a remote center of rotation can be established for a spherical mechanism with three-degree-of-rotations. In the chapter, the working principle of the shoulder mechanism is described. The kinematics of the mechanism is analyzed. Mechanism design and exoskeleton control are also presented.

Keywords: Assistive exoskeletons; shoulder mechanism; exoskeleton control; double parallelogram

5.1 Introduction

Exoskeletons can provide powers for assisting upper and lower body movements. Most developments with exoskeletons have been carried out for lower body systems. Compared with the lower limb, the movement at the upper limb is more sophisticated, due to the fact that a high level of range of motion and dexterity is needed, which poses a great challenge for the design of exoskeletons to be compatible with human motion [1–7].

A major challenge in upper-body exoskeleton design is the shoulder joint. For humans, the shoulder complex is one of the most anatomically complex areas in the human body, for which the total number of degrees-of-freedom is arguably

[1]Department of Mechanical and Manufacturing Engineering, Aalborg University, Denmark

up to five. The large number of degrees-of-freedom and also the range of motion increase the complexities of both design and control of the exoskeletons. There have been many different designs developed in the past, with aims to improve motion dexterity, flexibility, comfortability, etc. For powered motion assistance applications, a lightweight yet mechanically rigid design is desirable.

In this chapter, a novel design of an exoskeleton for the shoulder joint is described. The new design is a hybrid mechanism that consists of two revolute joints connected together via a double parallelogram linkage (DPL). By virtue of a DPL, a remote center (RC) of rotation can be established to design a mechanism with three degree-of-freedom rotations. In this chapter, the state-of-the-art of shoulder mechanisms is reviewed. The working principle is described. The kinematics of the mechanism is analyzed. The shoulder mechanism design is described. In connection with the shoulder exoskeleton control, different control techniques are described and discussed. Usability testing results are finally presented.

5.2 State-of-the-art in shoulder exoskeletons

The design challenges with power assistive upper-body exoskeletons are mainly related to the complexity of the human body, especially the shoulder complex, one of the most anatomically complex areas in the human body [8]. The shoulder complex is commonly described as two parts, the shoulder joint, or glenohumeral joint, and the shoulder griddle (see Figure 5.1). The glenohumeral joint can be described as a ball-and-socket joint with three degree-of-freedoms that describes the orientation of the humerus. The shoulder griddle is made up of the sternoclavicular joint, the acromioclavicular joint and the scapulothoracic joint. These three joints form a closed kinematic chain and are therefore unable to move independently [9]. The dominant motions of the shoulder girdle are the two rotational motions: elevation/depression and protraction/retraction. In total, the shoulder complex has five degrees-of-freedom. However, an exoskeleton that can power all degrees-of-freedom (DOFs) is often not practical, particularly not for portable/wearable exoskeletons. Instead, only the rotational degrees-of-freedom are considered.

Conventional designs of shoulder exoskeletons use a serial linkage system with 3-revolute (3R) joints [10] to generate the spherical motion of the human shoulder joint (see Figure 5.2(a)). A problem with a serial structure is its workspace limitations. The wearer of the exoskeleton can only raise the upper arm for a small angle in the frontal plane before the shoulder mechanism collides with their shoulder, neck, or head. To avoid this problem, some alternative designs have been proposed. The design in [10] minimized the effect of these problems by designing their exoskeletons so that the singular configurations and collision problem of the 3R mechanism occur at postures that are less likely for the wearer to reach. In an approach reported in [11], one of the links in the 3R mechanism is replaced with a circular guide to further avoid collision with the wearer (see Figure 5.2(b)). In [12], the circular guide, which often has a heavy and complicated construction, is avoided by moving one of the degrees-of-freedom of the shoulder to the elbow (see Figure 5.2(c)). A problem with this solution is the added mass at the elbow due to

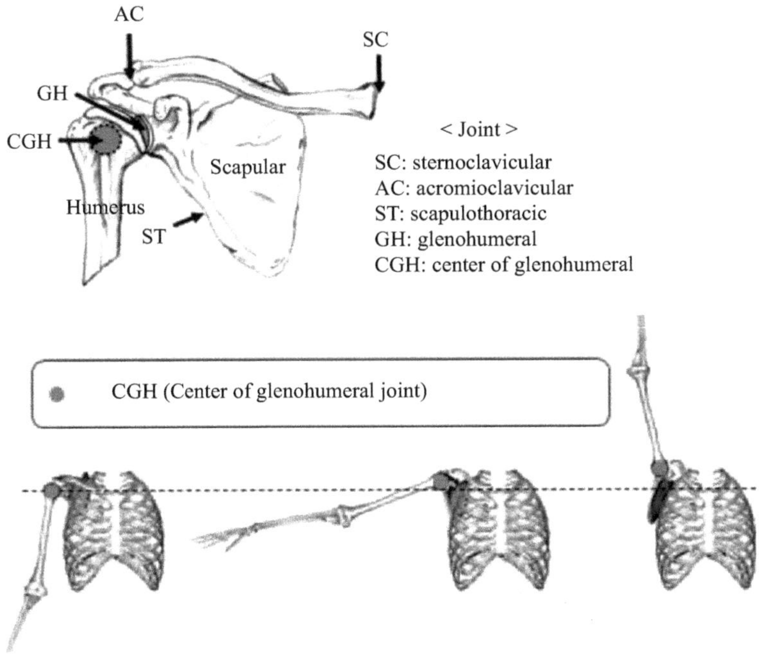

Figure 5.1 Structure of the shoulder complex and movement of the glenohumeral joint [9]

the additional actuator/joint. In [13], an extra link is added to the 3R mechanism making it a 4R mechanism (see Figure 5.2(d)). Because of the redundancy of the mechanism, the robot can avoid singular configurations and collisions with the wearer. For portable exoskeletons adding an active redundant degree-of-freedom causes a concern on both the weight and the occupied space of the exoskeleton.

The design of the shoulder mechanism described in this chapter adopts DPLs. This leads to a design of a compact structure, light weight yet rigid design, and a large range of motion free of singularities.

5.3 Kinematics of spherical shoulder exoskeleton

The proposed design is a hybrid mechanism that consists of two revolute joints connected together via four links that form a DPL, which under the given configuration form a RC of motion mechanism, as illustrated in Figure 5.3. The mechanism works kinematically equivalent to the 3R wrist mechanisms, meaning that it can rotate about three independent axes that all coincide in one point, namely, the RC.

A schematic drawing of the spherical shoulder mechanism is shown in Figure 5.4. The design parameters include four link lengths L_i, $i = 1, \ldots, 4$ for the two parallelograms A-B-C-D and D-E-F-G. In addition, two offset angles ϕ_1 and ϕ_2 are

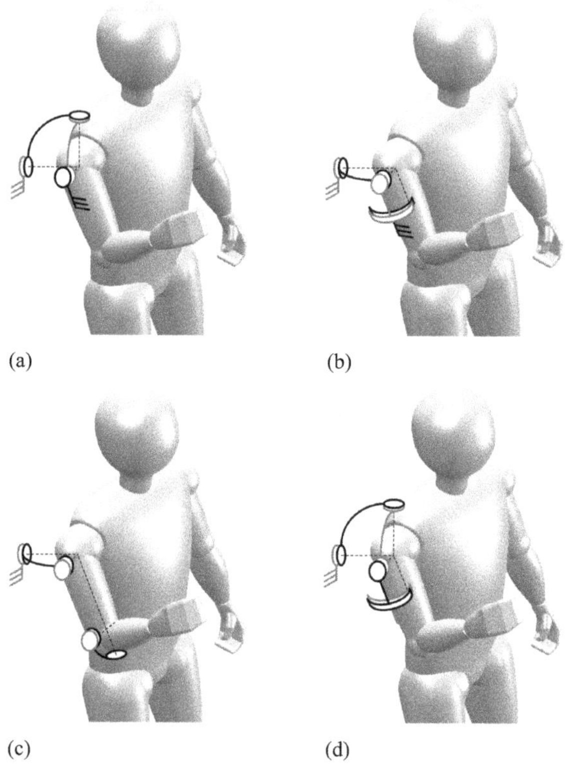

Figure 5.2 Principle of a shoulder joint mechanism using (a) three revolute joints, (b) two revolute joints and a circular guide, (c) three revolute joints with one placed at the elbow, and (d) a redundant joint

defined to account for the offset angle of the two parallelograms due to the size of actuators (motors). As the parallelograms are serially connected with each other, they form a third virtual parallelogram between points B-D-E-RC. The range of motion of the DPL is determined from this virtual parallelogram.

As the mechanism has a hybrid structure, namely, a planar linkage connecting two revolute joints for a spherical mechanism, the kinematic analysis is conducted for both the planar linkage and the spherical mechanism.

5.3.1 Planar kinematics of the DPL

In Figure 5.4, a coordinate system x_0, y_0, z_0 is fixed to the first revolute joint, where the z_0-axis is fixed along line $\mathcal{L}_1$. In the DPL, AC is the input link and EG is the output link. In general, we can formulate the planar kinematics of the DPL using three Cartesian coordinates per link; x and z coordinates of the center point of each link and an orientation angle of the link. Hence, a total of 15 equations are needed to describe the position and orientation of the links in the DPL

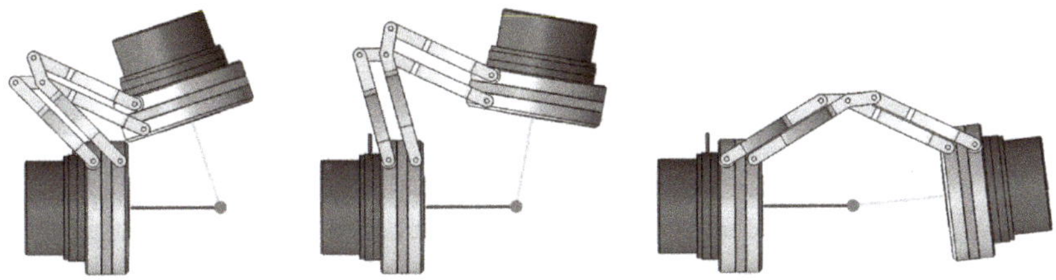

Figure 5.3 Working principle of double parallelogram linkage in the shoulder mechanism

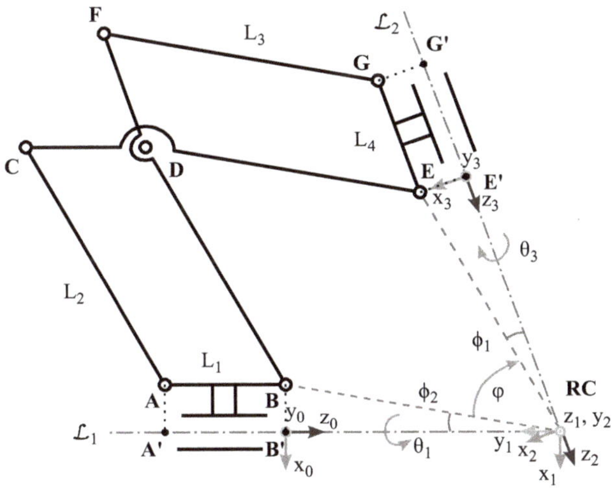

Figure 5.4 Kinematic model of the shoulder mechanism

$$\Phi(q(t), t) = 0 \tag{5.1}$$

where Φ is the set of constraint equations, q is the generalized coordinates of the links, and t is time. Differentiating the constraint equations yields

$$\dot{\Phi}(q(t)t) = \Phi_q \dot{q} + \Phi_t = 0 \tag{5.2}$$

where Φ_q is the constraint Jacobian, Φ_t is the partial derivative of Φ with respect to t, and $\dot{q}$ is the generalized coordinates velocities. The singular configuration is found from

$$\det(\Phi_q) = 0 \tag{5.3}$$

which yields

$$\psi = -\phi_2 \qquad \psi = \pi - \phi_2 \tag{5.4}$$

$$\psi = -\phi_1 \qquad \psi = \pi - \phi_1 \tag{5.5}$$

As seen from the result above, the DPL has four singular configurations, which are linked to the closed loop kinematic chain of the DPL. When one of the parallelograms is collapsed, i.e., ABCD or DEFG constitutes a straight line, the DPL gains a freedom and has two instantaneous degrees-of-freedom. At this configuration, an instantaneous change of ψ can lead to the parallelogram switching into an anti-parallelogram. As a result, the three axes of rotation of the DPL no longer coincide in a fixed RC. Therefore, to maximize the range of motion the DPL free of singularities, the two offset angles must be as close to each other in size as possible. The four singular configurations also reveal that the DPL cannot have a range of motion above 180°, within which the mechanism is free of singularity.

5.3.2 Kinematics of the shoulder mechanism

The kinematics of the shoulder mechanism is formulated based on Denavit–Hartenberg's convention, where Cartesian coordinate frames are attached to each link of a manipulator, as shown in Figure 5.4. The corresponding D–H parameters can be obtained as listed in Table 5.1. As the mechanism forms a spherical joint, the kinematics of the DPL can be expressed solely by rotations. The rotation matrix of the mechanism is readily obtained as

$$\mathbf{R} = \begin{bmatrix} c\theta_1 c\theta_2 c\theta_3 - s\theta_1 s\theta_3 & -c\theta_3 s\theta_1 - c\theta_1 c\theta_2 s\theta_3 & -c\theta_1 s\theta_2 \\ c\theta_1 s\theta_3 + c\theta_2 c\theta_3 s\theta_1 & c\theta_1 c\theta_3 - c\theta_2 s\theta_1 s\theta_3 & -s\theta_1 s\theta_2 \\ c\theta_3 s\theta_2 & -s\theta_2 s\theta_3 & c\theta_2 \end{bmatrix} \tag{5.6}$$

The Jacobian of the shoulder mechanism is the matrix that maps the joint velocities in the actuator space to the velocity state in the end-effector space:

$$\dot{\theta} = \mathbf{J}_\omega{}^{-1} \omega_\mathbf{e} \tag{5.7}$$

where $\dot{\theta} = \left[\dot{\theta}_1\ \dot{\theta}_2\ \dot{\theta}_3\right]^T$ is a vector with the joint angular velocities, $\mathbf{J}_\omega$ is the Jacobian and $\omega_\mathbf{e} = [\omega_x\ \omega_y\ \omega_z]^T$ is the end-effector angular velocities. The relation between the joint velocities and end-effector velocities is determined from the rotation matrix described in (5.6).

$$\omega_x = \dot{r}_{31} r_{21} + \dot{r}_{32} r_{22} + \dot{r}_{33} r_{23} \tag{5.8}$$

$$\omega_y = \dot{r}_{11} r_{31} + \dot{r}_{12} r_{32} + \dot{r}_{13} r_{33} \tag{5.9}$$

$$\omega_z = \dot{r}_{21} r_{11} + \dot{r}_{22} r_{12} + \dot{r}_{23} r_{13} \tag{5.10}$$

Table 5.1 Denavit–Hartenberg parameters of the proposed mechanism

Link, i	a_{i-1}	α_{i-1}	d_i	θ_i
1.	0	0	$L_3 \cos \phi_2$	θ_1
2.	0	90°	0	$(\psi + \phi_1 + \phi_2)$
3.	0	−90°	0	θ_3

where r_{ij} is the (i, j) entry of matrix **R**, and

$$\dot{r}_{ij} = \frac{dr_{ij}}{dt} = \frac{\partial r_{ij}}{\partial \theta_1}\dot{\theta}_1 + \frac{\partial r_{ij}}{\partial \theta_2}\dot{\theta}_2 + \frac{\partial r_{ij}}{\partial \theta_3}\dot{\theta}_3 \tag{5.11}$$

The Jacobian is found as

$$\mathbf{J}_\omega = \begin{bmatrix} 0 & s\theta_1 & -s\theta_2 c\theta_1 \\ 0 & -c\theta_1 & -s\theta_2 s\theta_1 \\ 1 & 0 & c\theta_2 \end{bmatrix} \tag{5.12}$$

Similar to the DPL, we determine the singularities of the shoulder mechanism with:

$$\det(\mathbf{J}) = -s\theta_2 = -s(\psi + \phi_1 + \phi_2) = 0 \tag{5.13}$$

which yields

$$\psi = -\phi_1 - \phi_2 \tag{5.14}$$

Equation (5.14) shows that the shoulder mechanism reaches a singular configuration where the joint axes constitute a common plane. Therefore, to maximize the range of motion that is free of singularities, the two offset angles must be as small as possible.

5.4 Shoulder mechanism design

In the construction of the shoulder mechanism, the challenge is to comfortably deliver the power assistance needed for moving collaboratively with the human limbs. To a large extent, this challenge is related to the method of actuation. For exoskeletons, the actuation has to drive the system safely and compliantly. A back-drivable transmission is thus needed to avoid the wearer being locked by the exoskeleton and enhance its safety and comfortability. Moreover, it also has to be lightweight and compact. Most active assistive exoskeletons are actuated by conventional electric motors or pneumatic muscle actuators (PMA). Electric motors are the most widely used actuation method due to their high precision and advanced motion control. The actuators are either placed locally at the joints or at the torso, where the power is transmitted via cables, like the design of CADEN-7 [10]. The PMA is analogy to human muscles in that they can perform one-way actuation. Their lightness, good power-to-weight ratio, and flexible structure make them desirable for portable exoskeleton. The flexible structure is also an advantage over the electric motors. The RUPERT (Robotic Upper Extremity Repetitive Trainer) is an active exoskeleton with four degrees-of-freedom that uses PMA [13]. However, a major disadvantage with PMAs is their highly nonlinear behavior, which makes them more complicated to control and in general less accurate than the electric motors.

Our shoulder mechanism is actuated by brushless DC-motors, which in combination with Harmonic Drives to transfer torques to the exoskeleton. The Harmonic Drive was selected for its high efficiency and back-drivability, which

Table 5.2 Specifications of the shoulder exoskeleton

Joint	Min. angle	Max. angle	Speed	Cont. torques
Shoulder rotation	−30°	60°	–	–
Shoulder flex/ext	−60°	170°	69.8 rpm	11.2 Nm
Shoulder abd/add	−10°	170°	43.6 rpm	18.0 Nm

Table 5.3 Joint construction in the exoskeleton

	Joint 1	Joint 3
Motor type	EC-60	EC-60
Gear type	LCS-17-80	CSD-25-50-2
Position sensor	Noncontact Encoder	Noncontact Encoder
Velocity sensor	Hall sensor in EC-60	Hall sensor in EC-60
Current sensor	EC-60 build-in sensor	EC-60 build-in sensor

allows the user to move even if the motors are powered off. In the mechanism, there are two active joints and one passive joint. The internal rotation, namely the rotation of θ_2, is selected to be passive in the consideration that this rotation requires less assistance. Table 5.2 lists the design specifications for the shoulder exoskeleton. The driving speed and torques are specified based on the simulation of arm motion assistance with 5-kg payload. In such a case, the exoskeleton provides a maximum of 50% of the needed physical assistance.

Using the power, force/torque, and velocity simulation results, different compositions of the motors and gears were investigated. The output torque, τ_{out}, and velocity, ω_{out}, were calculated using following equations:

$$\tau_{\text{out}} = \mu_T N_G \tau_{\text{in}}, \quad \omega_{\text{out}} = \frac{\omega_{\text{in}}}{N_G} \tag{5.15}$$

where μ_T is the total efficiency of the transmission, i.e., Harmonic Drive and bearings, N_G is the gear ratio of the Harmonic Drive and τ_{in} and ω_{in} are the input torque and velocity, respectively. The driving factor for selecting the motors and gears, besides the simulation results, is the total weight of the transmission system. Table 5.3 lists the selected motors and gears for the two active joints. For the passive joint, only noncontact encoders are used.

5.5 Control strategies of exoskeleton shoulders

5.5.1 State-of-the-art exoskeleton control

Safety and compliance are the challenging goals to accomplish for exoskeleton control. Several control strategies have been proposed to achieve these objectives.

Caignan *et al.* [15] developed an exoskeleton for shoulder rehabilitation. Impedance and admittance control strategies were implemented to perceive their performance for isolateral exercises around the shoulder complex. Admittance controller showed good tracking performance without relying on feedforward control, although to achieve a desired response, friction and gravity compensation torques are also required to implement the impedance controller. Intentional reaching direction (IRD)-based admittance control was proposed by Huang *et al.* [16] to control the upper limb power assist exoskeleton. Force sensing resistors (FSRs) based interface determines the IRD. Model free PID-based admittance control strategy was proposed by Yu *et al.* [17] for EXO-UL7 exoskeleton. Impedance control techniques, besides gravity compensation as feedforward, were proposed for an upper body exoskeleton [18,19]. The exoskeletons were developed for rehabilitation and shoulder pathology treatment. Rocon *et al.* [20] implemented an impedance control strategy for their WOTAS (Wearable Orthosis for Tremor Assessment and Suppression) exoskeleton, developed for the patients suffering from tremors. A notch-filter-based control scheme was implemented, whose frequency was set to the tremor frequency, which was found to be more effective reducing the tremor effect. MCS (muscles circumference sensors) and desired motion intention (DMI)-based adaptive impedance control is proposed by Khan *et al.* [21] for the control of a 7-DOF exoskeleton. MCS was developed to relate upper arm muscles circumference to joint forces, while DMI was computed through interaction forces using Radial Basis Function Neural Network algorithm.

Nef *et al.* [22] implemented PD control with gravity compensation technique on the ARMin exoskeleton, developed for arm mobilization therapy. Performance of gravity compensation, along with a constant feedforward torque, was evaluated for a new shoulder-elbow exoskeleton (NESM), developed by Crea *et al.* [23]. The method allows the exoskeleton to lift its own weight, besides the feedforward torques enable the exoskeleton to assist the user. Huang *et al.* [24] proposed trajectory planning-based adaptive PID control techniques to assist the user in normal routine tasks. Adaptive network-based fuzzy inference system was utilized to compute PID gains. Rahman *et al.* [25] developed ETS-MARSE (7-DOF exoskeleton) to rehabilitate upper limb functions. Sliding Mode Control (SMC) was implemented to track the desired position computed through force signals extracted from Electromyography (EMG) sensors.

5.5.2 Control algorithms

Design of sophisticated control strategies, to make the exoskeleton operation safe and sound, has always remained a challenging task. The structure of control algorithms for an exoskeleton is very different from the ones designed for conventional articulated robotic arms. In contrast with conventional robots, besides commanding a torque or position signal, the human is also the part of control system [26].

Various control strategies have been developed [27,28] for human-exoskeleton interaction. Their significance varies for each targeted application. Based on their applications, the control architectures are categorized into two domains, namely, *Trajectory-Based Control (TBC)* and *Interaction-Based Control*, as described in the following subsections.

5.5.3 Trajectory-based control

In a trajectory control approach, the exoskeleton is programmed to follow a predefined trajectory. Therapy exercises and monitoring of patients progress are the targeted applications of such control architectures [22,29]. The contribution of interaction forces is very limited. If the patient is off track or not able to complete the task, the trajectory generator will produce the respective response to help the patient follow the right trajectory and complete the task [22,29].

Predefined TBC strategies are mainly used in exoskeletons that are developed for medical purposes. Such rehabilitation devices mostly perform defined routines and exercises in order to help the patients regain his/her motor functions. Very sophisticated techniques have been developed to enable the robotic arm follow the desired trajectory. Such techniques can also be implemented on exoskeletons by considering the human inside the control loop.

The block diagram of TBC is shown in Figure 5.5.

The primary block of the TBC is a trajectory generator, which outputs smooth reference signals i.e., position, velocity, and accelerations. Tracking of the desired trajectories is ensured by a Trajectory Control block. Several well-established control strategies in linear [30–32] and nonlinear [33,34] domains have been reported. Within linear domain, PD and PID control techniques have been widely reported for position control of exoskeletons. In this case control law is given by

$$\tau_{\text{tbc}} = k_p e + k_d \dot{e} + k_i \int e \tag{5.16}$$

where τ_{tbc} is the output of the trajectory control block and k_p, k_d, and k_i are the proportional, derivative, and integral gains, respectively.

In the domain of nonlinear and robust control techniques, SMC is capable of providing better tracking response for the respective reference signal. The general form of SMC is given by

$$\tau_{\text{tbc}} = \tau_{\text{equ}} + \tau_{\text{dis}} \tag{5.17}$$

where τ_{equ} is computed by placing $\dot{s} = 0$, here s is the sliding surface that depends on the error dynamics and is given by

$$s = e + \lambda \dot{e} \tag{5.18}$$

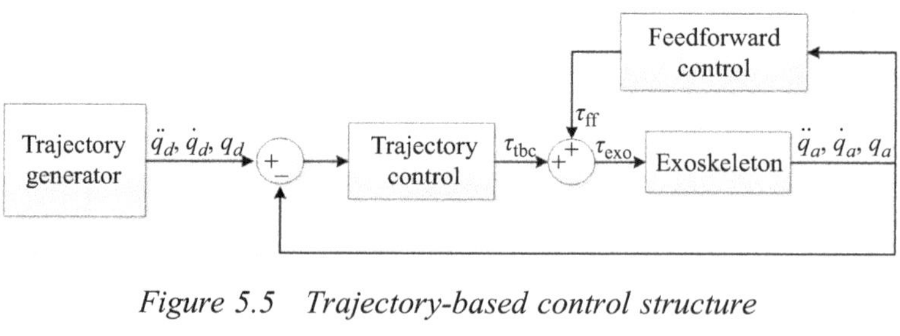

Figure 5.5 Trajectory-based control structure

Moreover, τ_{dis} is discontinuous control law given by

$$\tau_{dis} = k \cdot \text{sign}(s) \tag{5.19}$$

where λ and k are two gains. Moreover,

$$\text{sign}(s) = \begin{cases} +1 & s \geq 0 \\ -1 & s < 0 \end{cases}$$

sign(s) brings the discontinuity in the control action. It is the function of sliding surface.

SMC has two phases, i.e., a reaching phase and a sliding phase. The reaching phase is governed by the equivalent controller (τ_{equ}) which ensures that the system reaches the sliding surface. Once the system reaches there, τ_{dis}, defined by (5.19), becomes dominant and ensures that the system doesn't leave the sliding surface in future time. This discontinuous part adds the robustness in the designed control law against uncertainties and disturbances.

The control law for the feedforward control can be computed from the inverse dynamic model of a robotic manipulator that is given by

$$\tau = M(q)\ddot{q} + V(q, \dot{q}) + G(q) \tag{5.20}$$

where $M(q), V(q, \dot{q})$, and $G(q)$ represent inertia, coupling torques, and gravity terms, respectively.

Gravity and frictional forces both play a key role in making the exoskeleton operation smooth. Both forces can certainly increase the transparency between the human limb and exoskeleton motion, if handled properly. A feedforward block contributes to the control structure by compensating for the gravity $G(q)$ and Coulomb and viscous frictional forces [35]. Gravity compensation enables the exoskeleton to lift its own weight, hence it is not felt by the user. Moreover, if the user is weak and even unable to properly lift their arm, then his/her mass properties can also be considered and compensated.

5.5.4 Interaction-based control

Force interaction-based control strategies are implemented for the exoskeletons developed to rehabilitate [20] or assist [21] the humans in the normal routine works. These techniques can serve for the power assist exoskeletons [16] as well. The primary purpose of such control structures is to enable the exoskeleton to complement the human motion and augment the human strength and ability to remain active for longer period of time.

The interaction-based approach is very vital when exoskeletons are programmed to assist the user in daily routine tasks. The dexterity of human arm motion, in accomplishing a task, has to be matched by exoskeletons to fulfill the true meaning of assisting and rehabilitating the respective person. While interacting with the environment, namely, the human in the case of exoskeletons, the interaction behavior and the desired response, which are depicted mainly in terms of inertia, stiffness, and damping, need to be modeled perfectly. Pure motion control strategies don't work well when the desired conditions are not met, as there can be deviations in the desired trajectory if the system and environment models i.e., for the kinematics and dynamics, are not accurately defined [36]. The other possible techniques, such as impedance and admittance-based control, shown in Figures 5.6 and 5.7, respectively, have been widely investigated in area of interaction-based control to ensure the stability and compliance.

The general implementation of impedance-based control is shown in Figure 5.6.

The mechanical impedance at the interaction port, which computes the output force for the given velocity, is defined in the "Desired impedance" block. The relation is represented in term of a transfer function, that is given by

$$Z(s) = \frac{F(s)}{\dot{q}(s)} = Ms + \frac{K}{s} + B \tag{5.21}$$

where F is the force experienced by the human and $\dot{q}$ is the exoskeleton joint velocity. M, K, and B, representing inertia, stiffness, and damping parameters, define the interaction behavior, which can be varied by changing these three parameters.

Desired torque τ_{des}, computed through the mechanical impedance $Z(s)$ and the angular velocity $\dot{q}$, is compared with torque τ_{int} sensed by the feedback "Force sensor." The computed error torque τ_{err} along with the feedforward torque τ_{ff}, explained in the previous section, are then relayed to the exoskeleton.

The interaction between human and exoskeleton can also be expressed in terms of mechanical admittance. The block diagram of the control strategy is shown in Figure 5.7.

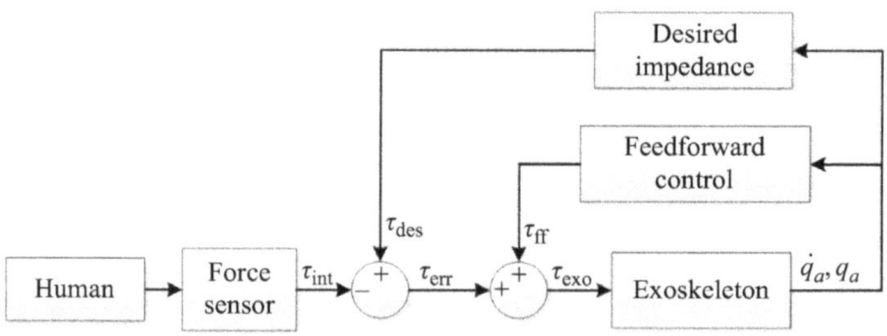

Figure 5.6 Impedance control

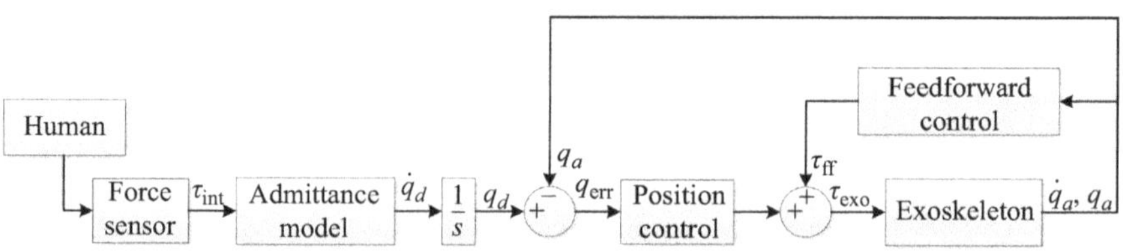

Figure 5.7 Diagram of admittance control

The mechanical admittance at the port computes the angular velocity for the given force. The relation is defined as

$$Y(s) = \frac{1}{Z(s)} = \frac{s}{Ms^2 + Bs + K} \tag{5.22}$$

It is the inverse of the mechanical impedance $Z(s)$ of Eq. (5.21). Force measured at the interaction port is passed to the admittance filter $Y(s)$, which computes the desired angular velocity $\dot{q}_d$. The admittance filter is followed by an integrator to compute the reference trajectory q_d which is tracked by implementing a position control strategy that generates the respective output torque. Feedforward torque τ_{ff}, can also be considered to eliminate the gravity pull on the exoskeleton.

5.6 Control of the shoulder mechanism

5.6.1 System description

The exoskeleton is developed for the assistance of elderly people to support movements including elbow flexion/extension and shoulder flexion/extension, adduction/abduction, and internal/external rotation. An admittance control scheme is implemented to control the elbow joint, while gravity compensation is implemented to control the shoulder joint. The control architecture is shown in Figure 5.8.

The port interaction forces, for admittance control, are measured through FSRs embedded inside the mechanical cuff (Figure 5.9). A group of eight FSRs are used to compute the desired motion intention.

The measured interaction forces are passed to the "Force model controller" to compute the direction intention. At first, a static force model is computed at rest position. This calibrates the FSRs to zero interaction torque τ_{int}. The change in port interaction forces during any movement is computed and compared with the static force model to compute the direction intention and interaction torque τ_{int}.

The interaction torque τ_{int} is then passed to admittance filter, to compute the desired angular velocity $\dot{q}_d$. The interaction behavior is defined by mechanical admittance $Y(s)$,

$$Y(s) = \frac{s}{Ms^2 + Bs + K} \tag{5.23}$$

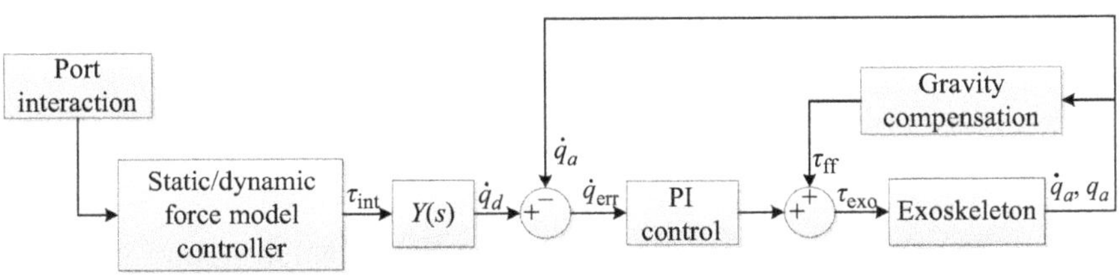

Figure 5.8 Diagram of admittance control

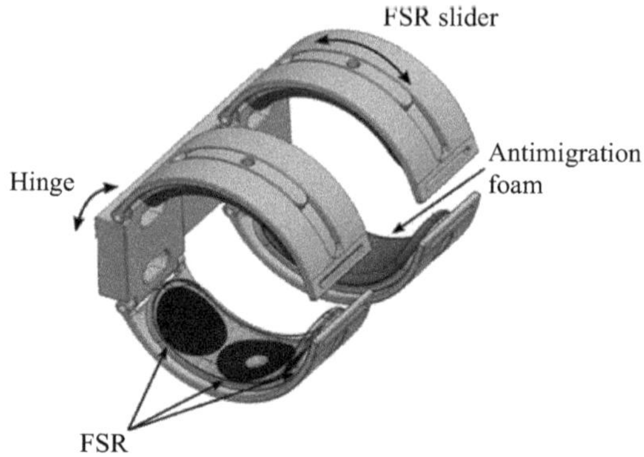

Figure 5.9 Port interaction force sensor built with FSRs

In the present study, instead of controlling the desired trajectory q_d, a PI control is implemented to track the desired angular velocity $\dot{q}_d$. The output torque computed by the PI controller is relayed to elbow joint motor, while "Gravity compensation" feedforward torque τ_{ff} is relayed to the shoulder complex, in order to operate the exoskeleton according to the user intention.

The final control law for the admittance control technique is given by

$$\tau_{exo} = k_p\dot{q}_e + k_i\int\dot{q}_e + \tau_{ff} \tag{5.24}$$

where τ_{exo} is the final torque relayed to the exoskeleton, k_p and k_i are the proportional and integral gains, respectively, and $\dot{q}_e$ is the angular velocity error.

Figure 5.10 shows the experimental results for the following set of admittance filter values $(M, B, K) = (0.075, 0.08, 0)$. The velocity feedback is measured through Maxon motors built in velocity sensors and an absolute magnetic position encoder (RFD 4000) has been installed outside the motor to measure the angular position. Interaction port force is measured through FSRs, where the positive sensor and negative sensor corresponds to the flexion and extension of the arm. Velocity reference signal, shown in Figure 5.10(b), relates to the desired mechanical behavior settled up through the admittance filter. The tracing performance with the admittance control is shown in Figure 5.10(b), where the lacking in desired trajectory is because of the static friction, which is not modeled in the current control scheme. Besides this the controller is able to follow the desired trajectory. The position response, shown in Figure 5.10(c), also verifies the anticipated performance of the implemented control strategy.

5.7 Shoulder joint usability test

A shoulder exoskeleton has been built at Aalborg University (AAU), Denmark, as shown in Figure 5.11(a) and (b). The shoulder joint has been used in an upper-arm exoskeleton (see Figure 5.10) that also includes an elbow joint, which gives a total

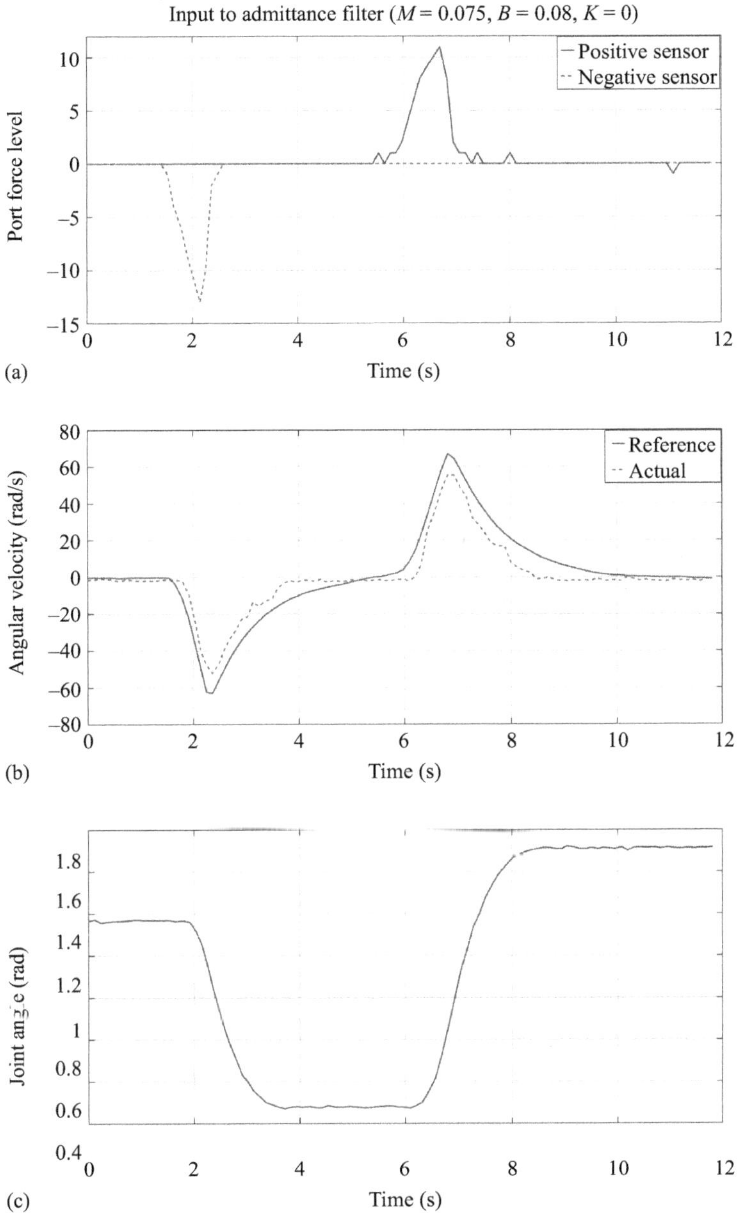

Figure 5.10 Interaction control on extension and flexion

of four degrees-of-freedom. The exoskeleton is strapped to the wearer through three attachments; (1) torso harness, (2) upper arm cuff, and (3) forearm cuff. The torso harness consists of a hard back (the shoulder exoskeleton base) with shoulder straps and snap buckle belt for rapidly fitting and easy tightening. The upper arm and

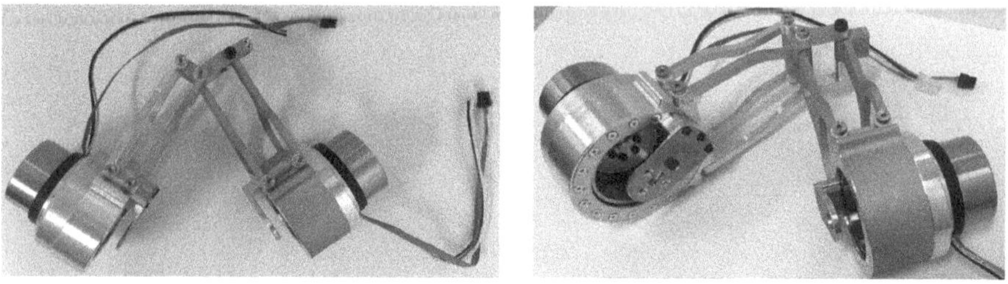

(a) (b)

Figure 5.11 A prototype of spherical shoulder mechanism

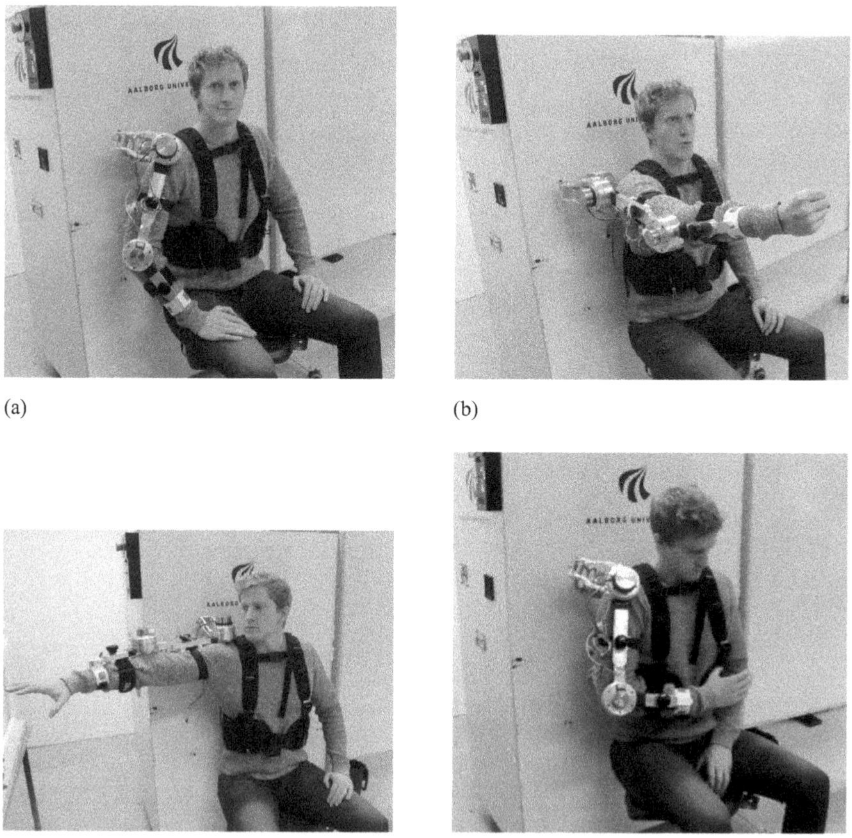

(a) (b)

(c) (d)

Figure 5.12 The shoulder joint mechanism in the AAU exoskeleton in (a) neutral/casual position, (b) 90° shoulder flexion, (c) 90° shoulder abduction, and (d) 90° shoulder internal rotation forearm cuffs both consist of a flexible plastics material and Velcro straps. Both cuffs are lined with foam and FSR sensors, which are used to measure the interaction forces between the exoskeleton and wearer, shown in Figure 5.9. The upper arm and the base of the shoulder joint are adjustable to fit the body size of the wearer.

The shoulder joint is able to realize the full range of motion of shoulder flexion/extension (see Figure 5.12(b)) without the exoskeleton colliding with the wearer. However, the upper-arm exoskeleton is supposed to match the movements of the shoulder griddle, which shifts the center of the glenohumeral joint. As a result, misalignment between the wearer and exoskeleton can occur when the flexion angle is above 90°. The shift of the glenohumeral joint is ignored, as it is assumed that any differences in the exoskeleton and human arm kinematics can be accommodated by passive compliance between the robot and the wearer, i.e., the torso harness. The shoulder joint is able to realize approx. 120° shoulder abduction before colliding with the wearer (see Figure 5.12(c)). Finally, the shoulder joint is able to realize 110° shoulder internal rotation, which is sufficient to enable the wearer to scratch him/herself on the opposite side of the abdomen (see Figure 5.12(d)). However, the shoulder joint is only able to realize 30° of shoulder external rotation. The prioritization of internal rotation of the shoulder is linked to the assumption that most of our lifts and other activities are done in front of our body.

5.8 Conclusions

This chapter is focused on the shoulder mechanism of exoskeletons. The design and control problems are reviewed and discussed. A new design of shoulder mechanism by virtue of double-parallelogram is described, for which the kinematics of the new shoulder joint is established and analyzed. In the chapter, different strategies of exoskeleton control are also described.

The new shoulder mechanism was constructed and developed. The shoulder mechanism has been applied to an upper-body exoskeleton for motion assistance. End-user usability testing on the shoulder mechanism shows that the new mechanism is able to produce the motion for the purpose of powered motion assistance.

References

[1] H. Lee, W. Kim, J. Han, and C. Han. The technical trend of the exoskeleton robot system for human power assistance. *International Journal of Precision Engineering and Manufacturing*, 13(8):1491–1497, 2012.

[2] Y. Mao and S. K. Agrawal. Design of a cable-driven arm exoskeleton (CAREX) for neural rehabilitation. *IEEE Transactions on Robotics*, 28(4):922–931, 2012.

[3] P. Garrec, J.-P. Friconneau, Y. Measson, and Y. Perrot. ABLE, an innovative transparent exoskeleton for the upper-limb. In *IEEE/RSJ Int. Conf. on Intelligent Robots and Systems, 2008*, 1483–1488, September 2008.

[4] J. F. Veneman, R. Kruidhof, E. Hekman, R. Ekkelenkamp, E. Van Assel donk, and H. Van Der Kooij. Design and evaluation of the LOPES exoskeleton robot for interactive gait rehabilitation. *IEEE Transactions on Neural Systems and Rehabilitation Engineering*, 15(3):379–386, 2007.

[5] T. Nef, M. Guidali, and R. Riener. ARMin III—arm therapy exoskeleton with an ergonomic shoulder actuation. *Applied Bionics and Biomechanics*, 6(2):127–142, April 2009.

[6] L. Zhou, S. Bai, M. S. Andersen, and J. Rasmussen. Modeling and design of a spring-loaded, cable-driven, wearable exoskeleton for the upper extremity. *Journal of Modeling, Identification and Control*, 36(3):167–177, 2015.

[7] L. Zhou, Y. Li, and S. Bai. A human-centered design optimization approach for robotic exoskeletons through biomechanical simulation. *Robotics and Autonomous Systems*, 91:337–347, 2017.

[8] A. E. Engin. On the biomechanics of the shoulder complex. *Journal of Biomechanics*, 13(7):575–590, 1980.

[9] D. Koo, P. H. Chang, M. K. Sohn, and J. Shin. Shoulder mechanism design of an exoskeleton robot for stroke patient rehabilitation. *IEEE International Conference on Rehabilitation Robotics*, 2011. pp. 1–6. doi: 10.1109/ICORR.2011.5975505

[10] D. Naidu, R. Stopforth, G. Bright, and S. Davrajh. A 7 DOF exoskeleton arm: Shoulder, elbow, wrist and hand mechanism for assistance to upper limb disabled individuals. In *IEEE AFRICON Conference*, pp. 13–15, 2011.

[11] J. C. Perry, J. Rosen, and S. Burns. Upper-limb powered exoskeleton design. *IEEE/ASME Transactions on Mechatronics*, 12(4):408–417, 2007.

[12] D. Chakarov, I. Veneva, M. Tsveov, and T. Tiankov. New exoskeleton arm concept design and actuation for haptic interaction with virtual objects. *Journal of Theoretical and Applied Mechanics*, 44(4):3–14, 2014.

[13] H. S. Lo and S. Xie. Optimization and analysis of a redundant 4R spherical wrist mechanism for a shoulder exoskeleton. *Robotica*, 32(8):1101–1211, 2014.

[14] T. G. Sugar, J. He, E. J. Koeneman, *et al.* Design and control of RUPERT: A device for robotic upper extremity repetitive therapy. *IEEE Transactions on Neural Systems and Rehabilitation Engineering*, 15(3):336–346, 2007.

[15] C. R. Caignan, M. P. Naylor, and S. N. Roderick. Controlling shoulder impedance in a rehabilitation arm exoskeleton. In *IEEE Inter. Conf. on Robotics and Automation (ICRA)*, pp. 2453–2458, 2008.

[16] J. Huang, W. Huo, W. Xu, S. Mohammed, and Y. Amirat. Control of upper-limb power-assist exoskeleton using a human-robot interface based on motion intention recognition. *IEEE Transactions on Automation Science and Engineering*, 12(4):1257–1270, 2015.

[17] W. Yu, J. Rosen, and X. Li, PID Admittance control for upper limb exoskeleton. In *IEEE, Proceedings of American Control Conference*, pp. 1124–1129, 2011.

[18] B. Kim and A. D. Deshpande. Controls for the shoulder mechanism of an upper-body exoskeleton for promoting scapulohumeral rhythm. In *IEEE Inter. Conf. on Rehabilitation Robotics (ICORR)*, pp. 538–542, 2015.

[19] C. Carignan and M. Liszka. Design of an arm exoskeleton with scapula motion for shoulder rehabilitation. In *IEEE Inter. Conf. on Advanced Robotics*, pp. 524–531, 2005.

[20] E. Rocon, J. M. Belda-Lois, A. F. Ruiz, M. Manto, J. C. Moreno, and J. L. Pons. Design and validation of a rehabilitation robotic exoskeleton for tremor assessment and suppression. *IEEE Transactions on Neural Systems and Rehabilitation Engineering*, 15(3):367–378, 2007.

[21] A. M. Khan, Deok-won Yun, M. A. Ali, J. Han, K. Shin and C. Han. Adaptive impedance control for upper limb assist exoskeleton. In *IEEE Inter. Conf. on Robotics and Automation (ICRA)*, pp. 4359–4366, 2015.

[22] T. Nef, M. Mihelj, and R. Riener. ARMin: a robot for patient-cooperative arm therapy. *Medical and Biological Engineering and Computing*, 45(9):887–900, 2007.

[23] S. Crea, M. Cempini, M. Moise, A. Baldoni, E. Trigili, and D. Marconi. Validation of a Gravity Compensation Algorithm for a Shoulder-Elbow Exoskeleton for Neurological Rehabilitation. *Converging Clinical and Engineering Research on Neurorehabilitation II, Springer*, Cham, 495–499, 2017.

[24] J. Huang, K. Young, and C. Ko. Effective control for an upper-body exoskeleton robot using ANFIS. In *IEEE Inter. Conf. on System Science and Engineering (ICSSE), 2016*, 1–4, 2016.

[25] M. Rahman, C. Ochoa-Luna, M. Saad, and P. Archambault. EMG based control of a robotic exoskeleton for shoulder and elbow motion assist. *Journal of Automation and Control Engineering*, 3(4):270–276, 2015.

[26] C. J. Yang, J. F. Zhang, Y. Chen, Y. M. Dong, and Y. Zhang. A review of exoskeleton-type systems and their key technologies. *Journal of Mechanical Engineering Science*, 222(8):1599–1612, 2008.

[27] K. Anam and A. A. Al-Jumaily. Active exoskeleton control systems: State of the art. *Procedia Engineering*, 41:988–994, 2012.

[28] J. M. P. Gunasekara, R. A. R. C. Gopura, T. S. S. Jayawardane, and S. W. H. M. T. D. Lalitharathne. Control methodologies for upper limb exoskeleton robots. In *IEEE/SICE Inter. Symposium on System Integration (SII), 2012*, 19–24, 2012.

[29] S. Balasubramanian and J. He. Adaptive control of a wearable exoskeleton for upper-extremity neurorehabilitation. *Applied Bionics and Biomechanics*, 9(1):99–115, 2012.

[30] K. H. Ang, G. Chong, and Y. Li. PID control system analysis, design, and technology. *IEEE Transactions on Control Systems Technology*, 13 (4):559–576, 2005.

[31] M. Khairudin, Z. Mohamed, and A. R. Husain. Dynamic model and robust control of flexible link robot manipulator. *Telecommunication Computing Electronics and Control*, 9(2):276–286, 2013.

[32] T. Qaisar and A. Mahmood. Robust control of a customized robotic arm with unstructured uncertainties. In *Inter. Conf. on Emerging Technologies (ICET), 2015*, 1–5, 2015.

[33] A. Bartoszewicz and R. J. Patton. Sliding mode control. *Journal of Adaptive Control and Signal Processing*, 21(8–9):635–637, 2007.

[34] R. H. Middleton and G. C. Goodwin. Adaptive computed torque control for rigid link manipulators. *Systems and Control Letters*, 10(1):9–16, 1988.

[35] J. González-Vargas, J. Ibáñez, J. L. Contreras-Vidal, H. van der Kooij, and J. L. Pons (eds). Wearable Robotics: Challenges and Trends. *Proceedings of the 2nd International Symposium on Wearable Robotics (WeRob)*, Segovia, Spain, October 18–21, 2016.

[36] B. Siciliano, and O. Khatib. *Springer Handbook of Robotics*. Springer-Verlag Berlin Heidelberg, 2008.

Chapter 6

Calibration platform for wearable 3D motion sensors

Bingfei Fan[1], Qingguo Li[2], Chao Wang[1], and Tao Liu[1]

Abstract

With the development of microelectromechanical system technologies, wearable motion sensors (WMSs) have played an increasingly significant role in the design of exoskeletons, where the WMSs are used for motion detection, orientation estimation, or position estimation. For these applications, the accuracy of the WMSs will affect the overall performance of the exoskeletons. The purpose of this chapter is to present an evaluation platform for assessing the accuracy of WMSs, assisting to develop or choose proper WMSs for exoskeletons. The presented evaluation platform is an instrumented gimbal with three rotation axes. Each axis is equipped with a DC motor and an absolute encoder. Thus, each axis can be controlled independently, and the rotation angle around each axis can be output accurately. In addition, each axis can rotate continuously via the equipped electrical slip rings. One of the major advantages of the instrumented gimbal is that it can be used for accurate motion analysis without needing any additional equipment. In order to verify the function of the platform, validation experiments were conducted, including a static accuracy test, and dynamic accuracy test with and without magnetic disturbances. Results show that the designed gimbal has good potential for evaluating the orientation of WMSs under different conditions.

Keywords: Inertial and magnetic sensors, instrumented gimbal, wearable motion sensors, coordinate frame alignment, orientation estimation

6.1 Introduction

Wearable motion sensor (WMS) usually refers to a microelectromechanical system-based inertial measurement unit (IMU), which consists of a triaxial

[1]State Key Laboratory of Fluid Power and Mechatronic Systems, School of Mechanical Engineering, Zhejiang University, China
[2]Department of Mechanical and Materials Engineering, Queen's University, Canada

Table 6.1 The functions of WMSs in the design of exoskeletons

Study	Exoskeletons	Functions of WMSs
Trkov *et al.* [5]	Robotic knee assistive device for slip-induced fall prevention	Slip detection
Mazumder *et al.* [6]	Lower limb active exoskeleton	Ambulatory activity classification together with a pressure sensor-based insole
Rocon *et al.* [7]	Upper limb and lower limb rehabilitation robotics	Monitoring tremor, gait or helping to evaluate rehabilitation progress
Song *et al.* [8]	Upper limb rehabilitation exoskeleton	Detect the rotation motion of upper limb
Mateos *et al.* [9]	Industrial assistive exoskeleton	Worked as subsidiary sensors, helping classify the upright position from inclined position

accelerometer and a triaxial gyroscope. In some cases, a triaxial magnetometer is also integrated into the IMU, forming a full three degrees of freedom (DOFs) drift-free orientation estimation unit. WMSs have been widely used in human motion analysis [1] for their advantages of low cost, small size, and low consumption, such as with gait analysis [2,3] and hand pose estimation [4]. More recently, WMSs have played an increasingly significant role in the design of exoskeletons, where the WMSs are used for intention recognition, pose estimation or position estimation. By reviewing the scientific literature, the functions of WMSs in the exoskeletons are listed in Table 6.1.

Sensor's orientations are required in these applications. Hence, the accuracy of the orientation estimation of the WMSs will affect the overall performance of the exoskeleton. For motion sensor-based orientation estimation, sensor fusion is always required to fully exploit the complementary properties of gyroscopes, accelerometers, and magnetometers [10]. Popular sensor fusion algorithms (SFAs) include the quaternion-based extended Kalman filter [11], nonlinear complementary filters [12], and gradient descent algorithm [13]. Many derived fusion algorithms still emerge continually. Through these developments, the accuracy of the orientation estimation has been improved gradually. During developing a WMS, an evaluation platform is needed to choose a proper SFA or tune parameters for existing SFAs. Stereophotogrammetry is one of the common evaluation tools. However, this method needs a set of expensive equipment and the evaluation is restricted to a special laboratory. Moreover, stereophotogrammetry also suffers from nonnegligible error [14]. In some cases, a fine tuned stereophotogrammetry system can cause a maximal inaccuracy of 0.5° [15].

A multiaxis rotation platform represents an alternative evaluation platform. With this platform, users only need to fix the WMS on the test area. Then, the motion of the platform can be controlled by using the controller. Compared with traditional stereophotogrammetry, the multiaxis rotation platform provides a simple and controllable WMSs evaluation method. Zhang *et al.* used a triaxial turntable to

Table 6.2 The expected features of the instrumented gimbal

No.	Expected features
1.	High accuracy and moderate cost
2.	Continuous rotation
3.	Free of magnetic disturbances
4.	Capable of outputting reference orientation
5.	Adjustable rotation speed

perform static and dynamic accuracy tests [16], where the turntable was able to provide the reference angle at a high accuracy. However, the ferromagnetic material made frames would generate magnetic disturbance, which indicates it not suitable for WMSs containing magnetometers. Moreover, it cannot rotate continuously due to the wire on the device. Lai *et al.* designed an aluminum gimbal that is simple and low cost [17]. However, the angular resolution of each axis is 0.29°, which is not high enough for evaluating the motion sensors, and the gimbal cannot rotate continuously. Lebel *et al.* designed a gimbal for assessing WMSs [18,19]. The distance between WMSs and ferromagnetic material was enlarged intentionally, so that the magnetic disturbances generated by the gimbal itself could be neglected. In addition, all three axes can rotate continuously so that the evaluation can cover a full range of motion (0°–360°). The major design deficiency is that the gimbal merely provides the power for rotation, so stereophotogrammetry is still needed to obtain the reference orientation.

In previous work, an instrumented gimbal was designed to quantify 3D joint angles measured by inertial sensors [20]. The orientation of the gimbal is measured by a potentiometer attached to the bearings of each axis. This gimbal can only be rotated manually due to the lack of motors and the rotation range cannot reach up to 360°. As an improvement to previous work, the objective of this chapter is to develop a general-purpose evaluation platform to assess the performance of WMSs used for exoskeletons. The evaluation platform is also designed as an instrumented gimbal, and the expected features are listed in Table 6.2. Users could analyze the accuracy of WMSs under various conditions with the instrumented gimbal, such as different time durations, different speeds, and with or without magnetic disturbances. Thus, the gimbal can be regarded as a useful tool for developing WMSs used in exoskeletons.

6.2 Design of instrumented gimbal

6.2.1 The mechanical structure of the instrumented gimbal

Euler angles with ZYX order are used to represent sensor orientation in this chapter. The rotations around Z, Y, and X axis are called yaw (ψ), pitch (θ), and roll angles (ϕ). Corresponding with the three axes of Euler angles, the gimbal should

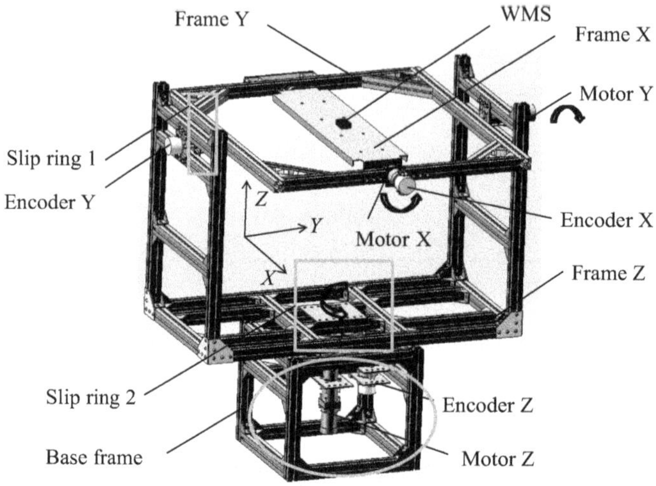

Figure 6.1 The model of designed three-axis instrumented gimbal

also contain three DOFs. As shown in Figure 6.1, the mechanical structure of the gimbal is a common gimbal structure. Each frame can rotate around the axis as indicated by the arrows. The gimbal consists of a base frame and three rotatable frames: frames X, Y, and Z. All the frames are assembled by aluminum sections. Each frame is equipped with a motor and an encoder. The motor is used to provide the power for rotation and the encoder is used to measure the angle. The motors are brushless DC motors (Times Brilliant Tech., China) with 24-V nominal voltage and 0.6-N-m max moment. Considering that the moments of inertia of frames Y and Z are larger than that of frame X, speed reducers are used for frames Y and Z with reduction ratios of 6 and 36, respectively. The absolute encoder of each axis is capable of outputting the angle related to a preset zero position. To avoid cable wrapping problems, both the power cable and the signal cable are passed through electrical slip rings. Therefore, each frame can rotate continuously. The slip ring 1 used for frame X contains 12 channels, and the slip ring 2 used for frames X and Y contains 24 channels. Moreover, each frame contains a mass adjustable module so as to keep balance when rotating.

During rotating, the maximum angular acceleration of each axis is calculated by using (6.1).

$$a = \frac{M \times i}{J} \tag{6.1}$$

where M is the moment of the motor, i denotes the reduction ratio, and J denotes the moment of inertia of each frame.

The maximum angular acceleration and other parameters of the designed gimbal are listed in Table 6.3.

Table 6.3 Parameters of the designed instrumented gimbal

Platform dimensions (length/width/height)	$1130 \times 1020 \times 1290$ mm^3
Degrees of freedom	3
Range of motion (*X*, *Y*, *Z*)	Continuously
Resolution (*X*, *Y*, *Z*)	0.09°, 0.09°, 0.09°
Maximum rate of motor (*X*, *Y*, *Z*)	600°/s, 360°/s, 180°/s
Maximum angular acceleration (*X*, *Y*, *Z*)	360°/s^2, 80°/s^2, 70°/s^2
Test section area	150×150 mm^2
Maximum load weight	1 kg

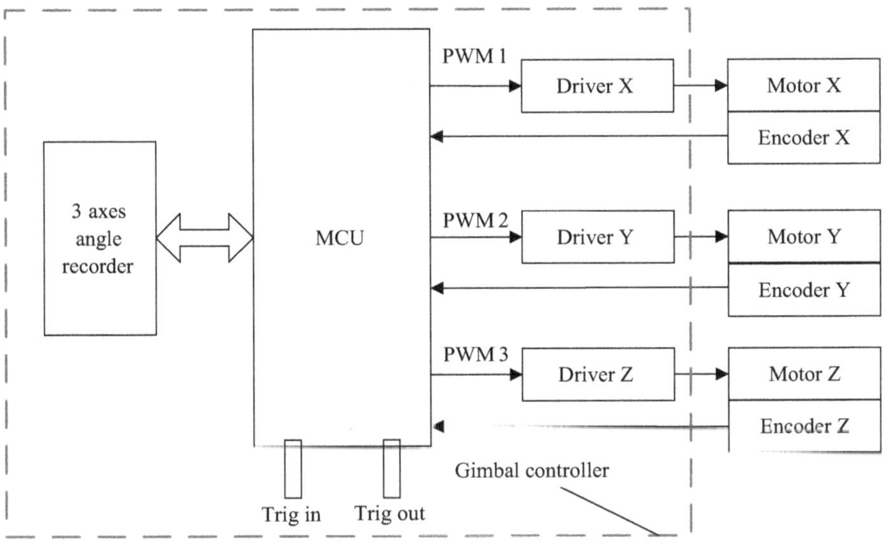

Figure 6.2 The controller of the instrumented gimbal

6.2.2 The controller of the designed gimbal

In order to generate various test patterns for WMSs, it is crucial to control the DC motor and record the angle of each frame. The controller of the instrumented gimbal consists of three parts: recorder, micro controller unit (MCU), and motor drivers. As shown in Figure 6.2, each motor is connected to a motor driver (Times Brilliant Tech., China). The MCU connects the motor drivers through pulse width modulation signals and other switching signals, so as to control the start-stop, direction, and the speed of the motors. The 12-bit encoder measures the rotation with a resolution of 0.09°, and an output frequency of 200 Hz. All three axes angles are send to the MCU for preprocessing and stored in the SD card in the recorder module, so that offline analysis can be done conveniently. The trig in and out interfaces are used to synchronize with the third-part systems.

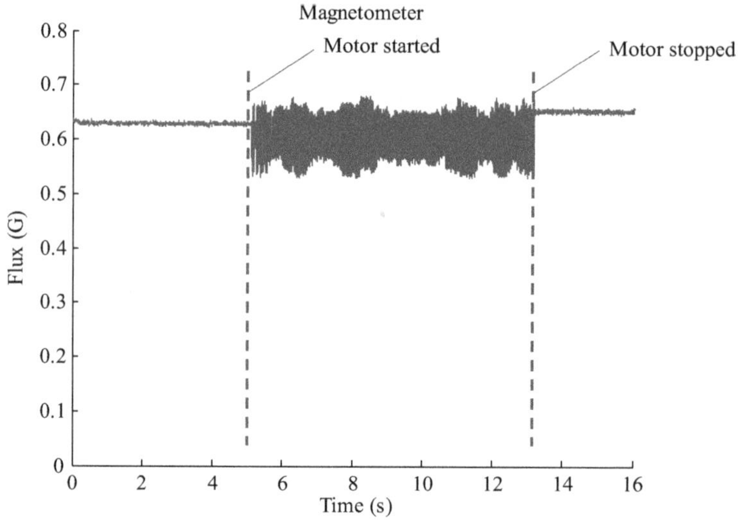

Figure 6.3 The measured magnetic field when the distance is 5 cm

6.2.3 The method for eliminating the magnetic disturbances

To be compatible with the magnetometers contained in some WMSs, the instrumented gimbal should ideally be a nonmagnetic device. However, the DC motor and the current-carrying wire would generate magnetic disturbances inevitably. An effective way to attenuate the magnetic disturbances is to increase the distance between the test area and the magnetic disturbance sources [21]. The magnetic field intensity is inversely proportional to the cube of the distance between test area and the magnetic interference sources [22]. However, it is hard to model the magnetic disturbances due to the complex distribution of magnetic sources, thus we determined the minimum distance between the test area and DC motors through test. In this test, the magnetometer (HMC5883L: Honeywell International, Inc.) contained in a WMS (x-IMU, x-io Technologies, UK) was adopted as the magnetic sensor. The distance from WMS to the DC motor was set to 5 cm first, and the result is shown in Figure 6.3. It can be seen that when the motor was started, the measured magnet intensity was drastically changed. When the distance was set to 30 cm, the DC motor has almost no effect on the measured value as shown by the results in Figure 6.4. As a result, the distances between the test area and any ferromagnetic material are set to at least 40 cm. The distance is also consistent with the recommended value presented in [21].

6.2.4 Calibration of the gimbal

Calibration was conducted to ensure the accuracy of the gimbal. The rotational accuracy of each axis was verified by a laser calibration device, which contains a laser receiver, a laser receiver and a controller. Figure 6.5 shows the calibration scenario of frame Z. The laser receiver is fixed on frame Z, and the transmitter is

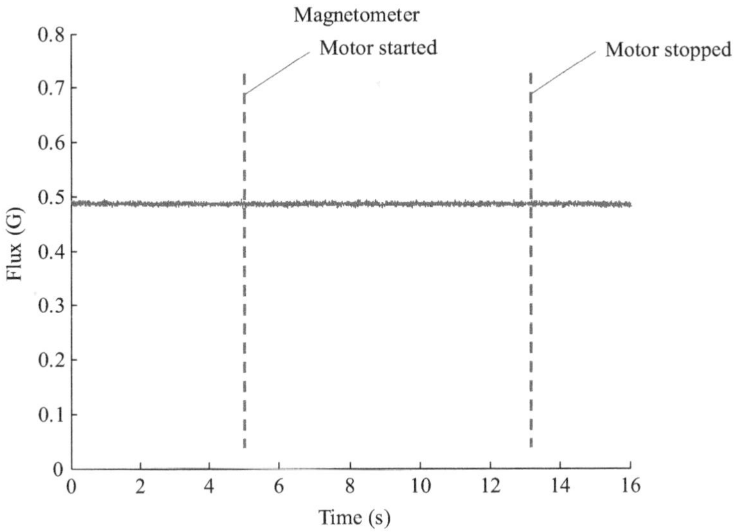

Figure 6.4 The measured magnetic field when the distance is 30 cm

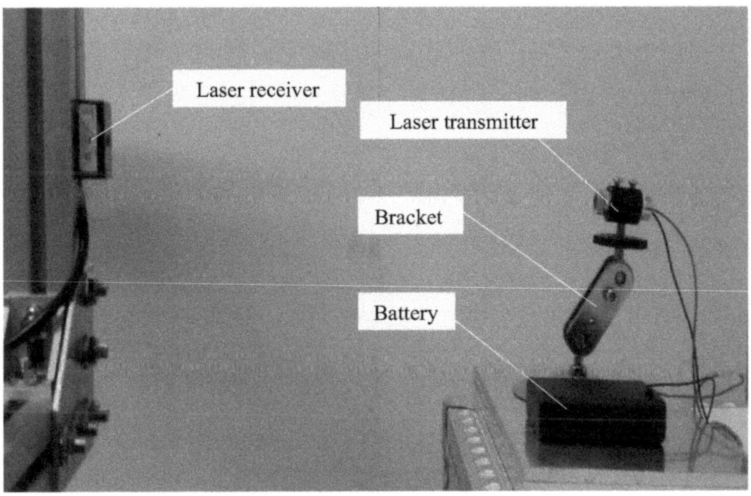

Figure 6.5 Laser setup for the gimbal calibration

supported by a bracket and fixed on a table on the ground. The battery provides the power for the transmitter. During the calibration procedure, once the laser was received by the receiver, a pulse was generated and sent to the controller, which then recorded the angle of the tested frame. Frame Z was rotated ten cycles, and the rotation angle of each cycle was checked to see if it was exactly the same as the

Table 6.4 The output angle at each cycle (degree)

Frame	Cycle										Mean error
	1	2	3	4	5	6	7	8	9	10	
X	360	360	360	359.912	360	359.912	360	359.912	359.912	360	0.0352
Y	360	360	360	360	360	360	360	359.912	359.912	359.912	0.0264
Z	360	359.912	360	360	359.912	359.912	359.912	360	359.912	359.912	0.0528

initial state. The angle of each axis is set to 360° in the first cycle. The same calibration was applied to frames X and Y. The output angles at each cycle of each frame are provided in Table 6.4.

From this table, it can be seen that the error of each cycle will not exceed the minimum resolution (0.09°), and the error would not accumulate over time. It means that the gimbal is well designed and has the expected accuracy.

6.3 Orientation evaluation with instrumented gimbal

6.3.1 Orientation error analysis using the instrumented gimbal

There are many methods to represent orientation: Euler angles, quaternion, axis angles, rotation matrix, etc. [23]. Among these representing methods, Euler angles and quaternion are the most commonly used methods. Euler angles represent a sequence of three elemental rotations. Although this method may lead to ambiguous results and suffer from singularity problems [24], it is visualized and easy to understand. In the designed gimbal, the angles of the frames are consistent with the ZYX order Euler angles except for the value range. The original output angle of each frame is from 0° to 360°, while the value range for roll and yaw is from −180° to +180°, for pitch is from −90° to +90°. Therefore, the output of the gimbal should be converted to Euler angles so that future comparison work can be done conveniently.

The angles of each frame can be converted to Euler angles using (6.2)–(6.4). Taking the Euler angles of the gimbal as the gold standard, the orientation error can be obtained. However, there is a special case when the pitch crosses ±90°. As shown in Figure 6.6, when the rectangle is rotated around the *Y*-axis from position A, to position B, the pitch value of the two positions is the same, but their orientation really has changed. In Euler angles representing method, roll and yaw would offset 180° suddenly when pitch crosses ±90°. This feature is reflected in the convention described in (6.4).

$$\phi = \begin{cases} \theta_x & \text{otherwise} \\ \theta_x - 180^\circ & \text{if } \theta_x > 180^\circ \end{cases} \tag{6.2}$$

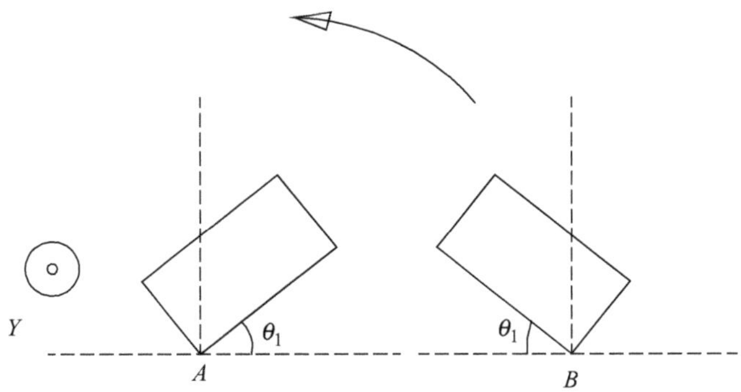

Figure 6.6 The roll and yaw offset diagram when pitch crosses ±90°

$$\psi = \begin{cases} \theta_z & \text{otherwise} \\ \theta_z - 180^\circ & \text{if } \theta_z > 180^\circ \end{cases} \tag{6.3}$$

$$\begin{cases} \theta = -\theta_y + 180^\circ & \text{if } \theta_y > 90^\circ \\ \phi = \phi + 180^\circ & \text{if } \theta_y > 90^\circ \\ \psi = \psi + 180^\circ & \text{if } \theta_y > 90^\circ \\ \theta = \theta_y - 360^\circ & \text{if } \theta_y > 270^\circ \end{cases} \tag{6.4}$$

where θ_x, θ_y, and θ_z represent the angle of each frame, and ϕ, ψ, and θ are the roll, yaw, and pitch angle, respectively.

Figure 6.7 shows the Euler angles of the gimbal when all frames were rotating anticlockwise continuously. As can be seen, jumps occur when pitch crosses ±90°. It is similar for the WMS (x-IMU, x-io Technologies, UK) attached on the test area of the gimbal, as shown in Figure 6.8. The difference is that the rate of change of the jump is slower than the gimbal. In this case, if Euler angles of these two systems were compared directly, a large comparison error would be induced. Thus, the Euler angles comparison method is improper for this case.

Compared with Euler angles, the quaternion method can avoid jump and singularity problems, and it can be converted to other representing methods easily. The quaternion of the gimbal can be converted from Euler angles by (6.5), while the quaternion of the WMS is output by its built in fusion algorithm [25].

$$q = \begin{bmatrix} q_0 \\ q_1 \\ q_2 \\ q_3 \end{bmatrix} = \begin{bmatrix} \cos\left(\frac{\varphi}{2}\right)\cos\left(\frac{\theta}{2}\right)\cos\left(\frac{\psi}{2}\right) + \sin\left(\frac{\varphi}{2}\right)\sin\left(\frac{\theta}{2}\right)\sin\left(\frac{\psi}{2}\right) \\ \sin\left(\frac{\varphi}{2}\right)\cos\left(\frac{\theta}{2}\right)\cos\left(\frac{\psi}{2}\right) - \cos\left(\frac{\varphi}{2}\right)\sin\left(\frac{\theta}{2}\right)\sin\left(\frac{\psi}{2}\right) \\ \cos\left(\frac{\varphi}{2}\right)\sin\left(\frac{\theta}{2}\right)\cos\left(\frac{\psi}{2}\right) + \sin\left(\frac{\varphi}{2}\right)\cos\left(\frac{\theta}{2}\right)\sin\left(\frac{\psi}{2}\right) \\ \cos\left(\frac{\varphi}{2}\right)\cos\left(\frac{\theta}{2}\right)\sin\left(\frac{\psi}{2}\right) - \sin\left(\frac{\varphi}{2}\right)\sin\left(\frac{\theta}{2}\right)\cos\left(\frac{\psi}{2}\right) \end{bmatrix} \tag{6.5}$$

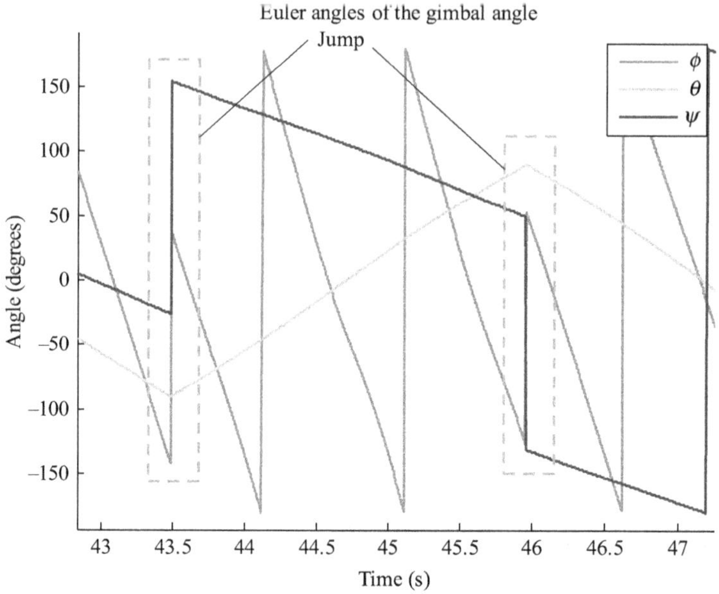

Figure 6.7 The Euler angles of the gimbal when rotated anticlockwise continuously

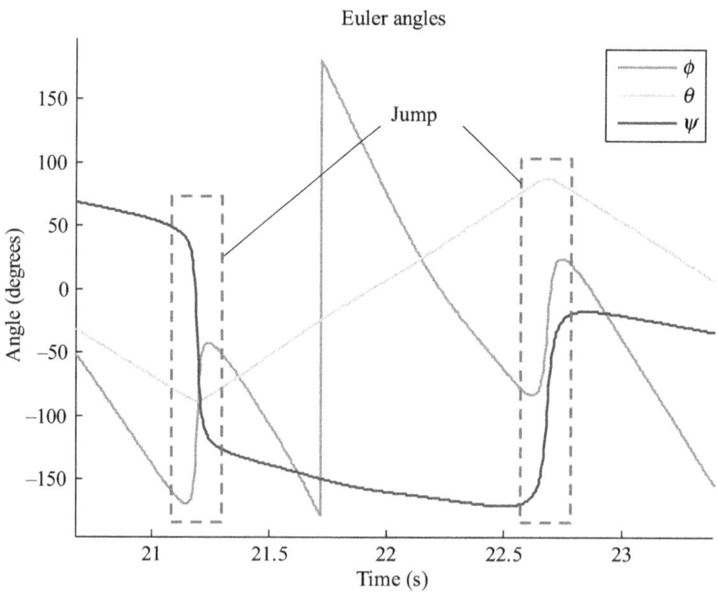

Figure 6.8 The Euler angles of the WMS when fixed on the gimbal rotated anticlockwise continuously

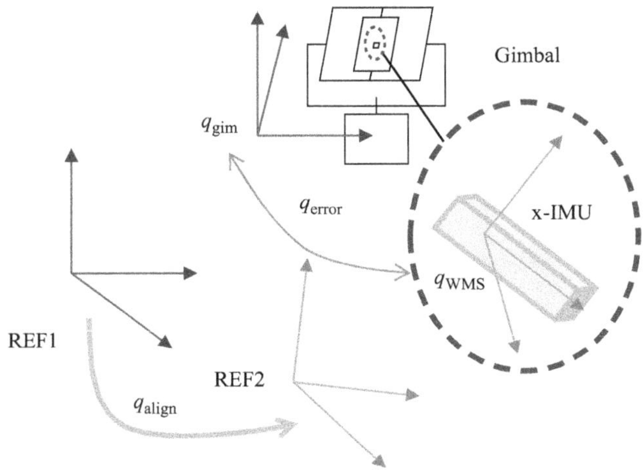

Figure 6.9 Coordinate frame alignment for WMS and the designed evaluation platform

Error analysis can be performed using quaternion comparison, and this method is also used in [26]. Quaternion multiplication and conjugation are the basic operations. In this chapter, $\otimes$ denotes a quaternion multiplication, and $q_B{}^* = [q_0, -q_1, -q_2, -q_3]$ denotes the conjugate quaternion of $q_B \cdot q_c = q_A \cdot q_B$ describes a vector first rotated at q_B, and then rotated at q_A. $q_c = q_B{}^*$ describes a vector inversely rotated at q_B. Thus, the quaternion error q_e between q_A and q_B can be described as (6.6).

$$q_e = q_A \otimes q_B^* \tag{6.6}$$

However, this equation is valid only under the condition that q_A and q_B are in the same coordinate system. In the designed evaluation platform, as shown in Figure 6.9, the reference coordinates of the gimbal and the WMS are REF1 and REF2, respectively. Thus, there must be a misalignment between these two coordinate frames. Let, q_{align} denotes the fixed misalignment between REF1 and REF2. Supposing that REF1 is the unified coordinate frame, the error between q_{gim} and q_{WMS} estimated by WMS can be calculated by the following equations:

$$^{W}_{R1}q = q_{\text{WMS}} \otimes q_{\text{align}} \tag{6.7}$$

$$q_{\text{error}} = q_{\text{gim}} \otimes {}^{W}_{R1}q^* \tag{6.8}$$

where $^{W}_{R1}q$ denotes the WMS's orientation relative to the REF1 coordinate frame.

Through coordinate frame alignment, the misalignment could be compensated. For the WMSs, the accuracy of the orientation estimation is affected by many factors [27], such as magnetic disturbances, and external movement acceleration. The accuracy of the estimated orientation is relatively high [13] in a magnetic disturbances free environment and slow motion state. Supposing that the error of the orientation is subject to Gaussian distribution, the mean error between q_{gim} and

q_{WMS} can be regarded as q_{align} when the WMS is fixed on the gimbal and rotated at a very low speed (quasi static state). In addition, it is necessary to convert q_{error} to Euler angles for easier understanding. Quaternion can be converted to Euler angles using (6.9)–(6.11) [23].

$$\phi = \operatorname{atan2}\left(2q_2q_3 + 2q_0q_1, q_3^2 - q_2^2 - q_1^2 + q_0^2\right) \tag{6.9}$$

$$\theta = -\operatorname{asin}(2q_1q_3 - 2q_0q_2) \tag{6.10}$$

$$\psi = \operatorname{atan2}\left(2q_1q_2 + 2q_0q_3, q_1^2 + q_0^2 - q_3^2 - q_2^2\right). \tag{6.11}$$

6.3.2 Selected wearable motion sensor

In this study, a commercially available WMS (x-IMU, x-io Technologies, UK) was selected as the analyzed object. This WMS contains a triaxial gyroscope, a triaxial accelerometer, and a triaxial magnetometer, and all sensors were calibrated before they were acquired [25]. The sensor orientation can also be estimated through the built in SFA. All the calibrated sensors' data and the estimated orientation can be stored in an SD card or transmitted to a computer through USB or Bluetooth.

6.3.3 Sensor configuration

In the gimbal, the angle output by the absolute encoder is relative to a preset zero position. Thus, before using the gimbal, zero positions should be set correctly. The REF2 of the WMS in Figure 6.9 is a north-west-up coordinate frame. In order to remain compliant with this coordinate frame, the same coordinate frame should be established for the gimbal. Hence, the zero position of *Z* should be set when *X* points to north, and the zero position of *X* and *Y* should be set when frames X and Y remain horizontal. An inclinometer (DXL360S, JINGYAN instrument, Dongguan, China) with a resolution of 0.05° was used to find the horizontal of frames X and Y, and a compass was used to find the zero position of *Z*. In this way, a north-west-up unified coordinate frame was established.

All three axes of the gimbal can be controlled independently. Thus, the gimbal can work on either single-axis mode or multiaxis mode. During the experiments, the WMS was mounted on the testing area of frame X with double-sided tape, and the *X*-axis of the sensor was aligned with that of the gimbal (Figure 6.10(a)). A magnetic disturbance generator was installed beside the WMS, which is used to simulate the magnetic disturbance in the surroundings. As shown in Figure 6.10(b), the generator consists of a square-paper tube and a circular permanent magnet (diameter: 15 mm, thickness: 2 mm). The tube was attached on frame X and the magnet was put into the tube. When frame Y rotates, the magnet reciprocated in the paper tube due to gravity, so that the magnetic disturbance changed with the distance between the magnet and the WMS. When testing, the sensor data of the WMS was sent to a computer through Bluetooth. All three axes angles of the instrumented gimbal were stored on an SD card first, and then converted to quaternion. The sampling rates of the gimbal and WMS were 200 and 256 Hz, respectively. For comparison, the quaternion from the WMS was down-sampled to 200 Hz. The two

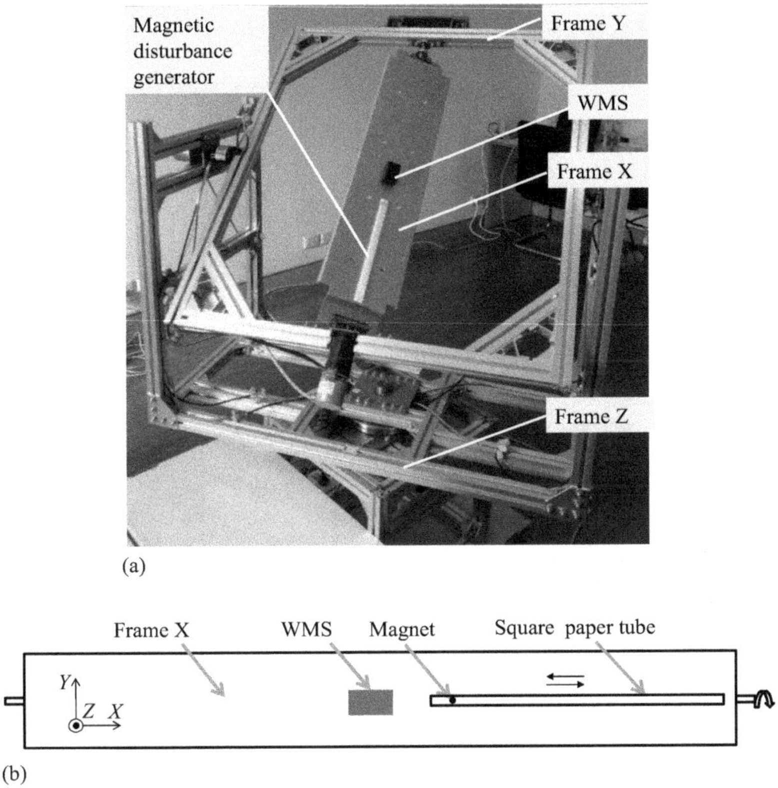

Figure 6.10 (a) WMS accuracy evaluation setup and (b) the detail of the magnetic disturbance generator

systems were time-synchronized manually, and all the comparisons were performed using MATLAB® (The MathWorks Inc., Natick, MA, USA).

6.3.4 Hard-iron calibration for magnetometer

The designed gimbal aims at evaluating WMSs containing magnetometers, which requires that the gimbal itself should not generate magnetic disturbance. This feature is guaranteed by enlarging the distance between any ferromagnetic-material and the test area of the gimbal. After finishing the design, it should also be verified. Similarly, x-IMU was used to measure the magnetic field in the test area. Before this test, the hard-iron bias was removed by the accompanying tool [25]. During the test, the x-IMU was fixed on the test area in frame X, and the x-IMU and gimbal were then started, so that they could rotate around three axes simultaneously. Finally, the magnetic field data was sent to a PC through Bluetooth. Figure 6.11 shows the original vector of the magnetic field and its magnitude, of which the mean value is 0.484 G and the standard deviation is 0.008 G. The results show that the magnitude of the magnetic field is almost a constant. So the test area of the

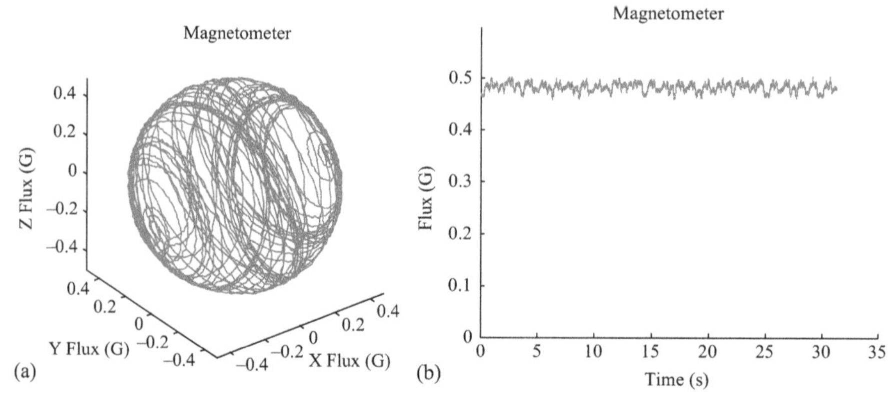

Figure 6.11 The magnetic field measured by x-IMU when the gimbal was rotating. (a) The measured magnetic field and (b) the magnitude of the magnetic field

gimbal can be regarded as a magnetic disturbance free area, and the magnetometer was well calibrated and ready for evaluating the accuracy of the sensor orientation.

6.3.5 Coordinate frame alignment (CFA) for WMS before experiments

When the WMS was attached on the test area of the gimbal, the misalignment error between the gimbal and the sensor coordinate systems must be compensated. The method of calculating the alignment error is described in Section 6.3.1. When the WMS is newly attached to the gimbal, or the zero positions of the gimbal are reconfigured, the coordinate frame alignment should be performed.

6.4 Experimental method

Following the preparation work, evaluation experiments were performed. Although the gimbal can simulate many different conditions, comprehensively listing all the experiments is impractical and unnecessary. So in this section, only experiments under typical conditions were carried out, including static accuracy test, dynamic accuracy test, and the effect of magnetic disturbances test. The detailed experimental protocol is described in the following sections.

6.4.1 Static accuracy test

The x-IMU was kept in the test area of the gimbal, and then the frames of gimbal were rotated manually. When one frame was rotated, the other frames were kept still, and 25 points in each circle were collected. In each static position, the frame was kept still for about 2–3 s. Both the data of the gimbal and the WMS were recorded for off-line analysis. The root mean square errors (RMSEs) and Bland–Altman plot method [28] are used to show the results.

6.4.2 *Dynamic accuracy test*

6.4.2.1 Single-axis rotation

In this test, only one axis was rotated, while the other two axes were kept still by a stopper. Each frame was rotated at both a low and high speed. Specifically, frame X was rotated at 320°/s and 520°/s; frame Y was rotated at 80°/s and 160°/s, and frame Z was rotated at 35°/s and 65°/s. The Euler angles of the WMS are compared with those of the gimbal directly. For convenience, the ranges of all the Euler angles were converted to ±90° to ensure the curves of the angles were continuous. The RMSEs of Euler angles were calculated for statistical analysis.

6.4.2.2 Multiaxis rotation

The major advantage of the designed gimbal is that all three axes can rotate continuously and simultaneously, which makes it possible to fully evaluate the WMSs. For example, the effect of time, velocity, and magnetic disturbances can be evaluated. In this chapter, a simple three-axis rotational experiment was performed to show the potential of the gimbal. Before the test, both the gimbal and the x-IMU were well configured. The gimbal rotated at a constant speed for about 50 s. The RMSEs of Euler angles converted from quaternion error were calculated.

6.4.2.3 The effect of magnetic disturbances

In this experiment, a circular permanent magnet (diameter: 15 mm, thickness: 2 mm) was mounted on the magnetic disturbance generator, so as to generate the magnetic disturbance. The gimbal was kept at zero position at the beginning of the experiment. The gimbal was then started by the remote control, all the frames rotated simultaneously and the magnet slid in the paper tube. The rotation speed of axes *Z*, *Y*, and *X* were about 40°/s, 70°/s, and 360°/s, respectively, and the duration of the trial was about 60 s. Error analysis was performed based on quaternion comparison. The quaternion error between the gold standard (Instrumented gimbal) and the estimated orientation was converted to Euler angles for interpretation. The RMSEs of the Euler angles were calculated and analyzed.

6.5 Results and discussion

6.5.1 *Static accuracy*

The static accuracy of the estimated Euler angles is shown in Figure 6.12. The horizontal axis shows the mean of the gimbal angle and the x-IMU angle, and the vertical axis shows the error of Euler angles. The RMSEs of roll, pitch and yaw angle are 0.98°, 0.73°, and 1.01°, respectively.

6.5.2 *Dynamic accuracy*

6.5.2.1 Single-axis rotation

The RMSEs of Euler angles are summarized in Table 6.5, and the detailed comparison of the Euler angles is shown in Figure 6.13. It can be seen that the accuracy

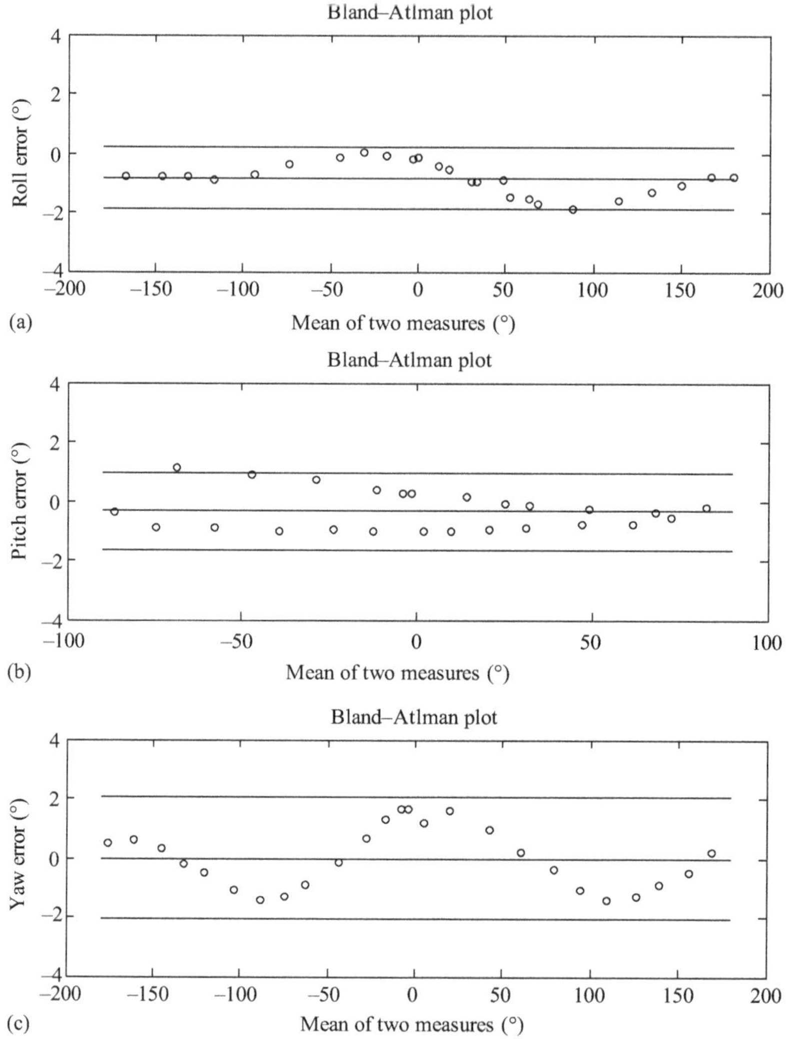

Figure 6.12 Static accuracy of Euler angles estimated by x-IMU. (a) The error of roll angle; (b) the error of pitch angle; and (c) the error of yaw angle

Table 6.5 Single axis evaluation for the orientation of WMS

Z speed	RMSE (yaw)	Y speed	RMSE (pitch)	X speed	RMSE (roll)
35°/s	0.58	80°/s	0.49	320°/s	0.82
65°/s	0.65	160°/s	0.56	520°/s	1.19

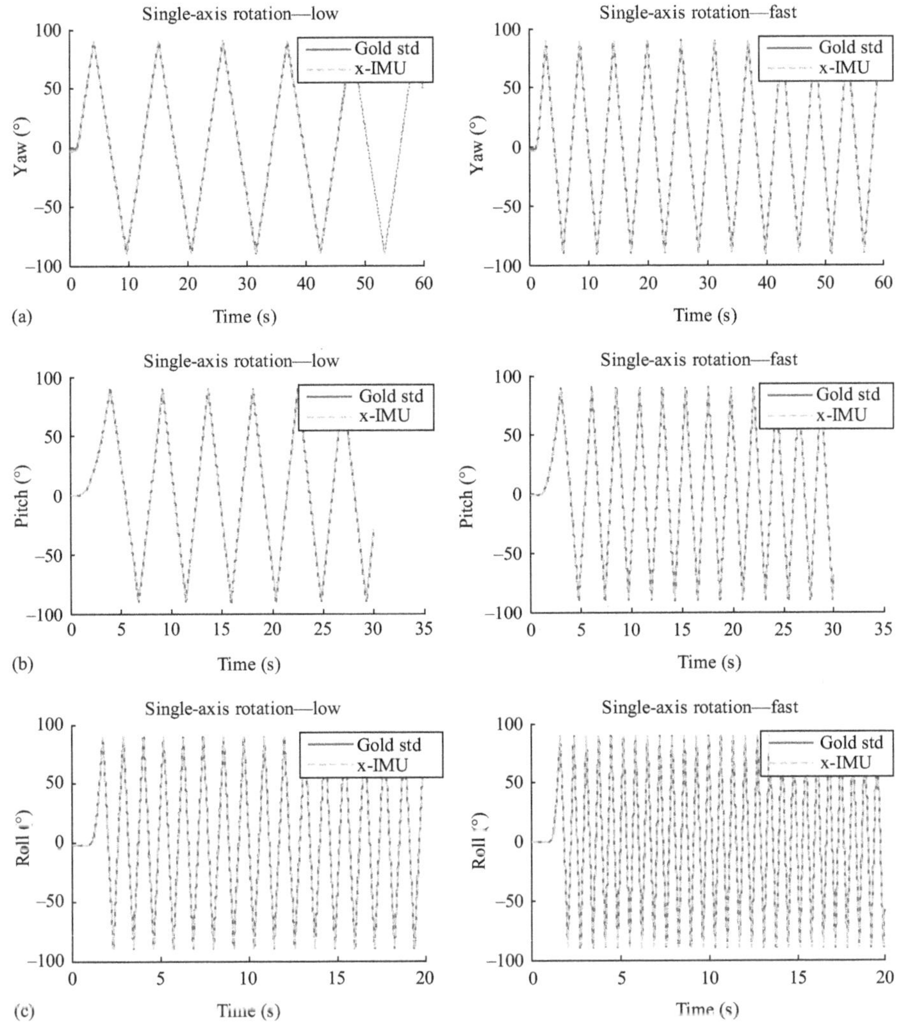

Figure 6.13 Dynamic accuracy of Euler angles estimated by x-IMU. (a) The error of roll angle; (b) the error of pitch angle; and (c) the error of yaw angle

in the single-axis rotation condition is higher than that in the static condition. The probable explanation for this is that when the sensor is in a stationary state, the orientation is calculated by the measured acceleration and magnetic field, and the gyroscopes do not contribute to the orientation estimation. While in the dynamic state, the angular velocities measured by the gyroscopes contribute to the orientation estimation, and hence improved the accuracy. In addition, it can be seen that the accuracy is higher at lower speeds. This can be interpreted that when the speed is high, the error between measured acceleration and the gravity is large,

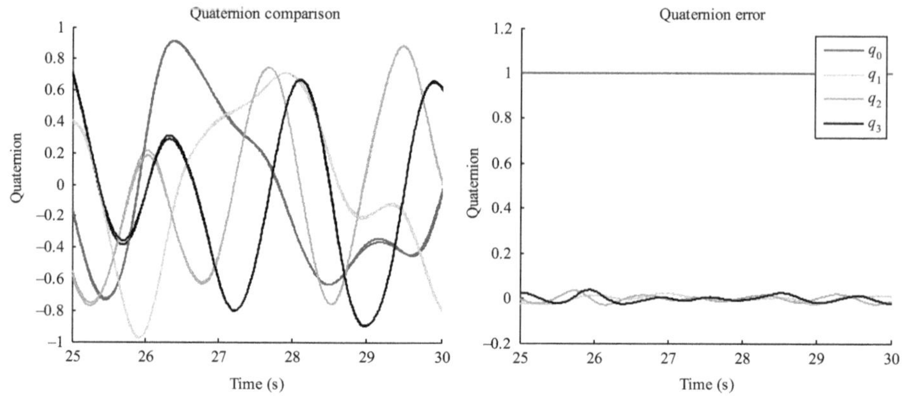

Figure 6.14 Quaternion comparison between the gimbal and the WMS

and although the fusion algorithm can improve the accuracy to some extent, the error would increase inevitably.

6.5.2.2 Multiaxis rotation

The results of multiaxis rotation are shown in Figures 6.14 and 6.15. Figure 6.14 shows the example of quaternion differences and the calculated q_{error} between the gimbal and the WMS. As the quaternion error in this figure is not easy to understand, it was converted to Euler angles for a better understanding. The converted Euler angles are shown in Figure 6.15. The RMSEs of roll, pitch, and yaw are 1.64°, 1.60°, and 1.70°, respectively. Although these angles are not relative to the Earth frame any more, the error trends are clear. From the results, it can be seen that the RMSEs of multiaxis rotation are larger than that of single axis rotation.

6.5.3 The effect of magnetic disturbances

During this experiment, the magnitude of the measured magnetic field is shown in Figure 6.16. The disturbed magnetic field is plotted as a red line. It can be seen that strong magnetic disturbance occurred periodically.

Figure 6.17 shows the Euler angles converted from quaternion error between the two measurements. The RMSEs of roll, pitch, and yaw are 3.62°, 2.84°, and 2.88°, respectively. It can be seen that the accuracy of WMS is seriously affected by the magnetic disturbance. For example, the pitch error increased by more than 10° sometimes. So it is recommended that any ferromagnetic material be avoided in the surroundings when using this WMS.

From all the results, it can be seen that the designed gimbal can be used for performance evaluation under various conditions. Hence, when developing WMSs or selecting commercially available WMSs for an exoskeleton, testing patterns can be simulated on the gimbal according to the working conditions of the designed exoskeleton. This will assist in developing a proper WMS for the exoskeletons. The example tests presented in this chapter have shown the merits of the instrumented

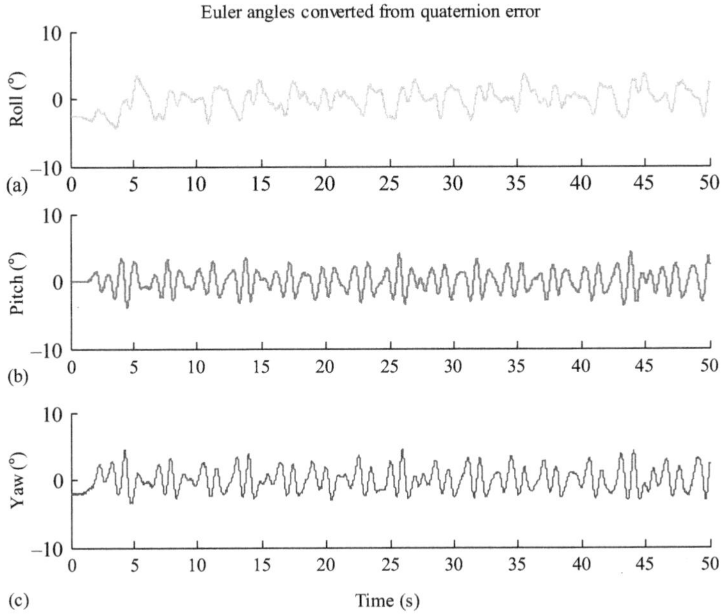

Figure 6.15 Euler angles error when all three axes rotated simultaneously. (a) The error of roll angle; (b) the error of pitch angle; and (c) the error of yaw angle

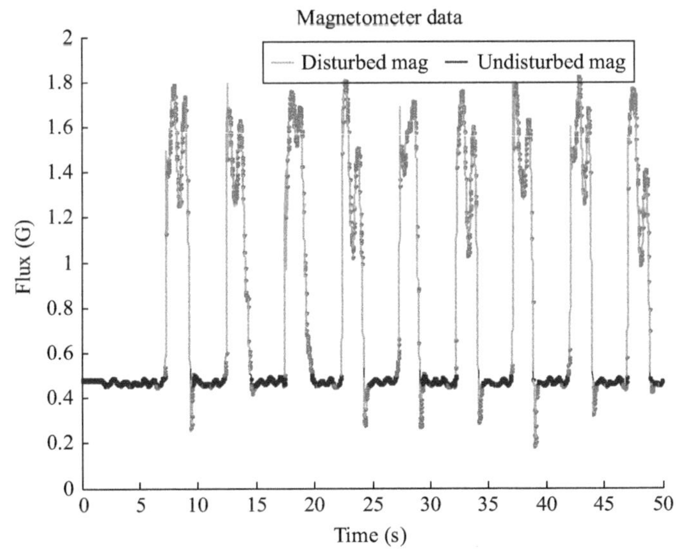

Figure 6.16 The magnitude of the measured magnetic field during the dynamic test

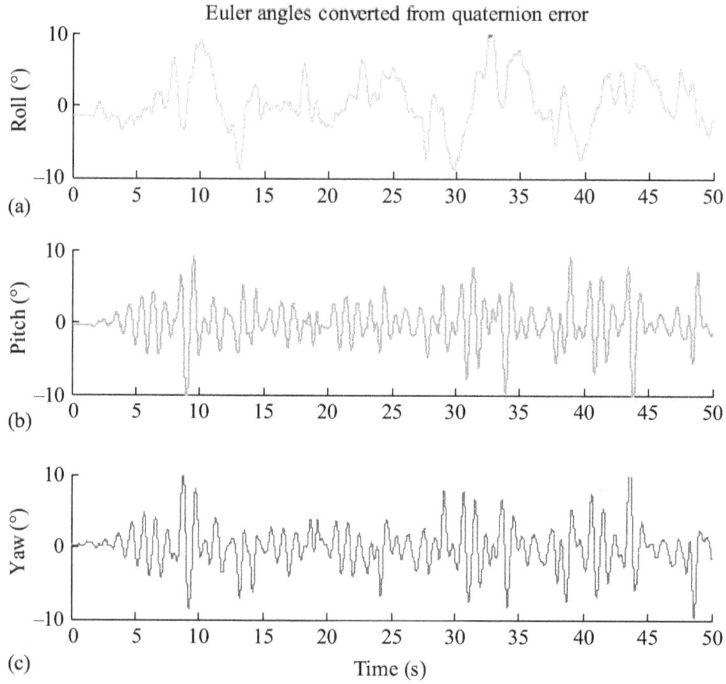

Figure 6.17 Euler angles converted from quaternion error between the two measures. (a) Roll angle error; (b) pitch angle error; and (c) yaw angle error

gimbal. Compared with the gimbals presented in previous literature [16–18], the designed instrumented gimbal offers the following advantages:

1. Each frame can rotate continuously, so that the test pattern can cover a full range of rotation (0°–360°) around each axis;
2. The gimbal is capable of directly measuring the angle of each axis, so that the error analysis for WMS can be performed without needing any additional equipment.
3. The test area of the gimbal is nearly free of magnetic disturbances, so that the gimbal can be used to evaluate WMSs containing magnetometers.

In our study, the instrumented gimbal is assembled by aluminum profiles. As a limitation, the machining precision of the parts, the stiffness and the mass unbalance of the frames influence the accuracy of the instrumented gimbal, which will induce error to the reference angle. Further design should enhance the stiffness of the frames, preferably using a welded frame instead of a bolted connection frame, and improve the mass unbalance problem of each frame carefully.

6.6 Conclusion

In this study, an instrumented gimbal was designed for evaluating the performance of WMSs used in exoskeletons. Compared with other similar evaluation devices, the advantages of the designed instrumented gimbal are that all three axes can rotate continuously and simultaneously. Proper distance was set between test area and its nearby ferromagnetic materials to ensure that the test area is nearly free of magnetic disturbances. So the gimbal is compatible with the magnetometers contained in some WMSs. Moreover, the gimbal is an independent device which can output the orientation with high accuracy. These features make it possible to fully evaluate the WMSs under controlled conditions: different time durations, different speeds, with or without magnetic disturbances. In order to validate the design, experiments were performed, including static accuracy tests, dynamic accuracy tests, and the effect of magnetic disturbances tests. Results show that the gimbal can be used to analyze the accuracy of the WMS under different conditions, so that can help to improve the accuracy of WMSs for the applications of exoskeletons. In the future, more conditions should be simulated for WMSs according to the working conditions of exoskeletons, and a standard test flow should be summarized to validate the WMSs.

Acknowledgment

This work was supported in part by the NSFC Grant No. 51775485 and U1613203; the Zhejiang Provincial Natural Science Foundation of China under Grant No. LR15E050002, the State Key Laboratory of Fluid Power and Mechatronic Systems under Grant GZKF-201702 and NSERC Discovery Grant to Q. Li.

References

[1] Iosa M, Picerno P, Paolucci S and Morone G 2016 Wearable inertial sensors for human movement analysis *Expert Review of Medical Devices* **13** 641–659.

[2] Tao W, Liu T, Zheng R and Feng H 2012 Gait analysis using wearable sensors *Sensors (Basel)* **12** 2255–2283.

[3] Tao W J, Zhang J Y, Li G Y, *et al.* 2016 A wearable sensor system for lower-limb rehabilitation evaluation using the GRF and CoP distributions *Measurement Science and Technology* **27** 025701.

[4] Kortier H G, Antonsson J, Schepers H M, Gustafsson F and Veltink P H 2015 Hand pose estimation by fusion of inertial and magnetic sensing aided by a permanent magnet *IEEE Transactions on Neural Systems and Rehabilitation Engineering* **23** 796–806.

[5] Trkov M 2016 Modeling, sensing, and control of human bipedal walking with foot slip. Ph.D. Thesis, Rutgers University, New Jersey, USA.

[6] Mazumder O, Kundu A S, Lenka P K and Bhaumik S 2016 Ambulatory activity classification with dendogram-based support vector machine: application in lower-limb active exoskeleton *Gait & Posture* **50** 53–59.

[7] Rocon E, Moreno J C, Ruiz A F, Brunetti F, Miranda J A and Pons J L 2007 Application of inertial sensors in rehabilitation robotics. In: *10th IEEE International Conference on Rehabilitation Robotics*, (Noordwijk, Netherlands) pp. 145–150.

[8] Song Z, Guo S and IEEE 2011 Development of a real-time upper limb's motion tracking exoskeleton device for active rehabilitation using an inertia sensor. In: *2011 9th World Congress on Intelligent Control and Automation (WCICA 2011)*, pp. 1206–1211.

[9] Mateos L A, Ortiz J, Toxiri S, Fernandez J, Masood J and Caldwell D G 2016 Exoshoe: a sensory system to measure foot pressure in industrial exoskeleton. In: *6th IEEE International Conference on Biomedical Robotics and Biomechatronics (BioRob)*, (Singapore) pp. 99–105.

[10] Sabatini A M 2011 Estimating three-dimensional orientation of human body parts by inertial/magnetic sensing *Sensors (Basel)* **11** 1489–1525.

[11] Sabatini A M 2006 Quaternion-based extended Kalman filter for determining orientation by inertial and magnetic sensing *IEEE Transactions on Biomedical Engineering* **53** 1346–1356.

[12] Mahony R, Hamel T and Pflimlin J-M 2008 Nonlinear complementary filters on the special orthogonal group *IEEE Transactions on Automatic Control* **53** 1203–1218.

[13] Madgwick S, Harrison A and Vaidyanathan R 2011 Estimation of IMU and MARG orientation using a gradient descent algorithm. In: *Proceedings of the 2011 IEEE International Conference on Rehabilitation Robotics (ICORR)*, (Zurich, Switzerland) 29 June–1 July 2011; pp. 1–7.

[14] Chiari L, Della Croce U, Leardini A and Cappozzo A 2005 Human movement analysis using stereophotogrammetry—Part 2: Instrumental errors *Gait & Posture* **21** 197–211.

[15] Ligorio G and Sabatini A M 2015 A novel Kalman filter for human motion tracking with an inertial-based dynamic inclinometer *IEEE Transactions on Biomedical Engineering* **62** 2033–2043.

[16] Zhang S, Shuai Y, Liu C, Yuan X and Sheng L 2016 A dual-linear Kalman Filter for real-time orientation determination system using low-cost MEMS sensors *Sensors* **16** 264.

[17] Lai Y C, Jan S S and Hsiao F B 2010 Development of a low-cost attitude and heading reference system using a three-axis rotating platform *Sensors (Basel)* **10** 2472–2491.

[18] Lebel K, Boissy P, Hamel M and Duval C 2013 *Inertial measures of motion for clinical biomechanics: comparative assessment of accuracy under controlled conditions—effect of velocity PLoS One* **8**(11): e79945. doi:10.1371/journal.pone.0079945.

[19] Lebel K, Boissy P, Hamel M and Duval C 2015 Inertial measures of motion for clinical biomechanics: comparative assessment of accuracy under

controlled conditions—changes in accuracy over time *PLoS One* **10**(3): e0118361. doi:10.1371/journal.pone.0118361.

[20] Brennan A, Zhang J, Deluzio K and Li Q 2011 Quantification of inertial sensor-based 3D joint angle measurement accuracy using an instrumented gimbal *Gait & Posture* **34** 320–3.

[21] de Vries W H K, Veeger H E J, Baten C T M and van der Helm F C T 2009 Magnetic distortion in motion labs, implications for validating inertial magnetic sensors *Gait & Posture* **29** 535–541.

[22] Willis D M, Gardiner A R, Davda V N and Bone V J 1997 Planar charged-particle trajectories in multipole magnetic fields *Annales Geophysicae-Atmospheres Hydrospheres and Space Sciences* **15** 197–210.

[23] Diebel J 2006 Representing Attitude: Euler Angles, Unit Quaternions, and Rotation Vectors; Stanford University: Stanford, CA, USA.

[24] Valenti R G, Dryanovski I and Xiao J Z 2015 Keeping a good attitude: a quaternion-based orientation filter for IMUs and MARGs *Sensors (Basel)* **15** 19302–19330.

[25] x-io Technologies Inc., x-IMU User Manual, version 5.2 ed., November 2013. Available online: http://www.x-io.co.uk/downloads/x-IMU-User-Manual-v5.2.pdf (accessed 20 August 2016).

[26] Koning B H W, Baten C T M and Koopman H F J M 2012 Three dimensional rotation offset correction. In: *Proceedings of the XII International Symposium on 3D Analysis of Human Movement*, no. 5, pp. 3–6.

[27] Ligorio G, Bergamini E, Pasciuto I, Vannozzi G, Cappozzo A and Sabatini A M 2016 Assessing the performance of sensor fusion methods: application to magnetic-inertial-based human body tracking *Sensors (Basel)* **16** 153.

[28] Bland J M and Altman D G 2010 Statistical methods for assessing agreement between two methods of clinical measurement *International Journal of Nursing Studies* **47** 931–936.

Chapter 7

Control and performance of upper- and lower extremity SEA-based exoskeletons

Crea Simona, Parri Andrea*, Trigili Emilio*, Baldoni Andrea*, Muscolo Marco*, Fantozzi Matteo*, Moisè Matteo*, Cortese Mario*, Giovacchini Francesco*, Carrozza Maria Chiara*, and Vitiello Nicola*,***

Abstract

In the last years compliant actuators have become extremely popular in the field of wearable robotics, due to their ability to realize safe human–robot interfaces. One of the most well-known example of a compliant actuator is the series elastic actuator (SEA), consisting of an elastic element (e.g. a spring) in series with a stiff actuator; such actuation architecture is considered to be the simplest design solution to realize compliant, compact, and light-weight actuators for wearable robots; in addition, from the control perspective, SEA architecture allows for simple force or torque control in addition to position control.

In this chapter, three wearable robots for upper- and lower-limb rehabilitation and assistance, developed at The BioRobotics Institute of Scuola Superiore Sant'Anna, are described. The three devices have similar SEA-based actuation units, integrating commercial electromagnetic motors and custom torsional springs, with constant stiffness and linear torque-deformation characteristics. Closed-loop torque control performance show that the systems can be highly transparent when controlled under zero-torque modality, i.e. the interaction with the human is minimal and the actuators do not hinder the user's movement, and bandwidths and output torques are compatible with the human movements to be assisted.

Keywords: series elastic actuators, torque control, compliance, elbow exoskeleton, shoulder-elbow exoskeleton, hip exoskeleton

*The BioRobotics Institute, Scuola Superiore Sant'Anna, Italy
**MARE Laboratory, Don Carlo Gnocchi Foundation, Italy

7.1 Compliant actuators with series elasticity for wearable robots

Traditional industrial robots have stiff actuators to allow precise and stable position control. However, most of the wearable robotics applications do not need precise trajectory tracking: they need to control close physical human–robot interaction which requires the robot to absorb unpredicted shocks and supply additional mechanical power to the user's limb; moreover, actuators with reduced reflected inertia and output impedance would guarantee the user's safety more than typical stiff actuators. In the last years, the development of novel actuation units capable of achieving those goals has turned out to be fundamental for wearable applications. Compliant actuators have become extremely popular in the field of wearable robotics. Their ability to absorb shocks due to unpredictable interaction with the user or the environment (such as in case of spastic muscle contractions which is a typical feature of poststroke movements or collisions with external objects), to store and release energy, and to realize safe human–robot interfaces make them one of the most preferred design solutions [1,2].

A compliant actuator is defined as a device that depending on the applied external force, can change its own equilibrium position, i.e., the position of the actuator that generates null output force or torque [3]. There are two main approaches to realize compliant actuators. On the one side, a desired compliance of the actuation unit can be artificially achieved through the implementation of an impedance or admittance control: the physical compliance of the system is that of typical stiff actuators and the compliant behavior fully depends on the controller and its programed output. On the other, compliance can be integrated directly into the actuation unit by means of passive elastic elements; these elastic elements in turn, can have fixed or variable impedance. In both the cases, a physical elasticity provides inherent compliance even when the system is turned off, regardless of the controller. In addition, the control strategy can further reduce the output compliance of the actuator by modifying the equilibrium position of the actuator.

Fixed-compliance (or fixed-stiffness) actuators incorporate a fixed-stiffness passive element in their structure, typically in between the stiff actuator and the output load. The most well-known example of a fixed-stiffness actuator is the series elastic actuator (SEA), consisting of an elastic element (e.g., a spring) in series with a stiff actuator [4,5]. The resulting physical compliance of the system can be approximated by the spring compliance, which is a fixed value, but can be virtually adjusted by dynamically changing the equilibrium position of the spring.

Variable-compliance actuators are capable of regulating their physical compliance, by means of the combination of passive and active elements. As an example, variable-compliance (or variable-impedance, VIA) actuators typically include two actuators to set the position and the stiffness of the joint [6,7]. Although the resulting performance of these actuators cannot be achieved with conventional stiff actuators (the range of stiffness and energy storage capability of VIAs is extremely high) and the integration of elastic elements in the actuation can potentially result in the development of inherently-safer actuation units, there are many drawbacks of this architecture for applications in wearable robotics. This architecture typically needs two motors to control one joint and the encumbrance and weight of the final design might not be optimal for applications where the robot should be portable and lightweight.

For the reasons of lightweight and portability but still the need for safety, fixed-compliance actuators are preferred. Fixed compliance in a series configuration is considered to be the simplest design solution with enough efficiency to realize compliant, compact, and light-weight actuators. As an example, from a biological view point, series elasticity has been naturally integrated into humans' articulations: in the muscle-tendon architecture, muscles generally contribute most to the active mechanical work, whereas tendons, which are in series with muscles, provide the majority of elastic energy savings [8].

The spring in the SEA is the key component of the mechanism. Custom springs, linear and torsional ones, are typically designed taking into account the desired maximum torque and stiffness to interface with the actuator and the human. Clearly, simple geometries with high structural efficiency and fatigue limits and ease of manufacture are preferred. Moreover, springs must have a highly linear and repeatable behavior in order to provide a precise measurement of the delivered torque (i.e., by measuring the compression of the compliant element, the force or torque on the load can be calculated by the Hooke's Law) and allow better controllability.

As mentioned, from the control view point, the classical position control has been widely used where the robot joints or end-effector is moved to track a predefined trajectory and the actuator must be as stiff as possible. However, in many applications, force or torque control is necessary strategy to comply with the user's movement. Indeed, the control strategies of wearable devices typically fall under two main categories, the robot-in-charge and patient-in-charge modes [9]. In the *robot-in-charge* mode, the robot should force the limb movement along predetermined trajectories: in this case, it is important that the robot has enough bandwidth and power to realize the desired positioning and trajectory tracking performance with the desired, relatively high, impedance. In the *patient-in-charge* mode, it is important that interaction forces between the robot (such as an exoskeleton) and the human are controlled toward zero; in other words, the perceived output impedance of the robot is low. Assistive force or torque can be provided if needed, or convergent or divergent force/torque fields can be applied to train the motor control functions of the user during the rehabilitation process.

SEA architecture typically allows for simple force or torque control in addition to position control [10]. A feedback loop calculates the error between the measured and the desired forces and the error signal is used in a controller that calculates the appropriate current to be applied by the motor to correct the error.

A simple model for a SEA is shown in Figure 7.1, where J_m is the motor mass, k_s is the spring stiffness, T_m is the force on the motor, T_l is the output force, and θ_m and θ_l the angles of the motor shaft and load. From the Newton's Laws, the T_m is a function of the load torque T_l as expressed by the following equations [11]:

$$\begin{aligned} T_m + k_s(\theta_l - \theta_m) &= J_m\ddot{\theta}_m \\ -k_s(\theta_l - \theta_m) &= T_l \end{aligned}$$

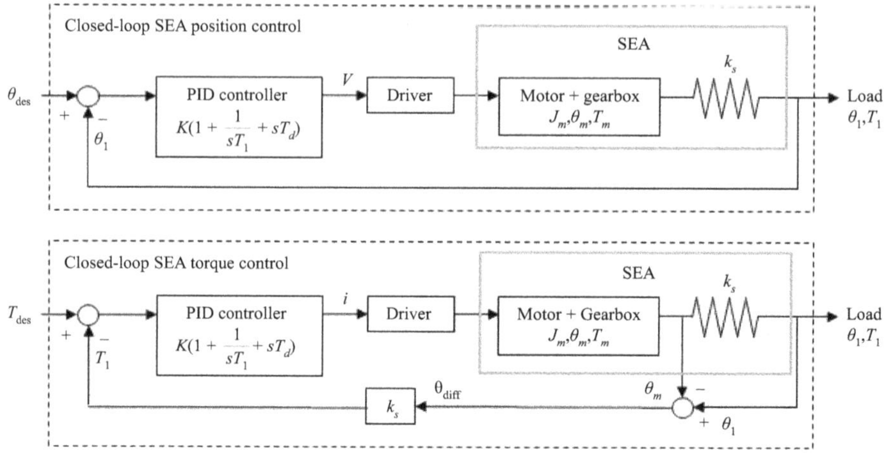

Figure 7.1 Closed-loop control of a SEA. Position control and torque control schemes are represented

After transforming the equations to the Laplace domain, the relationship between T_l and T_m is given:

$$T_m(s) = \left(1 + \frac{J_m}{k_s}s^2\right)T_l(s) + J_m s^2 \theta_l(s)$$

The motor torque T_m is the source torque to provide an output torque T_l when the actuator is moving.

The SEA model results in a second order equation; thus, classical proportional-integral-derivative (PID) controllers are suitable to control the output force or position. Classical PID controllers with the form $K(1 + \frac{1}{sT_i} + sT_d)$ can be used for closed-loop control, as described in [12] (Figure 7.1).

According to [12], the output mechanical impedance of a system can be defined as the ratio between force and position and has a frequency-dependent behavior. The resulting output impedance of a SEA system with the additional PID controller is

$$Z(s) = \frac{T_l(s)}{\theta_l(s)} = \frac{-s^2 J_m k_s}{s^2 J_m + k_s + K\left(1 + \frac{1}{sT_i} + sT_d\right)}$$

Studying the interaction impedance in the frequency domain determines whether a system is stable based on the rules set by Colgate and Hogan [13] and quantifies the transparency of the system in zero-torque conditions.

In this chapter, we provide the description of three SEA-based wearable robots developed at The BioRobotics Institute of Scuola Superiore Sant'Anna. Three devices with similar SEA-based actuation units have been designed for upper- and lower limb rehabilitation and assistance. Devices are presented in detail: particular attention is given to the SEA design and closed-loop torque and position control performance. Also, a comparison with similar state-of-the-art devices is provided.

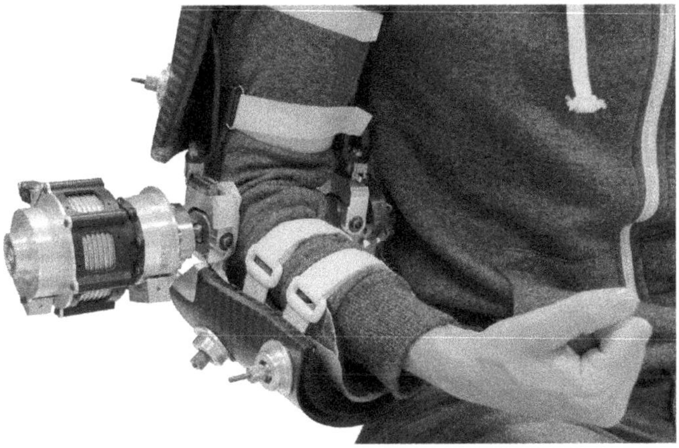

Figure 7.2 Overview of the NEEM

7.2 NEUROExos elbow module

NEUROExos elbow module (hereafter called NEEM, Figure 7.2) is a highly ergonomic powered exoskeleton for elbow rehabilitation.

Its first design [14] presented three main innovative features: (i) a tailored shell structure that realizes a comfortable physical human–robot interface area; (ii) a passive mechanism for self-alignment of joint axes; and (iii) a bioinspired antagonistic actuation based on nonlinear elastic elements.

In order to realize a safe human–robot interaction, i.e., to interact with the patient and compliantly interact to sudden movements of the person, a second version of the NEEM has been recently presented in [15]. The new design integrates a new actuation unit based on a SEA unit, with a custom torsional spring and an electromagnetic motor, and implements a novel control algorithm that mobilizes the user's limb and guarantees safety without forcing the subject with nonphysiological movements or excessive load. High-linearity and repeatability of the torque-deformation characteristic of the developed custom spring is the greatest advantage of this SEA with respect to other solutions from the state of the art.

The exoskeleton is composed of two shell frames corresponding to the arm and forearm. Each shell frame consists of a structural outer shell made of carbon-fiber and a thermoformed plastic inner shell that is customizable to the user's arm. The actuation rotates the lower frame with respect to the upper frame; the two frames are connected through a four degrees-of-freedom (DOFs) [14] passive mechanism to allow self-alignment of the joint axes. A number of mechanisms have been implemented to allow the necessary adjustments to wear the exoskeleton: for example, the upper frame hangs from an adjustable spherical gimbal, used to set the orientation of the exoskeleton. An extensible horizontal cantilever supported by the main vertical stand sets the horizontal and vertical position of the

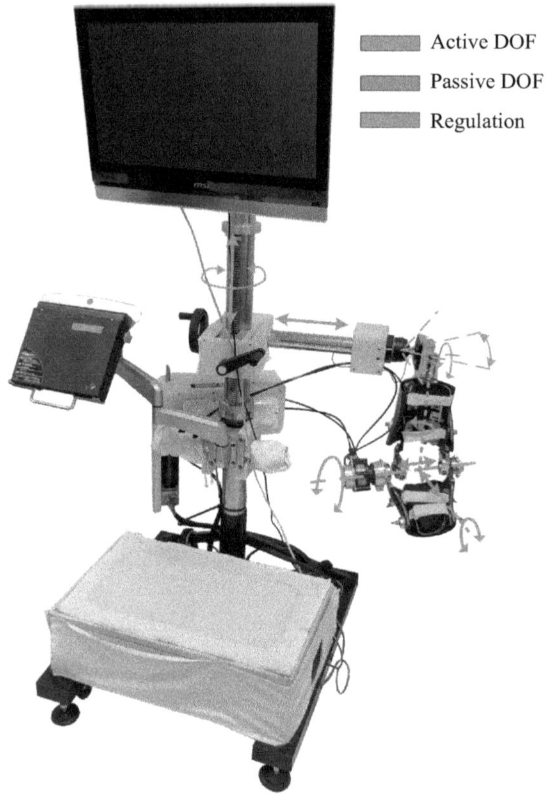

Figure 7.3 The NEEM system. Active and passive degrees of freedom and regulations are indicated by arrows

exoskeleton with respect to the main stand. The set of active and passive DOFs and size regulations are reported in Figure 7.3. The control and interface electronics are mounted on the vertical stand. The stand is mounted on a wheeled platform that makes the exoskeleton portable within the clinical setting and usable with seated or bedridden patients.

7.2.1 SEA architecture: mechanics and control

The actuation unit of the joint is located remotely, on the main vertical stand. The actuator includes a DC servomotor (Maxon Motor® EC60, 400 W), a compact reduction stage with high reduction ratio (Harmonic Drive® CPL-17A-080-2, transmission ratio 1:80), a safety torque limiter on the Harmonic Drive output (R+W SK1/15/D/20) that disengages when the applied torque overcomes a threshold (between 35 and 70 N m, settable depending on the application), and a grooved pulley.

Torque is transmitted from this pulley to the exoskeleton actuated joint through a Bowden-cable mechanism: two antagonist cables (stainless steel AISI 316 wire rope, outer diameter 1.0 mm, Carl Stahl® CG 719100), routed by bendable Bowden hoses (flat wire spiral with inner tube, inner diameter 2.2 mm, Carl Stahl® FKP224049) transmit the rotation from the motor pulley to the exoskeleton joint. The pulley assembly is built in Ergal aluminum alloy (7075-T6) and the Tecatron GF40. A custom patented torsion spring[1] couples the actuated pulley and the lower frame of the exoskeleton rotations: the coupling realizes a SEA architecture. The spring has a constant stiffness of 98.75 N m/rad and has a linear torque-deformation characteristic. The maximum allowed deflection is 17 deg which corresponds to a peak torque of 30 N m; if the deformation is higher than the threshold, the stiffness drastically increases with a nonlinear behavior due to the spring coils touching. Structurally the maximum applicable torque that can be sustained is about 54 N m. As a consequence, the spring shows an almost infinite lifetime to alternated load cycles in the range ±30 N m.

The sensory system is composed of two 32-bit absolute encoders (Renishaw® RESOLUTE™, RESA30USA052B ring plus RA32BAA052B30F read head), one for each extremity of the spring; the more proximal to the human joint is labeled as "joint position encoder" and provides the elbow joint angular value; the difference of the two encoder readings gives the spring deformation, thus the torque transmitted through the spring. Absolute encoders avoid the calibration phase, which is usually performed to correct the bias on the torque reading at platform start and can introduce undesired errors if the calibration procedure is not performed correctly.

Two closed-loop control strategies have been implemented: *joint position* and *joint torque* control.

- **Joint position control.** The joint position control moves the lower link of the exoskeleton along a desired reference position. The closed-loop position control uses a Proportional Integrative (PI) regulator operating on the error between the desired angle value and measured angle. The PI then outputs a desired voltage in the range ±48 V. A commercial servo amplifier (EPOS2 70/10, Maxon Motor®) controls the output voltage which is mapped by the motor-gear to ±230 deg/s of maximum joint speed.

  PI parameters were tuned empirically in order to have a bandwidth up to 1 Hz. The performance of the closed-loop position control was characterized by the analysis of the step and chirp responses in unloaded conditions.

  The results of the step response in extension and flexion (steps of 60 deg) were assessed in terms of rise time, settling time, and overshoot. Rise time was lower than 0.3 s, and the settling time was estimated to be around 0.4 s, and the overshoot was less than 0.05 deg. For the chirp response, the angular position was moved along a linear chirp with a frequency ranging from 0.3 to 4 Hz and with an amplitude of 20 deg. The estimated—3-dB bandwidth of the resulting amplitude Bode diagram was found to be 1.55 Hz.

[1]Giovacchini F., Cempini M., Vitiello N., Carrozza M.C. *Torsional transmission element with elastic response.* WO2015001469, International Publication Date: January 8, 2015

- **Joint torque control.** The joint torque compensator controls the amount of torque delivered to the user's joint. The closed-loop torque control still implements a PI regulator, whose input is the error between the desired torque value and the measured one and outputs a desired current in the range ±10 A. The servo amplifier controls the output current that corresponds to an output torque range of ±40 N m with the corresponding motor/gearbox. The applied torque is estimated by measuring the deformation of the spring (i.e., the difference of the joint position encoder and the encoder measuring the rotation of the motor-gear, multiplied by the stiffness of the torsion spring). Again, the PI parameters were tuned empirically in order to have the minimum output impedance in the transparent mode and the performance of the closed-loop torque control was assessed by the analysis of the step and chirp responses in unloaded conditions.

  The results of the step response in increasing and decreasing torque (steps of 2 N m) were evaluated in terms of rise time, settling time, and overshoot. In this case, the rise time was lower than 0.02 s, settling time was lower than 0.3 s, but the overshoot was definitely too high (i.e., over 0.6 N m).

7.2.2 High-level control

Position and torque controllers, respectively, allow the system to perform the so-called robot-in-charge and patient-in-charge rehabilitation paradigms.

In the robot-in-charge paradigm, the robot executes the movement; this rehabilitation paradigm is typical in the initial rehabilitation phase when the patient is more severely impaired.

In the patient-in-charge paradigm, the patient leads the movement while the robot can aid and assist in the accomplishment of the movement, supplying additional torque when necessary and setting the desired torque to zero (thus making the exoskeleton completely transparent to the user actions) when the user is able to perform the task autonomously. Transparency of the exoskeleton can be also useful to make the robot compliant to sudden spasticity of the mobilized joint, without forcing undesired trajectories or excessive load.

The possibility to switch between the two control modes allowed the implementation of rehabilitation exercises that take advantages of both strategies: for example, the smart passive joint mobilization described in [15] is based on the joint moving along a predetermined trajectory, where the detection of excessive resistance offered by the user to a the preset movement (i.e., excessive torque) results into the automatic switch to the transparent mode.

7.3 NEUROExos shoulder–elbow module

The NEUROExos shoulder–elbow module (NESM, Figure 7.4) is a SEA-based upper-limb exoskeleton designed for the physical rehabilitation of poststroke subjects. Four SEA actuation units address the shoulder abduction/adduction (sA/A), flexion/extension (sF/E), and internal/external rotation (sI/E), as well as the elbow flexion/extension (eF/E).

Figure 7.4 Overview of the NESM

The device has been designed to comply with the complexity of the human shoulder articulation, optimizing the human–robot kinematic coupling for a safe and comfortable interaction with the user. A simple approximation as a spherical joint with three intersecting axes is not sufficient to convey the full range of motion (ROM) of the shoulder complex, which emerges from the combined action of the glenohumeral, acromioclavicular, sternoclavicular, and scapulothoracic joints [16]. For this reason, the NESM kinematic chain has been endowed with additional passive DOFs, to confer translational movements to the active rotational joints and ensure their proper alignment with the corresponding anatomical joint axes. Three of the passive DOFs are embedded within the exoskeleton support structure, to follow the shoulder elevation/depression, the trunk rotation around the vertical axis and its translation along the forward/backward direction. The exoskeleton is connected to the user's trunk with a bar linkage having a rotational joint on the robot side and a ball joint on the human side, which reduces the constraints on the scapular movement during the sA/A. At the elbow level, the self-aligning kinematic chain of the NEEM has been replaced with a slider enabling the translation of the elbow axis along the forward/backward direction. Finally, a set of lockable sliding and pin joints have been designed to adapt the exoskeleton mechanical structure to users with different anthropometric sizes. In particular, the axis of rotation of the eF/E joint can be tilted and translated in the anterior/posterior direction and the position of the arm cuff can be regulated to fit different humerus lengths. Additional adjustable joints have been included within the support structure, so that they permit the height and orientation of the exoskeleton to be adjusted to fit semi-laid or wheelchair users. The whole system and all its active and passive degrees of freedom and size regulations are shown in Figure 7.5.

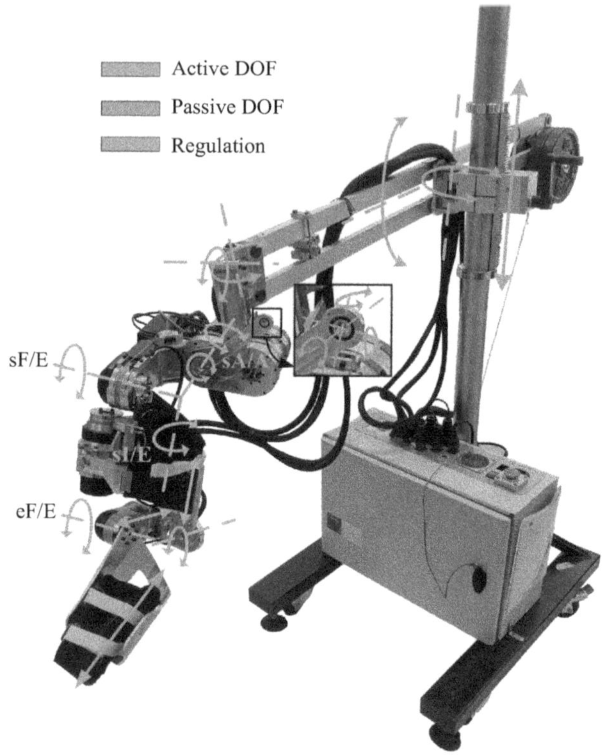

Figure 7.5 The NESM system. Active and passive degrees of freedom and regulations are indicated by arrows

The exoskeleton has been set up on a wheeled platform with a vertical stand to be easily moved inside a clinical facility. The platform also contains the electronics and control box.

7.3.1 SEA architecture: mechanics and control

The NESM has a modular mechanical structure with three main blocks, each bearing its own actuation unit. The first, namely, the shoulder section, is composed of an L-shape aluminum flange encompassing the user's shoulder and comprising two identical SEA units for sA/A and sF/E joints. In order to optimize inertia distribution, the two actuation units are placed next to each other on the rear of the flange. The actuators employ a 90 W DC motor (Maxon Motor® EC 90 Flat, 90 W, Sachseln, Switzerland), a reduction stage with 1:160 gear ratio (Harmonic Drive® CSD-25-160-2A) and a servo driver (Elmo Motion Control® Gold Solo Whistle). While the sA/A unit is coaxial with the addressed anatomical axis, a cable-pulley stage allows the transmission of sF/E rotation. In particular, two idle pulleys at the corner of the flange route the cables from the sF/E driver to the sF/E actuation pulley, placed on the lateral side. The elastic element is a custom torsional spring with constant stiffness of 165.8 N m/rad, able to withstand a maximum peak torque of 60 N m. Finally, the sensing elements are two 32-bit absolute optical encoders (Renishaw® RESOLUTE™ RA32BAA075B350AF), one for each side of the spring.

The second assembly, the arm section, is dedicated to the sI/E joint. The core structure comprises two coaxial hemicylindrical shells rotating around their common axis, perpendicular to both sF/E and sA/A axes and aligned with the humerus direction. The sI/E motion is achieved through an external parallel chain composed of two articulated parallelograms realizing a remote center of rotation. This chain is actuated by means of two capstan pulleys, one driven by the sI/E SEA unit and the other fixed with the moving shell. Finally, the elbow section includes the elbow joint. The actuator lies perpendicular to the elbow axis and the motion is achieved by means of two bevel gears with perpendicular axes. The SEA units for the sI/E and eF/E joint are identical and employ a 100-W DC motor (Maxon Motor® EC 60 Flat, 100 W), a 1:120 gear ratio reduction stage (Harmonic Drive® CPL-17A-120-2A) and a commercial servo driver (Elmo Motion Control® Gold Solo Whistle) for the actuator side. The spring is capable to withstand a maximum torque of 35 N m before yielding, with a stiffness equal to 98.7 N m/rad.

The position controller is based on a PID regulator, whose parameters are tuned empirically. The controller regulates its output depending on the error between the reference joint angle and the measured one, enabling the SEA unit to provide the torque necessary to move the joint along a desired trajectory, overcoming external applied loads (e.g., gravity). A preliminary experimental validation was carried out to evaluate the performance of the position controller. Two weights of 1.5 kg were attached to the arm and forearm cuffs, to simulate the average inertia of a human limb and 40 sinusoidal inputs of 0.1-Hz frequency and 10-deg amplitude were commanded simultaneously as a reference position to each joint. The root mean square error was 1.23 deg for sA/A, 1.06 deg for sF/E, 0.73 deg for sI/E, and 0.98 deg for eF/E.

The torque controller is based on PID regulators as well and operates on the error between the desired torque and the torque estimated by the deformation of the spring. A null reference torque can be set to the controller, making the robot transparent to the user's movements. In addition, a gravity compensation algorithm has been implemented and validated in [17] to counteract the effect of the weight of the device. The torque due to gravity is calculated in static conditions according to the joint configuration, so that the SEA units can provide the compensatory torque to cancel it.

The performance of the closed-loop low-level torque controllers has been assessed by an experimental characterization of its dynamic behavior. With two identical SEA units, the characterization has been conducted on one joint for each, namely, sA/A and eF/E. In particular, the step response was analyzed in static conditions by pushing the joint toward its mechanical stop with a preload torque and applying torque steps of 9 N m. Both increasing and decreasing steps were repeated 50 times. In both joints, the desired value is reached with an overshoot lower than 0.1 N m. The mean rising and falling times were, respectively, 0.38 and

0.37 s for the sA/A joint, while for the eF/E joint the mean values were 0.29 and 0.30 s, respectively.

7.3.2 High-level control

The two control modalities are employed similarly to the NEEM, the position control being more suitable in the early stages of the rehabilitation, while the torque control is more appropriate at later stages.

For the position mode, an inverse kinematic algorithm has been designed to reach different positions within the workspace of the robot, reproducing functional tasks typical of activities of daily living with the aim of improving cortical reorganization and motor recovery [18]. Smooth trajectories with null velocities at the beginning and at the end of the movement are provided and passive mobilization can be performed limiting the specific ROM of the subject. Abnormal torque profiles can be detected to immediately switch into the transparent mode [19]. This is extremely useful in the treatment of poststroke subjects when dealing with sudden, unpredictable spastic contractions.

More related to the traditional rehabilitation protocols, a set of exercises has been designed to act selectively on a single joint and the muscular groups involved in its mobilization. Repetitive movements can be executed in presence of diverging and converging torque fields, whereas assistive or resistive torques can be provided to train specifically the agonist or antagonist muscles involved in the movement, during the concentric and/or the eccentric phase.

7.4 Active pelvis orthosis

The Active Pelvis Orthosis (α-APO, Figure 7.6(a)) is a novel wearable robotic device for lower limb movement assistance. It is lightweight and designed for assisting hip flexion/extension (hF/E) during walking [20]. It has been conceived for providing active hip assistance to people affected by lower limb impairments who preserve sufficient mobility in daily living activities.

In order to satisfy ergonomic requirements, the design was based on the extensive use of lightweight composite materials like carbon-fiber. The orthosis can fit a wide range of user sizes thanks to several manual adjustments for fine regulations and passive DOFs which follow the gait motions out of the F/E plane (thigh abduction/adduction). These adjustments were embedded in the mechanical frame. As a result, the device ensures kinematic compatibility, enhancing the comfort of the human–robot physical interaction and addressing the match of intra- and intersubject anthropometric variability.

Even if highly ergonomic, as a laboratory prototype, the α-version has the main limitation of a wired power supply and communication with the remote desktop of the operator. To be usable and effective in assisting people in activities of daily living, a wearable robot should be portable and battery operated. With the ultimate goal of achieving a totally portable device, the β-version of the APO, namely, β-APO (Figure 7.6(b)), was designed and developed.

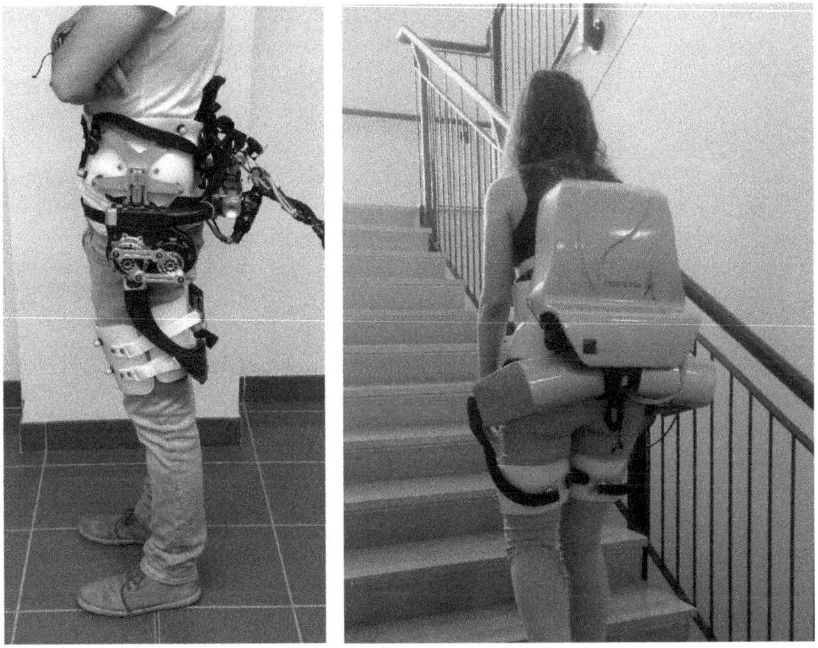

(a) (b)

Figure 7.6 Overview of the α-APO (a) and β-APO (b)

7.4.1 SEA architecture: mechanics and control

α-APO is endowed with two SEA units, one for each hF/E joint, mounted on the lateral arms of a C-shaped frame surrounding the user's trunk. The motor units have been designed taking as reference the hip angle and torque profiles reported by the Winter dataset for level-ground walking [21]. The target amount of assistance was set to 50 percent of the human torque required during level-ground walking assuming a natural cadence of 105 steps/min and a user weight of 80 kg [21]; hence, the actuator was designed in order to provide a maximum peak torque of 35 N m. The SEA series elasticity is obtained by means of the custom patented torsional spring with a stiffness of 100 N m/rad—a value comparable with the human hip average stiffness during ground level walking [22]—and bears the torsion stress without yielding, nor presenting hysteretic or nonlinear behavior.

Each actuation unit is deployed around two parallel axes. One is the axis of the 100-W DC motor (Maxon Motor®, EC60, 100 W) equipped with an incremental encoder (1024 ppr, MILE, Maxon Motor®) and coupled with a 80:1 Harmonic Drive (Harmonic Drive®, CPL-17A-080-2A) reduction stage. On the other axis, collocated with the human hF/E axis, there is the torsional spring in series with a 32-bit absolute encoder (Renishaw®, RESOLUTE™, ring: RESA30USA052B, read head: RA32BAA052B30), which measures the absolute hip joint angle. Each actuation unit has a weight of 1.2 kg. The transmission between the two parallel axes is based on a 4-bar mechanism, with a ROM between −30 and 110 deg, limited by emergency mechanical stops. Although this design is relatively small, the solution has one limitation as it partly prevents the user from swinging their arms.

The closed-loop control architecture is based on a PID regulator running at 1 kHz. The PID regulator returns an electrical current provided to the motor, within a saturation interval of ±3.2 A—corresponding to a torque range of ±35 N m. The motor current is controlled by means of a commercial servo amplifier (EPOS2 70/10, Maxon Motor®). As explained in [11], the bandwidth of a SEA system controlled by means of a PID compensator can be limited by design. Thus, PID regulator coefficients were tuned manually to achieve the widest closed-loop bandwidth. Both the high- and low-level layers run at 1 kHz. The APO control system implements a safety loop that switches off the actuation when the measured torque is higher than 30 N m, or the joint speed is greater than 400 deg/s. In addition, both the experimenter and the user can turn off the apparatus by means of a red, emergency button.

With the ultimate goal of achieving full portability, a relevant advancement from the α-APO to the β-APO[2] was the design optimization of the SEA unit grounded on the reduction of its weight and size. As a result, the β-APO SEA units were placed on the rear part of the device and deployed on a single-axis configuration in order to minimize the lateral size thereby allowing natural arm swinging. Each SEA consists of a 70-W DC motor (Maxon Motor®, EC60 Flat, 70 W) coupled with a 100:1 Harmonic Drive (Harmonic Drive®, CPL-14A-100-2A). The Harmonic Drive is connected to the torsional spring, whose deformation is directly measured by an absolute 17-bit rotary electric encoder (Netzer Precision Motion Sensors Ltd, DS-37™). The torque is transmitted to the hip joint through a pulley wrapped with a steel cable. A 17-bit absolute rotary electric encode (Netzer Precision Motion Sensors Ltd, DS-25™) placed on the hip axis measures the actual hip angle. Each actuation unit weighs slightly more than 1 kg, namely, 15 percent lower than the α-APO unit.

The low-level controller the α-APO was validated from three points of view: step response to different torque amplitudes, chirp response to a linear chirp torque and output impedance in zero-torque mode. The step-response experiments pointed out that the tracking capabilities of the device actuator are fast and prompt enough. The step response is mainly a property of the chosen motors and their PID tuning. The rise time, settling time, and the overshoot were 0.07 s, 0.18 s, and 2.51 N m, respectively, for the maximal step amplitude of 8 N m. The chirp response analysis tested the low-level controller bandwidth and a −3-dB control bandwidth of 15.5 Hz was determined, thus broadly enclosing the typical frequencies of human gait. Under the action of the zero-torque control, the output impedance for a

[2]Vitiello N., Giovacchini F., Cempini M., Fantozzi M., Moisè M., Muscolo M., Cortese M. *Actuation system for hip orthosis*. WO2016128877, International Publication Date: August 18, 2016

frequency of 1 Hz was lowered by −38 dB to 1.3 N m/rad, compared with the inherent passive impedance of 98.7 N m/rad.

Low-level, closed-loop, torque controllers of the β-APO (running on FPGA level at 1 kHz) are independent for each actuation unit and rely on a 2-pole–2-zero compensator designed to achieve the minimum joint output impedance (≈1 N m/rad at 0.5 Hz motion and ≈5 N m/rad at 3.0 Hz) and a relatively high closed-loop bandwidth (12.2 Hz) [23].

7.4.2 High-level control

In order to effectively assist human locomotion-related activities, inherently exhibiting intra- and intersubject variability, the assistive controller design was grounded on two main requirements: (i) prompt adaptiveness to the user intentions, e.g., variation of the walking cadence or step length; (ii) high reliability in decoding the ambulatory activities, i.e., walking, stairs climbing/descending, by means of algorithms requiring no cognitive participation from the user except the volitional movement itself.

To satisfy adaptiveness requirements, the high-level controller of the APO was designed to provide phase-invariant assistance thanks to an intuitive, not invasive, interface estimating the phase of ambulatory tasks in an ecological fashion. The proposed methodology is based on mathematical tools known as adaptive oscillators, able to converge to the features (i.e., the amplitude, frequency, and offset) of a time-continuous quasiperiodical signal [24], e.g., gait-related biomechanical signals. Phase-locked assistive torque, adaptive to the minor variations of the gait pattern, allows for seamlessly keeping the synchronization with the user's intended motion [25]. With the integration of external wearable sensory apparatus, namely, a pair of sensitive insoles [26], the portable β-APO enabled the possibility of designing several experimental activities with the goal of testing the feasibility of intention decoding algorithm for locomotion-related activities assistance in semi-structured environment—i.e., not only treadmill walking but also ground-level walking and stair climbing [27].

7.5 Performance, strengths, and challenges of SEAs in wearable robotics

Exoskeletons entail a close physical human–robot interaction that requires stringent design requirements to take into account human kinematic and kinetic properties. When designing a SEA unit, there are three relevant considerations that should be taken into account:

1. The spring stiffness. Having a spring with stiffness comparable to that of the human joint would ensure safer human–robot interaction. Although in biological systems joint impedance is not fixed (indeed, muscle impedance properties are modulated to create impedance matching between the load and

Table 7.1 Summary of the SEA units features

	Spring stiffness (N m/rad)	Max peak torque (N m)	Motor power (W)	SEA unit weight (kg)	Power density (N m/kg)	Bandwidth of torque controller (Hz)	Output impedance at 1 Hz (N m/rad)
NEEM	98.7	54	400	3.0	18.0	8.7	0.2
NESM eF/E Joint	98.7	30	100	1.3	23.1	7.0	1.7
NESM sA/A Joint	165.8	60	90	2.1	28.6	5.0	2.6
α-APO	98.7	35	100	1.2	28.0	15.5	1.3
β-APO	99.8	25	70	1.0	25.0	12.2	1.0

muscles [28]), it is important that the fixed spring stiffness of the SEA is in a range comparable to the physiological one. Moreover, from a control viewpoint, low stiffness reduces the torque bandwidth, whereas high stiffness increases the impedance and worsens the torque output resolution.

2. The bandwidth of the actuator. The bandwidth achieved by the actuator should be compatible with the one of the human movement. For example, a human arm has an achievable position bandwidth in the order of 2–5 Hz [29] and a torque control bandwidth of approximately 7 Hz, and reflex properties of the arm up to 20 Hz [30].
3. The output torque. The output maximum torque (without considering the gravity compensation) should be compatible with that of the human movement to be assisted.

Our SEA-based exoskeletons have been designed taking into account the abovementioned requirements and results of validation tests in unloaded conditions (without the human-in-the-loop) are reported in previous sections. Table 7.1 summarizes the main performance features of the actuators.

The key element of the SEA-based exoskeleton developed at The BioRobotics Institute is the torsional spring. Coupled with 17-bit rotary encoders they can provide a resolution of the output torque of 0.005 N m. The spring reproducibility on different scales allows for the achievement of the similar stiffness values in all the prototypes. The only exceptions are the NESM abduction/adduction and flexion/extension joints, in which the stiffness is higher (165.8 N m/rad) due to different functional requirements: indeed, those joints have to sustain higher torques for assisting the movement of the limb while also compensating for most of the gravity of the robot.

A compact design requires the employment of light-weight actuators with high power density. Although upper-limb SEAs are slightly heavier, this is acceptable since portability is not a functional requirement for their employment in rehabilitation scenarios. On the other hand, the weight for the lower limb SEAs must be lower than 1.2 kg in order to facilitate the portability of the system.

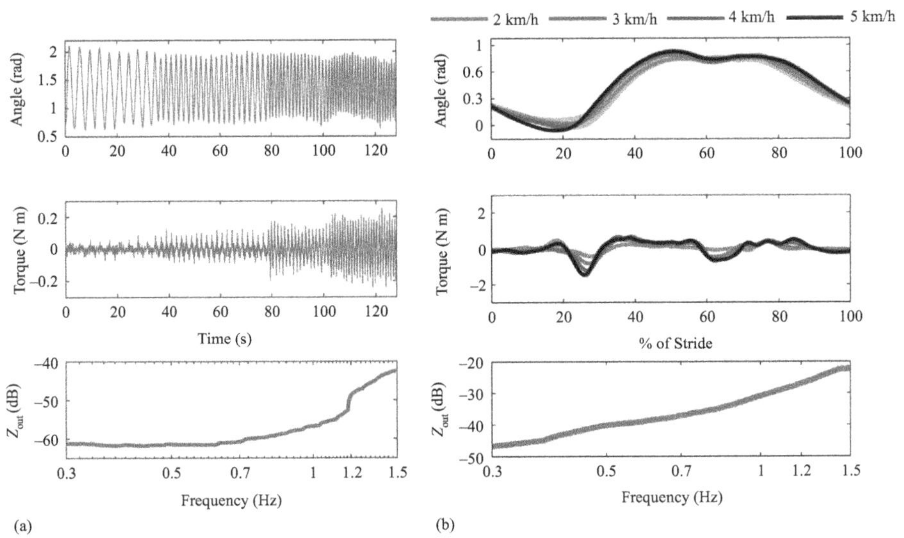

Figure 7.7 Performance of the NEEM (a) and β-APO (b) SEA units under zero-torque control in real human-in-the-loop application scenarios. The first row shows the flexion-extension movement of the articulation at different velocities. In the second row, interaction torques exerted at the human–robot interface are represented. The last row reports the output impedance normalized with respect to the inherent spring stiffness of the SEA unit

Closed loop bandwidth of the robotic joints is largely in the range of human movements that has to be assisted for all the platforms. When controlled under zero-torque modality, their interaction with the human counterpart is minimal and they do not hinder the user's motion. Furthermore, in order to assess the transparency of the system with the human in the loop, some validation tests have been carried out in application scenarios, i.e., elbow rehabilitation and gait training. In fact, the performance of output impedance as reported in the table is collected from manual movement displacement of the powered robotic joints. The measurement can be different when the human is interacting with the robot due to two main reasons: (i) inertia of the moving body segments, e.g., the movement of the forearm in NEEM and (ii) interaction with the environment, e.g., the foot strike when walking with the APO.

The transparency of the NEEM was assessed in terms of output impedance perceived by the user when performing repetitive movements with the exoskeleton in torque control, setting a zero-torque reference. In order to span the range of frequencies of common eF/E movements in rehabilitation scenarios (up to 1 Hz [14]), a metronome was used to provide the user with an acoustic reference at increasing frequencies. The results for a time window of 130 s with frequencies up to 1.5 Hz are shown in Figure 7.7(a). The perceived stiffness has a frequency-dependent behavior, ranging from 0.08 N m/rad at 0.3 Hz to 0.75 N m/rad at 1.5 Hz, corresponding to peaks of parasitic torques of 0.06 and 0.24 N m, respectively. As expected, the maximum parasitic torque is detected in correspondence of the transition from flexion to extension motion, where inertial effects are most prominent especially at higher velocities.

In order to quantify the output impedance of the portable β-APO orthosis in a real usability scenario, we performed a preliminary evaluation walking on a treadmill at four different speeds, i.e., 2, 3, 4, and 5 km/h, with the low-level closed loop torque controller set on a desired zero-torque reference. The results are shown in Figure 7.7(b). As expected, the parasitic stiffness is dependent on the frequency motion of the task; it increased with the walking speed. As a result, the residual torque applied at the human–robot interface was higher for the maximal walking speed. The increment was from 0.99 N m/rad at 2 km/h up to 2.01 N m/rad at 5 km/h for a resulting parasitic torque of 0.43 and 1.47 N m contrasting the free-motion of the user. The maximum parasitic torque is shown at the transition from the late swing to the early stance phases of gait, when the limb is propelled in the plane of progression and the hip joint redirects the limb movement. A second peak torque can be identified at the moment of the heel strike. In this case the instantaneous stress applied on the torsional element is represented by the shock absorption of the foot impacting with the ground.

In literature, there are quite a few examples of SEA-based exoskeletons for the upper- (elbow and/or shoulder) and lower (hip and/or knee and/or ankle) limb movement rehabilitation and assistance.

One of the most famous SEA-based upper-limb exoskeleton is the LIMPACT [31], a revised version of the device presented in [30]. In the first device, the identified frequency response function had a −3-dB gain bandwidth of 35 Hz, whereas the new version presents an amplitude-dependent torque bandwidth of 43–102 Hz, which is extremely high, compared to most exoskeletons in literature. The device employs four hydraulic actuators that ensure high torques.

Chien [32] recently presented a shoulder exoskeleton with two DOFs and linear SEAs. Step motors are connected to springs with stiffness of 40 N/mm. However, no tests have been carried out so far and information about performance of the system is not available yet.

The Wilmington Robotic Exoskeleton (WREX) is a passive upper-limb orthosis for children with muscular weakness, with four DOFs to allow sufficient range of movement and a gravity balance mechanism based on elastic bands. In the work by [33], the SEAs were introduced in the WREX. In that case, the controller was able to provide torque control with a bandwidth up to 3.18 Hz in the absence of disturbances. Spring stiffness was 2.51 N m/rad. Wolbrecht *et al.* [34] have actuated the WREX using pneumatic actuators, which are inherently compliant as well, reaching similar performance in torque control (3.5-Hz bandwidth).

In the framework of lower limb SEA-based powered orthoses, the SEA joints of the exoskeleton described in [35] are capable of delivering up to 60 N m of peak torque with a 300-W motor and a spring stiffness of 230 N m/rad. The closed-loop bandwidth of the torque controller is 6.5 Hz.

The linear SEA presented in [36] uses a 400-W DC motor and a linear spring characterized by a nonlinear elasticity, with a total weight of 1.4 kg. It provides up to 2600 N with a maximum bandwidth of 5 Hz.

In [37], authors developed a knee rotational SEA weighing 2.5 kg with a 150-W DC motor and a torsional spring with a stiffness of 100 N m/rad. The bandwidth reaches 9.6 Hz and output stiffness for a walking speed of 3.6 km/h is 10 N m/rad.

The LOwer-extremity Powered ExoSkeleton (LOPES) platform [38] presented a bandwidth of 12 Hz and resistive torque equal to 5 N m for a frequency motion of 1 Hz in zero-torque control.

The higher bandwidth is possessed by the SEA-based joint of the MIND-WALKER exoskeleton [39] with a stiffness of 800 N m/rad and a peak torque of 100 N m; its closed-loop torque controller reached 20 Hz for an applied torque of 2 N m. Similar performance were obtained by the knee orthotic system presented in [40].

In the specific scenario of hF/E assistance, a SEA-based assistive device named Hip exoskeleton for Superior Assistance (HeSA) was developed [41]. The actuation unit of the HeSA exoskeleton reaches 15 N m of peak torque for assisting walking and running pace, thus suitably covering gait-related walking frequencies.

Overall, the performance requirements for realizing safe, lightweight, and powerful torque sources were met for most of the presented SEA-based exoskeletons.

Different bandwidths resulted from different exoskeletons as a consequence of different target applications, from human gait assistance to upper-limb rehabilitation, and on target users' residual movement capabilities, from users with complete limb paralyses (such as complete spinal cord injury patients) to users with moderate movement capabilities (such as older adults, people following stroke events). In all the cases, the transparent behavior (zero torque or zero-force control mode) was achieved with a resulting low output impedance, which turns out to be fundamental for letting the user move freely when the assistance is not required or necessary.

It is worth mentioning that the evaluation of the actuator performance is one of the key features for validating exoskeletons with or without the human in the loop. Other factors, such as the kinematic coupling and assistive strategies, also have a fundamental role for realizing wearable devices that can be used effectively in activities of daily living or rehabilitation scenarios.

References

[1] Veneman J. F., Ekkelenkamp R., Kruidhof R., Van der Helm F. C., Van der Kooij H. 'A series elastic- and bowden-cable-based actuation system for use as torque actuator in exoskeleton-type robots'. *The International Journal of Robotics Research*. 2006:25(3), 261–281.

[2] Braun D. J., Petit F., Huber F., *et al.* 'Robots driven by compliant actuators: optimal control under actuation constraints'. *IEEE Transactions on Robotics*. 2013:29(5), 1085–1101.

[3] Van Ham R., Sugar T. G., Vanderborght B., Hollander K. W., Lefeber D. 'Compliant actuator designs'. *IEEE Robotics and Automation Magazine*. 2009:16(3), 81–94.
[4] Robinson D. W., Pratt J. E., Paluska D. J., Pratt G. A. 'Series elastic actuator development for a biomimetic walking robot'. *Proceedings of the 1999 IEEE/ASME International Conference on Advanced Intelligent Mechatronics*; Atlanta, GA, USA, Sep 1999. pp. 561–568.
[5] Pratt J., Krupp B., Morse C. 'Series elastic actuators for high fidelity force control'. *Industrial Robot: An International Journal*. 2002:29(3), 234–241.
[6] Vanderborght B., Van Ham R., Lefeber D., Sugar T. G., Hollander K. W. 'Comparison of mechanical design and energy consumption of adaptable, passive-compliant actuators'. *The International Journal of Robotics Research*. 2009:28(1), 90–103.
[7] Vanderborght B., Albu-Schäffer A., Bicchi A., *et al.* 'Variable impedance actuators: a review'. *Robotics and Autonomous Systems*. 2013:61(12), 1601–1614.
[8] Maganaris C. N., Paul J. P. 'In vivo human tendon mechanical properties'. *The Journal of Physiology*, 1999:521(1), 307–313.
[9] Veneman J. F., Kruidhof R., Hekman E. E., Ekkelenkamp R., Van Asseldonk E. H., Van Der Kooij H. 'Design and evaluation of the LOPES exoskeleton robot for interactive gait rehabilitation'. *IEEE Transactions on Neural Systems and Rehabilitation Engineering*. 2007:15(3), 379–386.
[10] Pratt J. E., Krupp B. T. 'Series elastic actuators for legged robots'. *Proceeding of SPIE 5422*; Orlando, FL, USA, Apr 2004. Unmanned Ground Vehicle Technology; 2004. pp. 135–144.
[11] Pratt G. A., Williamson M. M. 'Series elastic actuators'. *Proceedings of 1995 IEEE/RSJ International Conference on Intelligent Robots and Systems. Human Robot interaction and Cooperative Robots*; Pittsburg, PA, USA, Aug 1995. vol. 1, pp. 399–406.
[12] Williamson M. M. 'Series Elastic Actuators' (No. AI-TR-1524). Massachusetts Institute of Technology, Artificial Intelligence Lab, 1995.
[13] Colgate E., Hogan N. 'An analysis of contact instability in terms of passive physical equivalents'. *Proceedings of 1989 International Conference on Robotics and Automation*; Scottsdale, AZ, USA, May 1989. pp. 404–409.
[14] Vitiello N., Lenzi T., Roccella S., *et al.* Giovacchini 'NEUROExos: a powered elbow exoskeleton for physical rehabilitation'. *IEEE Transactions on Robotics*. 2013:29(1), 220–235.
[15] Vitiello N., Cempini M., Crea S., *et al.* 'Functional design of a powered elbow orthosis towards its clinical employment'. *IEEE/ASME Transactions on Mechatronics*. 2016:21(4), 1880–1891.
[16] Amabile C., Bull A. M. J., Kedgley A. E. 'The centre of rotation of the shoulder complex and the effect of normalisation'. *Journal of Biomechanics*. 2016:49(9), 1938–1943.
[17] Crea S., Cempini M., Moisè M., *et al.* 'Validation of a gravity compensation algorithm for a shoulder-elbow exoskeleton for neurological rehabilitation'.

Proceedings of the 3rd International Conference on NeuroRehabilitation; Segovia, Spain, Oct 2016. Springer; 2017. pp. 495–499.

[18] Takeuchi N., Izumi S. 'Rehabilitation with poststroke motor recovery: a review with a focus on neural plasticity'. *Stroke Research and Treatment*. 2013:2013, 1–13.

[19] Crea S., Cempini M., Moisè M., *et al.* 'A novel shoulder-elbow exoskeleton with series elastic actuators'. *Proceedings of VI EEE RAS/EMBS International Conference on Biomedical Robotics and Biomechatronics*; Singapore, Singapore, Jun 2016. pp. 1248–1253.

[20] Giovacchini F., Vannetti F., Fantozzi M., *et al.* 'A light-weight active orthosis for hip movement assistance'. *Robotics and Autonomous System*. 2015:73, 123–134.

[21] Winter D. A. *Biomechanics and Motor Control of Human Movement*. New York: John Wiley & Sons; 2009. p. 370.

[22] Walsh C. J., Endo K., Herr H. 'A quasi-passive legacy exoskeleton for load-carrying augmentation'. *International Journal of Humanoid Robotics*. 2007:4(3), 487–506.

[23] Parri A., Yan T., Giovacchini F., *et al.* 'A portable active pelvis orthosis for ambulatory movement assistance'. *Proceedings of the 2nd International Symposium on Wearable Robotics*; Segovia, Spain, Oct 2016. Springer; 2017. pp. 75–80

[24] Ronsse R., Lenzi T., Vitiello N., *et al.* 'Oscillator-based assistance of cyclical movements: model-based and model-free approaches.' *Medical & Biological Engineering & Computing*. 2011:49(10), 1173–1185.

[25] Yan T., Parri A., Ruiz Garate V., Cempini M., Ronsse R., Vitiello N. 'An oscillator-based smooth real-time estimate of gait phase for wearable robotics'. *Autonomous Robots*. 2016:41(3), 759–774.

[26] Crea S., Donati M., De Rossi S. M. M., Oddo C. M., Vitiello N. 'A wireless flexible sensorized insole for gait analysis'. *Sensors*. 2014:14(1), 1073–1093.

[27] Yuan K., Parri A., Yan T., *et al.* 'A realtime locomotion mode recognition method for an active pelvis orthosis. *Proceedings of 2015 IEEE/RSJ International Conference on Intelligent Robots and Systems*; Hamburg, Germany, Sep–Oct 2015. pp. 3893–3896.

[28] Farahat W. A., Herr H. M. 'Optimal workloop energetics of muscle-actuated systems: an impedance matching view'. *PLoS Computational Biology*. 2010:6(6), e1000795.

[29] Brooks T. L. 'Telerobotic response requirements'. *Proceedings of 1990 IEEE International Conference on Systems, Man and Cybernetics*; Los Angeles, CA, USA, Nov 1990. pp. 113–120.

[30] Stienen A. H., Hekman E. E., Ter Braak H., Aalsma A. M., Van der Helm F. C., Van der Kooij H. 'Design of a rotational hydro-elastic actuator for an active upper-extremity rehabilitation exoskeleton'. *Proceedings of 2nd IEEE RAS & EMBS International Conference on Biomedical Robotics and Biomechatronics*; Scottsdale, AZ, USA, Oct 2008. pp. 881–888.

[31] Otten A., Voort C., Stienen A., Aarts R., Van Asseldonk E., Van der Kooij H. 'LIMPACT: a hydraulically powered self-aligning upper limb exoskeleton'. *IEEE/ASME Transactions on Mechatronics*. 2015:20(5), 2285–2298.

[32] Chien L., Chen D. F., Lan C. C. 'Design of an adaptive exoskeleton for safe robotic shoulder rehabilitation'. *Proceedings of 2016 IEEE International Conference on Advanced Intelligent Mechatronics*; Banff, AB, Canada, Jul 2016. pp. 282–287.

[33] Ragonesi D., Agrawal S., Sample W., Rahman T. 'Series elastic actuator control of a powered exoskeleton'. *Proceedings of 33rd Annual International Conference of the IEEE Engineering in Medicine and Biology Society*; Boston, MA, USA, Sep 2011. pp. 3515–3518.

[34] Wolbrecht E. T., Leavitt J., Reinkensmeyer D. J., Bobrow J. E. 'Control of a pneumatic orthosis for upper extremity stroke rehabilitation'. *Proceedings of 28th Annual International Conference of the IEEE Engineering in Medicine and Biology Society*; New York City, NY, USA, Aug 2006. pp. 2687–2693.

[35] Accoto D., Carpino G., Sergi F., Tagliamonte N. L., Zollo L., Guglielmelli E. 'Design and characterization of a novel high-power series elastic actuator for a lower limb robotic orthosis'. *International Journal of Advanced Robotic Systems*. 2013:10(359), 1–12.

[36] Beil J., Perner G., Asfour T. 'Design and control of the lower limb exoskeleton KIT-EXO-1'. *Proceedings of 2015 IEEE International Conference on Rehabilitation Robotics;* Singapore, Singapore, Aug 2015. pp. 119–124.

[37] Dos Santos W. M., Caurin G. A. P., Siqueira A. A. G. 'Design and control of an active knee orthosis driven by a rotary series elastic actuator'. *Control Engineering Practice*. 2015:58, 307–318.

[38] Vallery H., Ekkelenkamp R., Van Der Kooij H., Buss M. 'Passive and accurate torque control of series elastic actuators'. *Proceedings of 2007 IEEE/RSJ International Conference on Intelligent Robots and Systems*; San Diego, CA, USA, Oct–Nov 2007. pp. 3534–3538.

[39] Wang S., Wang L., Meijneke C., *et al.* 'Design and control of the MINDWALKER exoskeleton'. *IEEE Transactions on Neural Systems and Rehabilitation Engineering*. 2015:23(2), 277–286.

[40] Kong K., Bae J., Tomizuka M. 'A compact rotary series elastic actuator for human assistive systems'. *IEEE/ASME Transactions on Mechatronics*. 2012:17(2), 288–297.

[41] Sugar T. G., Fernandez E., Kinney D., Hollander K. W., Redkar S. 'HeSA, hip exoskeleton for superior assistance'. *Proceedings of the 2nd International Symposium on Wearable Robotics*; Segovia, Spain, Oct 2016. Springer; 2017. pp. 319–323.

Chapter 8

Gait-event-based synchronization and control of a compact portable knee–ankle–foot exoskeleton robot for gait rehabilitation

Zhao Guo[1,2], Gong Chen[1], and Haoyong Yu[1]

Abstract

This chapter presents the mechanical design and control of a knee-ankle-foot exoskeleton robot, which is compact, modular and portable for stroke patients to carry out overground gait training. A novel compact series elastic actuator (SEA) is developed for safe human–robot interactions. In order to control this portable knee-ankle-foot robot, a novel human–robot synchronization method using gait event information is proposed. This method includes two steps. Firstly, seven gait events in one gait cycle are detected in real time with a hidden Markov model (HMM); secondly, an adaptive oscillator is utilized to estimate the stride percentage of human gait using any one of the gait events. Synchronous reference trajectories for the robot are then generated with the estimated stride percentage. The proposed synchronization method is implemented in the robot and tested in 15 healthy subjects. The results of the experiments reveal that our approach is efficient in achieving human–robot synchronization. It shows that this method has the advantages of simple structure, flexible selection of gait events and fast adaptation.

Keywords: gait event, rehabilitation robotics, knee-ankle-foot robot, series elastic actuator (SEA)

8.1 Introduction

Rehabilitation robots have been introduced into the earlier phases of recovery after stroke to overcome the major limitations of traditional manual therapy [1]. High-quality rehabilitation therapies can be delivered by robots with good quantitative measurements and consistency. Over the years, various gait rehabilitation robotic

[1]Department of Biomedical Engineering, National University of Singapore, Singapore
[2]School of Power and Mechanical Engineering, Wuhan University, China

devices have been developed based on different concepts [2,3]. However, most of existing robotic gait training systems, such as Lokomat [4], ReoAmbulator [5], LOwer-extremity Powered ExoSkeleton (LOPES) [6], and Active Leg Exoskeleton (ALEX) [7], integrate treadmills and fixed platforms [8]. They are meant for acute patients at big rehabilitation centers or hospitals; and rehabilitation progress of which is inferior to over ground gait training [9,10]. Therefore, there is a great need for portable wearable robotic systems for chronic stroke patients at community rehabilitation centers or home settings. Such portable devices should be lightweight, safe, and easy to don and doff.

A number of gait rehabilitation devices targeted at specifically either the ankle joint [11–13] or the knee joint [14–16] have been developed. However, the weight remains to be the main challenge to achieve portability. The ankle robot developed at Massachusetts Institute of Technology [11] weighs 3.1 kg for just the ankle joint. The commercialized knee assistive device developed by Tibion [15] weighs more than 4.5 kg for just the knee joint. A quasi-passive compliant stance control orthosis for knee joint assistance recently developed by Shamaei *et al.* weighs 3 kg [16]. Besides, portable devices aimed at providing active assistive torque to both the knee and ankle joints are very rare due to ineffective and bulky actuation design.

In this chapter, we present a compact and modular knee–ankle–foot exoskeleton robot aiming for subacute and chronic stroke patients to conduct gait rehabilitation at outpatient rehabilitation centers or at home. The system can be easily reconfigured into either a knee robot or an ankle robot to suit different symptoms, such as knee hyper-extension and drop foot. In order for the device to be able to provide assistance during human gait, the actuator must be lightweight, compact and have a large range of output force. Thus we developed a novel compact compliant series elastic actuator (SEA) [17], and corresponding linkage mechanism to achieve portable and modular robot design.

In terms of robot control for gait rehabilitation, the synchronization between the motion of the robots and the actual human gait is very important. For example, in impedance-control-based strategies, the robotic assistance is specified based on the deviation of the actual position of the lower limb joints from the reference trajectories of the robots [18,19]. The common concern in these strategies is that if the robot trajectory is not synchronized with the human gait, the robot may resist the human walking, and may even cause injury [20].

Motivated by the need for human–robot synchronization, researchers have developed various control strategies. Jezernik *et al.* developed an adaptation algorithm on Lokomat that can synchronize the robot trajectories with the human joints by minimizing the human–machine interaction torque [21]. Aoyagi *et al.* developed an algorithm on Pneumatically Operated Gait Orthosis (POGO) and Pelvic Assist Manipulator (PAM) that can adjust the replay timing of the reference trajectories according to a state, which is an 18-dimensional vector of position and velocity signals of the robot and human joints [20]. This algorithm is effective in synchronizing the user's gait and their own recorded gait trajectories; however, the applicability for synchronization between normal gait references and subjects or patients with abnormal gait patterns is not ensured [21].

More recently, various adaptive oscillators have been implemented in rehabilitation robotics in order to provide synchronous assistance [22–24]. Adaptive oscillators as a type of mathematical tool can synchronize with an external periodic signal and extract the frequency or/and phase information. Complex sensing or complicated adaptation rules are not required to achieve synchronization. Ronsse *et al.* implemented an adaptive oscillator in NEUROExos to provide sinusoidal assistance in cyclical elbow flexion/extension movements [22]. In order to generate non-sinusoidal trajectories for gait training, several adaptive oscillators need to be implemented [23]. The methods have been used in LOPES for walking assistance for hip joints [22], and in ALEX II to provide hip support during treadmill training [24]. To summarize, the existing oscillator-based algorithms are effective in synchronizing the robot with arbitrary periodic input trajectory; however, these methods are limited in the following aspects. First, several oscillators are needed for the generation of a non-sinusoid gait trajectory, which is relatively computationally expensive. Second, the adaptation to achieve synchronization is relatively slow (e.g., in [25], it takes around ten cycles to achieve synchronization) and the overall system may diverge if the gain of the input trajectory is set to be large to achieve faster adaptation. Third, the current methods focus on reconstructing the waveform of the input gait trajectories. However, for the purposes of gait rehabilitation, it is more important to provide a normal reference trajectory that is synchronous with the abnormal gait of the patients.

The phenomenon of synchronization can be found in the biological world. For example, a specific species of firefly, Pteroptyx malaccae, can achieve both frequency and phase synchronization in flashing, and their synchronization mechanism is modeled by an adaptive oscillator [26]. The oscillator will alter its frequency based on its phase difference from the stimulus (other firefly's flash) and eventually achieve synchronization. To achieve human–robot synchronization, we adopt this adaptive oscillator into gait rehabilitation robotics. The adaptive oscillator represents the estimated gait percentage and the actual gait event during walking is regarded as the external impulsive stimulus. The phase difference of the estimated gait percentage and the detected gait event will drive the frequency adaptation of the oscillator, and eventually the gait percentage can be accurately estimated. In the proposed method, wearable inertia measurement unit (IMU) sensors are utilized to measure the seven gait events are detected in real time with a hidden Markov model (HMM). The adaptive oscillator is employed to extract the percentage of the human stride based on the gait events. Any one of the seven gait events is adequate for synchronization; and two or more gait events in one gait cycle is beneficial for faster adaptation. The synchronous reference trajectories are generated with a lookup table (LUT) according to the estimated gait percentage. The oscillator-based synchronization and assistive strategy is implemented in our portable knee–ankle–foot robot. Experiments on 15 healthy subjects are conducted to evaluate the performance of the proposed method and verify the mechanical performance of the robot.

The rest of this chapter is organized as follows. Section 8.2 characterizes the compact robot, including the biomechanics analysis, robot design, and compliant actuator design. Section 8.3 introduces the human robot synchronization control, including gait-event detection, development of the adaptive oscillator, and assistive controller. Section 8.4 presents the experimental protocol. Section 8.5 gives the experimental results in different conditions. This chapter ends with the conclusion in Section 8.6.

8.2 Mechanical design of the knee–ankle–foot robot

8.2.1 Design specifications

Clinical Gait Analysis data are utilized to guide the design of the exoskeleton and actuator [27]. Figure 8.1 illustrates the biomechanics when a healthy subject (70 kg, 0.9 m leg-length) walking at 1.0 m/s. The peak knee torque reaches 30 N m during stance phase. It is worth noting that knee torque remains under 10 N m (30 percent of peak torque) for most of a gait cycle, and goes beyond this value in a small period only (thick dashed line in Figure 8.1(a)). Correspondingly, the ankle peak torque is 60 N m when the angle is close to 15° (42 percent gait cycle, blue circle in Figure 8.1(a)) during stance phase (Figure 8.1(b)). Similarly, ankle torque stays below 20 N m (30 percent of peak torque) for more than half of the gait cycle (thin dashed line in Figure 8.1(a)). This character is a driving factor for our novel SEA design. On the other hand, the power requirement of each joint is a driving factor for the motor selection. Thus, an average mechanical power between 100 and 150 W is required for the knee and ankle joints (Figure 8.1(c)).

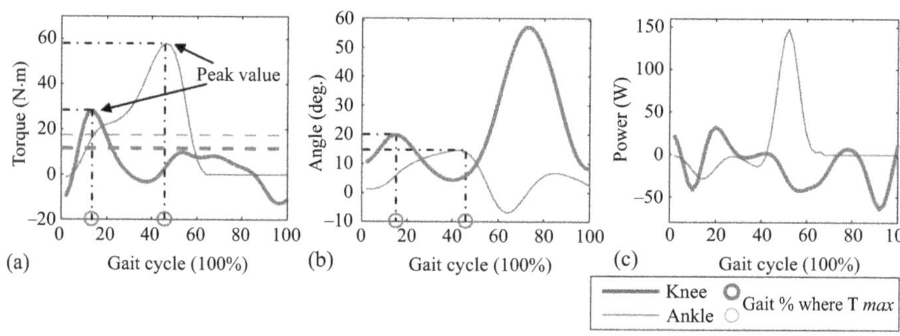

Figure 8.1 Biomechanics of human ankle and knee joints for a 70-kg healthy subject during normal gait cycle with 1.0-m/s speed, including (a) joint torque, (b) joint angle, and (c) joint power. Thick and thin lines represent knee and ankle joints respectively. Thin and thick circle is the gait percentage where joint torque reaches its maximum. Dot dash line in both (a) and (b) illustrates the joint angles and torques where joint torque is maximum. Dashed line in (a) is 30 percent of peak torque that covers the most part of a gait cycle

8.2.2 Mechanical structure design of the robot

Figure 8.2 shows the overall design of the knee–ankle–foot exoskeleton robot. Ball bearings are employed as the knee and ankle joint of the robot, which align with the knee and ankle joint of human (Figure 8.2(a)). Rotatory potentiometers are installed inside the bearing to determine the joint angles (Figure 8.2(c)). The modular design of a knee module and an ankle module (Figure 8.2(b)) satisfies the requirements of patients under different conditions, which can also be implemented separately. When using both modules together as a whole for the current prototype, a connecting linkage is employed to adjust the length of shank from 40 to 47 cm adapting to a range of different user heights. The actuation mechanism in this design is particular, which is a novel compact linear SEA through a four-bar linkage

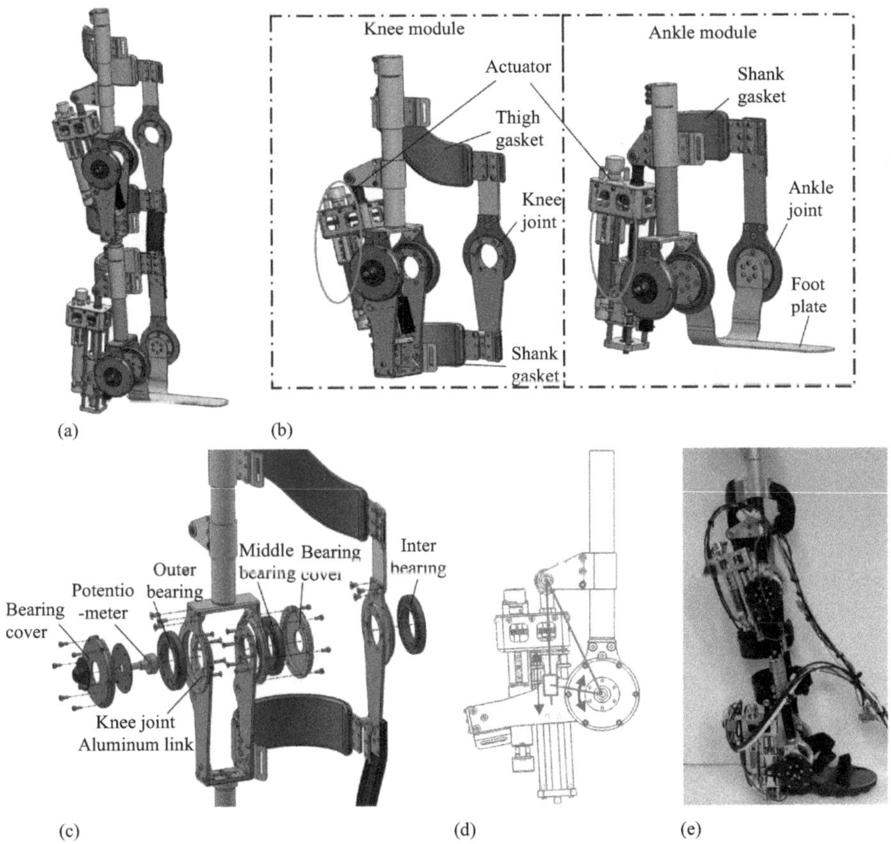

Figure 8.2 (a) CAD model of the knee–ankle–foot robot; (b) CAD models of the knee module and ankle–foot module; (c) explosive views of the robotic joint; (d) slider-crank linkage as the actuation mechanism; (e) a prototype of the robotic system

(Figure 8.2(d)). Figure 8.2(e) shows a prototype of the robot. The anthropomorphic structure is built with lightweight carbon fiber reinforced plastic composite material. The robot is easy to don and doff. The total mechanical weight of the prototype weighs about 3.5 kg, which makes it portable.

8.2.3 Compliant actuator design

Most current SEAs [12,13] adopted high stiffness springs in order to derive the high force transmission, which leads to the compromised force control performance, intrinsic compliance, and back drivability. The gait biomechanics indicates that the average joint torque in human lower limb joints during normal gait is much lower than the peak value. It implies that the actuator for gait assistive robots only needs to provide a relatively small torque for most periods of a gait cycle, and this can be handled by a soft spring. As for the higher peak torque, a stiffer spring is then required. On the basis of the above analysis, we develop a novel SEA design which overcomes the limitations of the conventional SEA designs and makes it more suitable for the gait rehabilitation robot. The novelty of this actuator design is the utilization of two springs in series with vastly different stiffness values to handle different force ranges. A soft translational spring handles low-force range, which enables the actuator to achieve the high force fidelity, low output impedance and high intrinsic compliance. A torsion spring with high effective stiffness, handles high-force operation when the soft spring is fully compressed. The torsion spring extends the output force into a larger range and improves the control bandwidth of the actuation system.

Figure 8.3(a) illustrates the composition of the compliant actuator. The actuator consists of an electrical motor that drives a ball screw through two sets of spring assemblies. Two unidirectional torsion springs are fitted into a custom-made coupler, which becomes a bidirectional torsion spring. Two rotary encoders are located at both sides of the motor and the difference of the two encoders' readings will be used to determine the deflection of the torsion spring. A pair of spur gears with 1:1 transmission ratio transfers the force to another parallel shaft of the ball screw, which helps reduce the total length of the actuator. Then, two linear springs are placed within an output carriage and pushed by the ball screw nut. Besides, a linear potentiometer is installed to measure the deflection of the linear spring, thus deriving the output force of the actuator based on Hook's law. Figure 8.3(c) further illustrates the design of the actuator in an exploded view. And the directions of the motion transmission are indicated as well.

In the current prototype of the actuator, please refer to Figure 8.3(b), a Maxon DC brushless motor (EC-4pole 120 W, 36 V) is adopted as the power source. A linear potentiometer from ETI is applied as the linear spring deflection sensor. Two incremental encoders with 1,024 tics/rev are used to measure the deflection of the torsion spring. The ball screw is chosen to be 140 mm in length with 2 mm/rev. The stroke of the actuator is 60 mm. The total weight of the SEA prototype is 0.85 kg.

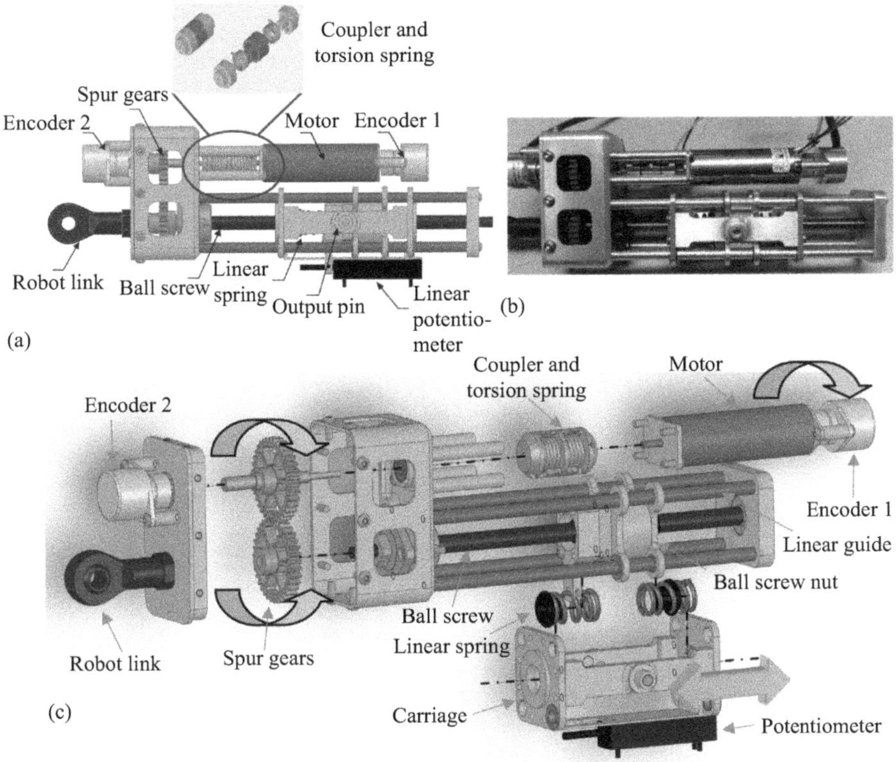

Figure 8.3 *(a) CAD model and (b) the prototype of the compliant force controllable SEA and (c) exploded views of the actuator and directions of motion transmission*

8.3 Human–robot synchronization control

In this section, we introduce the synchronization method for assistive control of the robot. The idea is to generate a synchronous reference trajectory for the robot by estimating the stride percentage of the human gait based on the gait event information and apply the appropriate assistive force to the subject.

8.3.1 Gait pattern of human walking

Gait describes the pattern of human walking. A gait cycle can be subdivided into seven gait phases, including loading response, mid-stance, terminal stance, pre-swing, initial swing, mid-swing, and terminal swing (Figure 8.4) [28]. The beginning of each gait phase is denoted as a gait event. Thus, correspondingly, there are seven gait events: Initial Contact, Opposite Toe Off, Heel Rise, Opposite Initial

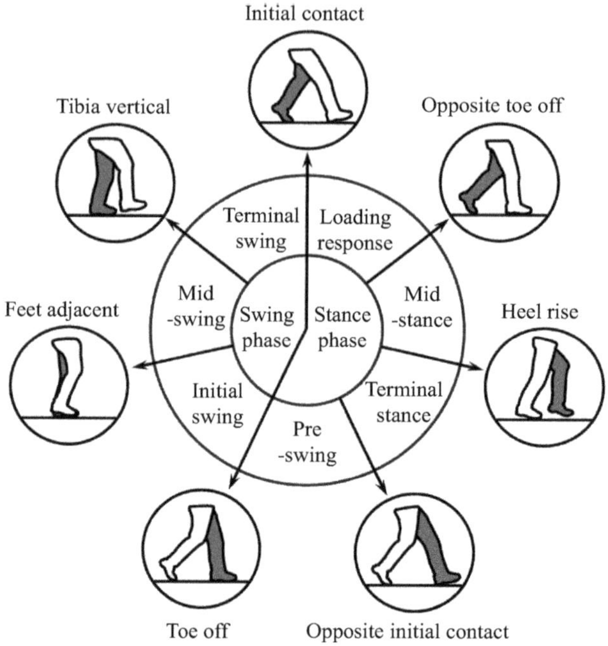

Figure 8.4 Gait events and gait phases in one gait cycle

Contact, Toe Off, Feet Adjacent, and Tibia Vertical [28]. The sequential occurrences of these gait events represent the transition of the gait phases, which propels the human body forward. Gait events follow a specific sequence and occur at specific periods within a gait cycle during normal walking.

The process of the gait-event-based synchronization method is illustrated in Figure 8.5. The method can be divided into two steps: gait-event detection with a HMM and stride percentage estimation with an adaptive oscillator for synchronization. Then, a synchronously assistive walking control method is implemented for the robot.

8.3.2 Gait events detection using HMM

In this subsection, the real-time gait event detection method is introduced. In our previous work, we have developed a HMM-based algorithm to detect the seven gait phases during walking; the gait events are then detected with the transitions of the gait phases [29,30]. IMU sensors are employed to collect the kinematics of the gait. Here it is defined as

$$z_t = [\omega_R, \alpha_R, \omega_L, \alpha_L, \theta_{R\text{knee}}, \theta_{R\text{shank}}] \tag{8.1}$$

where z_t is a six-dimensional vector denoting the observation features of the HMM at time t; ω_R, ω_L are the angular rates of both feet and α_R, α_L are their first-order derivatives; $\theta_{R\text{knee}}$ and $\theta_{R\text{shank}}$ are the angles of the right knee and the right shank, respectively. These selected features are capable of representing the gait phases.

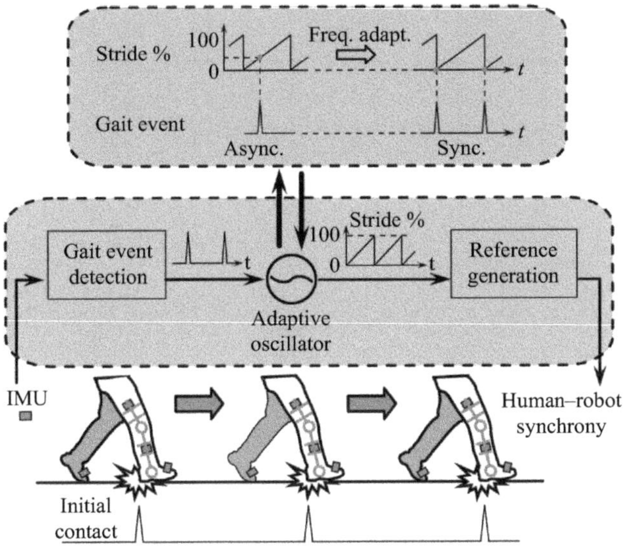

Figure 8.5 Flowchart of the gait-event-based synchronization method using an adaptive oscillator. The gait event of Initial Contact as an example is utilized for synchronization, which is described with a pulse signal. Both the frequency and the phase of the adaptive oscillator synchronize with the human gait based on the detected gait event. Synchronous trajectories are then generated for the robot

A HMM with seven states corresponding to seven gait phases is built as

$$S = \{LR, MSt, Tst, PSw, ISw, MSw, TSw\} \tag{8.2}$$

To implement the HMM for gait-event detection, three steps need to be carried out consecutively, including initialization, training, and decoding. HMM is defined as a statistical Markov model in which its unobservable state sequence $q = q_1, q_2, \ldots, q_T$ can be estimated through an observation sequence $Z = z_1, z_2, \ldots, z_T$, where T is the length of the observation sequence [31]. An HMM can be described by a 5-tuple (S, z_t, π, A, B), where $S = \{s_1, s_2, \ldots, s_N\}$ is a set of the N hidden states; π is the prior state probability vector $\pi = \{\pi_i | \pi_i = P[q_0 = s_i], i = 1, \ldots, N\}$; A is the state transition probability distribution matrix $A = \{a_{ij} | a_{ij} = P[q_t = s_i | q_{t+1} = s_j], i, j = 1, \ldots, N\}$; and B is the observation probability distribution matrix $B = \{b_i(z_t) | b_i(z_t) = P[z_t | q_t = s_i], i = 1, \ldots, N\}$.

For HMM initialization and training, subjects are required to conduct walking trials for the collection of the observation feature data. The benchmark seven phases are first annotated based on an adaptive threshold method [29]. With the observation data and the labeled gait phases, the parameter set of HMM $\lambda = (\pi, A, B)$ can be initialized. The initialization is based on the statistical results of the duration of each gait phases and the sequence of their transition. The HMM λ is further trained using the Baum–Welch algorithm for better performance [29]. The

Table 8.1 Online state decoding algorithm

Online Viterbi

for $i = 1, \ldots, 7$ **do**
    $\delta_1(i) = \pi_i \cdot b_{i,z_1}$
end for
for $2 \leq n \leq 300$ **do**
    for $1 \leq j \leq 7$ **do**
$\delta_n(j) = \max\limits_{1 \leq i \leq 7} \left(\delta_{n-1}(i) \cdot a_{ij}\right) \cdot b_{j,z_n}$
    end for
end for
return $q = \arg\max\limits_{1 \leq i \leq 7} \delta_{300}(i)$

The i and j indicate the gait phase, δ is a variable denoting likelihood, and q represents the decoded gait phase.

initialization and training sessions are done using MATLAB® and can be completed in a few minutes. Details about these sessions can be found in [29,31].

With the derived HMM, the most likely gait phase during walking is decoded. The procedure is described with pseudo-code in Table 8.1. In our application, an online Viterbi algorithm [31] with a moving window is implemented. Specifically, an observation sequence with 300 observations is employed for decoding, which is a sequence combining the current observation z_t and prior history observations. The first state of the decoded path is then regarded as the current gait phase. The occurrences of the gait events are then detected in the transitions of the gait phases.

8.3.3 Adaptive oscillator

As adopted from [26], a self-sustained oscillator is built to estimate the percentage of stride (*stride percent*):

$$d\phi/dt = \omega \tag{8.3}$$

where $\phi \in \mathbb{S}^1$ is the phase angle of the oscillator and the set $\mathbb{S}^1$ denotes the *unit circle*, t is time, and $\omega \in \mathbb{R}$ is the time-dependent variable frequency of the oscillator. The phase angle ϕ varies between 0 and 2π and grows uniformly within one cycle (Figure 8.5). The stride percentage is then derived as

$$\text{stride } \% = \frac{\phi}{2\pi} \times 100 \tag{8.4}$$

Considering that one of the seven gait events is utilized for synchronization, the occurrence of the gait event can be described with a periodic Dirac delta function $\delta(t/\tau)$, where τ is the period of the gait cycle. In order to achieve synchronization, the oscillator will adapt its frequency based on the incoming gait event as

$$d\omega/dt = \varepsilon(\Omega - \omega) + \delta(t/\tau) \cdot P \cdot G(\omega, \Delta\phi) \tag{8.5}$$

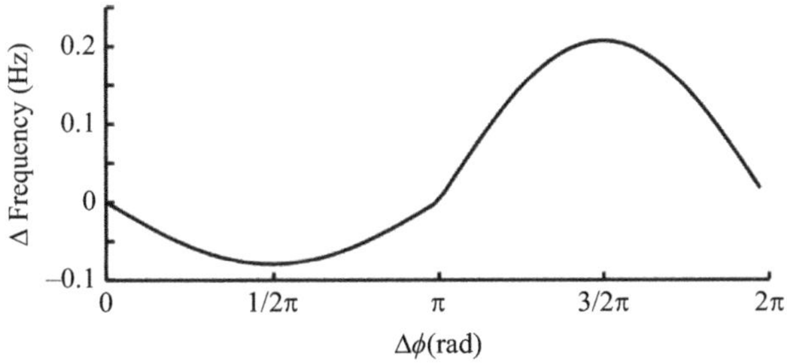

Figure 8.6 Phase response curve (PRC) G$(\omega,\Delta\phi)$*, with example of* $\omega = 0.7$*,* $\omega_{min} = 0.2$*,* $\omega_{max} = 2$

where ε denotes the adaptation rate of the oscillator frequency returning to the human gait frequency, Ω is the frequency of the gait cycle, and $\Omega = 2\pi/\tau$, P is a positive value that is related to the amount of the frequency change, $\Delta\phi$ is the phase difference between the adaptive oscillator and the actual gait, $\Delta\phi = (\phi - \mathbf{\Phi}) \bmod(2\pi)$, where $\mathbf{\Phi}$ is the predefined phase of the gait event, and $G(\omega, \Delta\phi)$ is the phase response curve employed to vary the frequency according to the detected gait event, which is specified as

$$G(\omega, \Delta\phi) = g^{+}(\Delta\phi)(\omega_{\min} - \omega) + g^{-}(\Delta\phi)(\omega_{\max} - \omega) \tag{8.6}$$

where $\omega_{\min}$, $\omega_{\max}$ are the allowed minimum and maximum frequencies, and $g^{+}(\Delta\phi)$, $g^{-}(\Delta\phi)$ are defined as

$$\begin{aligned} g^{+}(\Delta\phi) &= \max\left(\frac{1}{2\pi}\sin(\Delta\phi), 0\right) \\ g^{-}(\Delta\phi) &= -\min\left(\frac{1}{2\pi}\sin(\Delta\phi), 0\right) \end{aligned} \tag{8.7}$$

The phase response curve $G(\omega, \Delta\phi)$ is illustrated with Figure 8.6. In the above model, when $0 < \Delta\phi < \pi$, i.e., the oscillator is leading the gait event, the oscillator frequency will be decreased. Conversely, when $\pi < \Delta\phi < 2\pi$, i.e., the oscillator is lagging behind the gait event, the frequency will be increased.

The period of the gait is estimated by using the duration between two successive gait events:

$$\tau = 2\pi/\Omega = t_N - t_{N-1} \tag{8.8}$$

where t_N and t_{N-1} represent the time of the Nth and (N–1)th occurrences of the gait event.

A firefly only has one flash in one cycle, however, in our application there are $m(m = 1, 2, \ldots, 7)$ gait events in one gait cycle which can be utilized for faster synchronization. Hence, we extend the adaptive oscillator as

$$\begin{aligned} d\phi/dt &= \omega \\ d\omega/dt &= \varepsilon(\Omega - \omega) + \sum_{n=1}^{m} \delta(t_n/\tau) \cdot P \cdot G(\omega, \Delta\phi_n) \end{aligned} \tag{8.9}$$

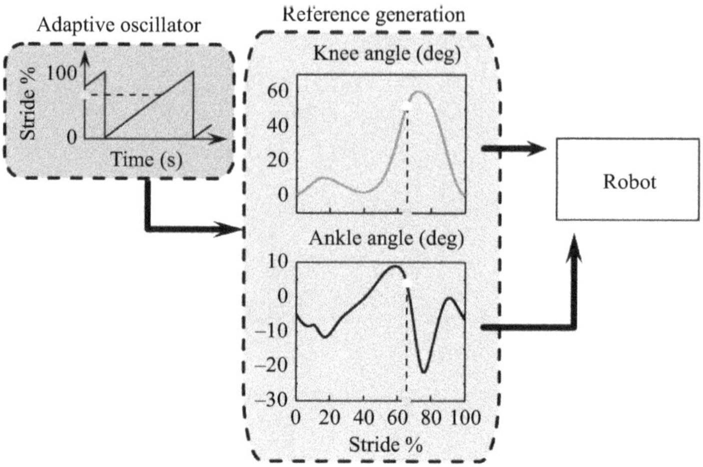

Figure 8.7 The reference trajectory for the robot is generated with a LUT based on the estimated stride percentage from the adaptive oscillator. Reference trajectories for multiple robotic joints can be generated from the same stride percentage

where $\delta(t_n/\tau)$ is the periodic Dirac delta function denoting the nth ($n = 1, 2, \ldots, m$) gait-event occurrence, and $\Delta\phi_n$ is the phase difference between the adaptive oscillator and the nth gait event, $\Delta\phi_n = (\phi - \mathbf{\Phi}_n)\text{mod}(2\pi)$, where $\mathbf{\Phi}_n$ is the pre-defined phase of the nth gait event. With the modified adaptive oscillator, more gait events in one gait cycle can be used to detect the phase difference between the oscillator and the actual gait, and thus synchronization can be achieved faster.

The period of the gait is estimated by using the duration between two successive gait events:

$$\tau = 2\pi/\Omega = \frac{1}{m}\sum_{n=1}^{m}\left(t_{n_N} - t_{n_{N-1}}\right) \tag{8.10}$$

where t_{n_N} and $t_{n_{N-1}}$ represent the time of the Nth and (N–1)th occurrences of the nth gait event.

The reference trajectories for the robot are generated using the estimated stride percentage based on a LUT. An example is illustrated in Figure 8.7, where a 2D-LUT is shown, corresponding to the robotic knee and ankle joints. The reference trajectories are recorded from a healthy subject during free walking (FW), and are normalized based on the stride percentage [7]. The phase of each gait event is also collected and averaged as the reference phase $\mathbf{\Phi}_n$ during synchronization. In our research, we use $\mathbf{\Phi}_1 = 0\%$, $\mathbf{\Phi}_2 = 7\%$, $\mathbf{\Phi}_3 = 48\%$, $\mathbf{\Phi}_4 = 50\%$, $\mathbf{\Phi}_5 = 60\%$, $\mathbf{\Phi}_6 = 77\%$, $\mathbf{\Phi}_7 = 86\%$. It is worth mentioning that the trajectories for different robot joints can be easily generated from the same oscillator without implementing other oscillators.

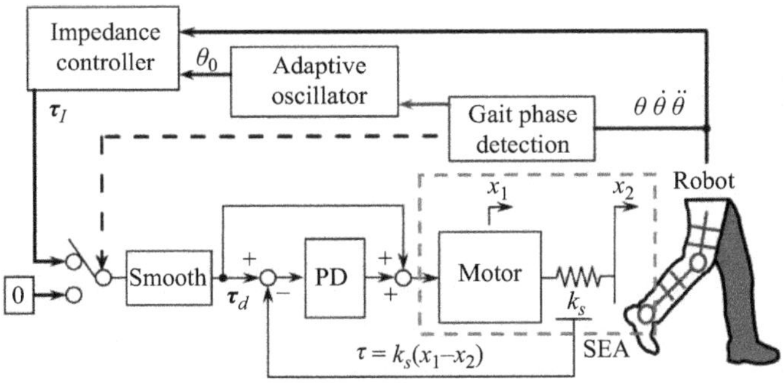

Figure 8.8 Control diagram of the knee–ankle–foot robot, where $\theta, \dot{\theta}, \ddot{\theta}$ represent the kinematics of the knee and ankle joints; θ_0 is the reference trajectory from the adaptive oscillator; τ_I is the output of the impedance controller; τ_d is the desired assistive torque; the dashed box represents the series elastic actuator (SEA), in which τ is the feedback torque, x_1, x_2 are the motor position and robot position; and k_S is the spring stiffness in the actuator

8.3.4 Assistive control of the robot

Based on the reference trajectory that is synchronous to the human gait, an assistive method is implemented for the robot with n joints to provide assistance during walking (Figure 8.8). The gait-event information is transmitted to the adaptive oscillator (Figure 8.8, solid arrow) to achieve synchronization between the robot and the human walking. The robot starts in zero-assistive control mode and switches to impedance control mode after five steps (Figure 8.8, dashed arrow). The step number is counted with the gait event of Initial Contact. Furthermore, with regard to potential safety issues, if synchronization is not guaranteed, i.e., the phase difference between the trajectory and the actual human gait is large, the robot switches to the zero-assistive mode, so that no assistance is given. The impedance controller is implemented in the robot to determine the assistive torque based on the reference trajectory, which is described as

$$\tau_I = \mathbf{K}_V(\boldsymbol{\theta} - \boldsymbol{\theta}_0) \tag{8.11}$$

where $\boldsymbol{\theta} = [\theta_1, \ldots, \theta_n] \in \mathbb{R}^n$ is the angles of n joints, $\boldsymbol{\theta}_0 \in \mathbb{R}^n$ is the joint angle reference, $\mathbf{K}_V = \mathrm{diag}(k_{v1}, \ldots, k_{vn}) \in \mathbb{R}^{n\times n}$ is a virtual stiffness matrix, where $k_{vi} \in \mathbb{R}$ is the virtual stiffness of the controller for the ith joint, and $\tau_I \in \mathbb{R}^n$ is the desired assistive torque from the controller.

To avoid a sudden change, the assistive torque is smoothed using an exponential function [24]:

$$\boldsymbol{\tau}_{dt} = \mathbf{A}\boldsymbol{\tau}_{It} + (\mathbf{I} - \mathbf{A})\boldsymbol{\tau}_{d(t-1)} \tag{8.12}$$

where $\tau_{It} \in \mathbb{R}^n$ is the desired torque τ_I at time point t, $\tau_{dt} \in \mathbb{R}^n$ is the smoothed desired torque τ_d at time t, and $\mathbf{A} = \mathrm{diag}(\alpha_1, \ldots, \alpha_n) \in \mathbb{R}^{n\times n}$ is a smoothing factor matrix, where $0 < \alpha_i < 1$ is the smoothing factor for the ith joint (shown in Figure 8.8 as the block labeled "smooth").

Proportional–derivative (PD) controllers plus feedforward term as the inner control loop to control the output force of the actuator, which are given as

$$\boldsymbol{u} = \mathbf{K}_P(\tau_d - \tau) + \mathbf{K}_D(\dot{\tau}_d - \dot{\tau}) + \tau_d \tag{8.13}$$

where $\tau_d \in \mathbb{R}^n$ and $\dot{\tau}_d \in \mathbb{R}^n$ are the desired force and its first-order derivative, $\tau \in \mathbb{R}^n$ and $\dot{\tau} \in \mathbb{R}^n$ are the force feedback and its derivative, $\boldsymbol{u} \in \mathbb{R}^n$ is the control output, and $\mathbf{K}_P = \text{diag}(k_{p1}, \ldots, k_{pn}) \in \mathbb{R}^{n\times n}$ and $\mathbf{K}_D = \text{diag}(k_{d1}, \ldots, k_{dn}) \in \mathbb{R}^{n\times n}$ are the matrices representing the PD control gains.

In our robotic system, the exoskeleton is driven by SEA. The force feedback of each actuator is estimated by the deflection of the spring based on Hooke's law as $\tau = k_S(x_1 - x_2)$, where k_S is the stiffness of the spring in the actuator, and x_1, x_2 are the motor position and robot joint position.

8.4 Experimental protocol

Experiments are conducted to evaluate the effectiveness of the synchronization method, and the feasibility of application in exoskeleton robot. Different gait events are used for synchronization to prove the flexibility of the proposed method.

8.4.1 Experimental setup

The proposed control strategy was tested with the exoskeleton for gait rehabilitation (Figure 8.9(a)). A wearable sensing system with seven IMU comprising commercial sensor chips (ADIS16405, Analog Devices, Inc.) is placed on the waist and two legs, including thighs, shanks, and feet, to measure gait kinematics

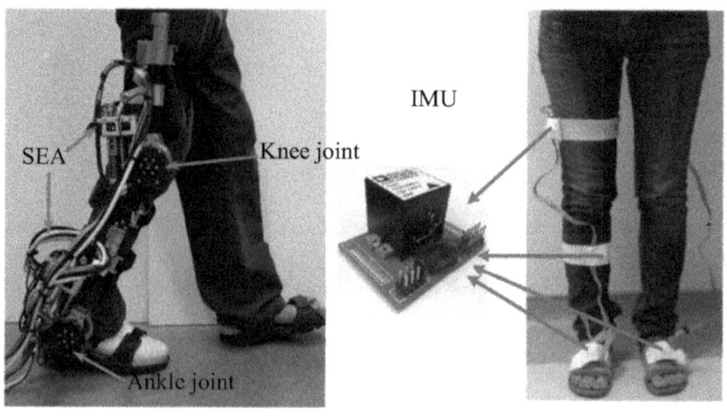

(a) (b)

Figure 8.9 (a) Prototype of a knee–ankle–foot exoskeleton robot; (b) motion capture system with IMU sensors

(Figure 8.9(b)) [29]. The controllers are implemented in the National Instrument CompactRIO 9074 embedded system. The sampling frequency is 2 kHz.

In our experiment, an abnormal gait pattern is simulated to resemble common pathological conditions of stroke patients, such as the stiff-knee gait [32]. An elastic bandage is wrapped around the knee and ankle joints of the exoskeleton to induce resistive torque. Flexion motions of the knee joint, and the ankle joint are hindered. The experiment aims to investigate the robustness of the synchronization method when an abnormal gait pattern is involved.

8.4.2 Experimental protocol

The experiment is tested on 15 healthy subjects (25.5 $\pm$ 5.3 years old, 174.9 $\pm$ 6.1 cm in height, 70.2 $\pm$ 10.7 kg in weight) to evaluate the effectiveness of the synchronization method. The subjects were informed of the protocol of the experiment and signed a consent form, which was approved by the Institutional Review Board of the National University of Singapore.

The subjects were required to wear the exoskeleton on their right leg, except in the "FW" condition. The subjects wear the IMU sensors on both legs in all conditions. Several trials served as practice to make the subjects familiar with the exoskeleton and the test scenario. At the beginning of each test, the subjects were asked to stand upright. The subjects were then asked to perform walking in a straight line at a preferred speed for about 30 m and stop in standing-up posture. There are five types of scenario, which are as follows:

1. **Free walking (FW):** subjects perform walking with the IMU sensors but without the robot. The average trajectories of the knee and ankle joints are recorded as the robot reference for each subject, and will be used in other conditions.
2. **Zero-assistive walking (ZA):** subjects wear the robot on their right leg. The robot functions in zero-assistive mode and tries to be transparent to the subjects.
3. **Simulated abnormal walking (SAW):** the subjects wear the robot on the right leg, which functions in zero-assistive mode. Elastic bandages are involved to provide resistive torque on the joints.
4. **Walking with low assistance (ASL):** the subjects wear the robot on the right leg, with elastic bandages on the joints. The impedance controller is implemented in both knee and ankle joints to provide assistance. The gait events of Opposite Initial Contact and Tibia Vertical are used for synchronization. A set of parameters were selected for the adaptive oscillator: $\omega_{\min} = 0.2$, $\omega_{\max} = 2$, $\varepsilon = 0.02$, $P = 11$. The impedance controller as described in (8.11) with $k_v = 0.2$ N m/deg was implemented in the knee and ankle joints to provide assistance. The smoothing factor in (8.12) was chosen to be $\alpha = 0.04$.
5. **Walking with high assistance (ASH):** this scenario is similar to the ASL scenario, except with $k_v = 0.4$ N m/deg for the impedance controller to provide higher assistive torque during walking.

8.4.3 Data analysis

The kinematics of the knee and ankle joints and the results of gait-phase detection were recorded. Each walking trial was segmented into gait cycles with the gait events of initial contact. Knee and ankle joint angles of ten gait cycles were taken and averaged in each condition. The corresponding assistive torque was also collected and segmented. The statistical significance of the changes in different conditions was evaluated with repeated measures analysis of variance (ANOVA). When a significant effect was found, Tukey's post hoc test was performed to contrast differences among the experimental conditions, with a p factor of 0.05. To enable better understanding, the phase error is normalized from $[0, 2\pi)$ to $[-\pi, \pi)$. Hence, a negative phase error represents a delay of the reference trajectory, and vice versa. In order to quantitatively evaluate how fast the adaption is, we define a value of ± 0.5 rad. When a step is reached after which the phase errors in the following five cycles are below this value, the number of this step is recorded. This number can be further compared among different experimental conditions to evaluate the robustness of the method.

8.5 Experimental results

8.5.1 Evaluation of synchronization

In this section, experimental results of a representative subject are given to show the details of the synchronization process, including the reference and actual gait trajectories, the adaptation of the frequency and the estimated phase error.

8.5.1.1 FW test

Figure 8.10 shows the experimental results of the FW test, in which the gait event of Initial Contact ($\Phi = 0\%$) is used for synchronization. The results include knee and ankle joint angles (Figure 8.10(a) and (b)), the phase difference between the adaptive oscillator and the actual gait (Figure 8.10(c)), the frequency of the adaptive oscillator and the estimated frequency of the actual gait (Figure 8.10(d)), and the phase angle of the adaptive oscillator (Figure 8.10(e)). It should be noted that in this configuration, where Initial Contact is used for synchronization, the phase angle equals the phase difference between the adaptive oscillator and the actual gait when Initial Contact is detected (Figure 8.10(c) and (e)). The figure shows that the test began with the subjects standing still. The subjects started walking at a random time point. The gait frequency is around 0.7 Hz (Figure 8.10(d)). The gait event of Initial Contact was detected in real time (circle in Figure 8.10(b)). The frequency of the adaptive oscillator was adapted according to (8.5) (Figure 8.10(d)) and the phase angle grows according to the frequency (Figure 8.10(e)). The phase error between the adaptive oscillator and the actual gait was reduced gradually (Figure 8.10(c)). In the last six steps in this figure, the averaged phase error is -0.053 rad, which shows a good performance. It was found that the reference trajectory was synchronized with human motion ($|\Delta\phi| < 0.5$ rad; shaded bar in Figure 8.10(c)) at the fourth step and onwards.

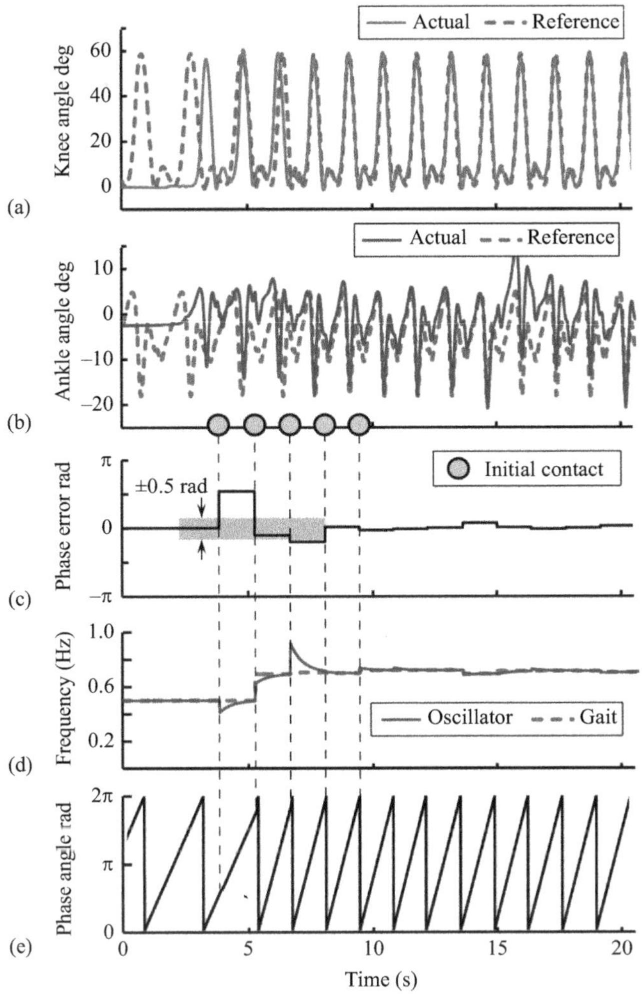

Figure 8.10 Experiment results of FW using Initial Contact for synchronization, including the actual knee (a) and ankle (b) joint angles, and their corresponding oscillating reference trajectory; (c) phase difference between the adaptive oscillator and the actual gait; (d) frequency of the adaptive oscillator and the estimated frequency of the actual gait; (e) phase angle of the adaptive oscillator

8.5.1.2 ZA test

This part describes the results of ZA mode to show the performance of the proposed algorithm with different gait events and experimental conditions. Without loss of generality, two gait events, Opposite Initial Contact ($\Phi_1 = 50\%$) and Tibia Vertical ($\Phi_2 = 87\%$), are used for synchronization.

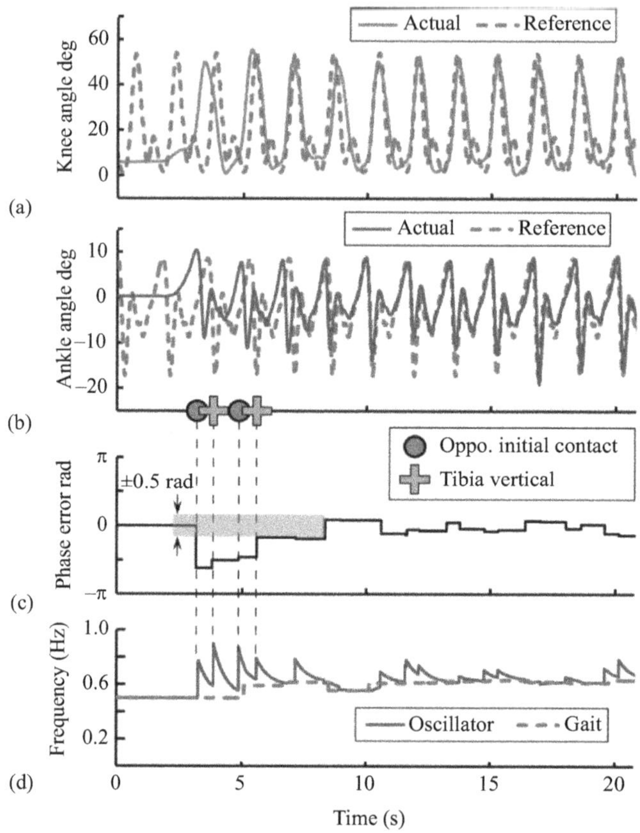

Figure 8.11 Experimental results of ZA using Opposite Initial Contact and Tibia Vertical for synchronization, including the actual knee (a) and ankle (b) joint angles, and their corresponding oscillating reference trajectory; (c) phase difference between the adaptive oscillator and the actual gait; (d) frequency of the adaptive oscillator and estimated frequency of the actual gait

As shown in Figure 8.11, the joint angles of the knee and ankle were slightly altered compared to those in the FW test (Figure 8.11(a) and (b)). The first peak in the knee joint angle was suppressed and the plantar flexion of the ankle joint was slightly reduced. However, the gait events were accurately detected in this condition (circle and cross in Figure 8.11(c)) and synchronization was achieved with the proposed method. In addition, when compared to the FW test, the phase error adaptation was faster with more gait events detected with the same parameters (Figure 8.11(c)). The synchronization was achieved after the third step, which was faster than the test where only one gait event was used for synchronization. The average phase error in the last six steps is 0.049 rad.

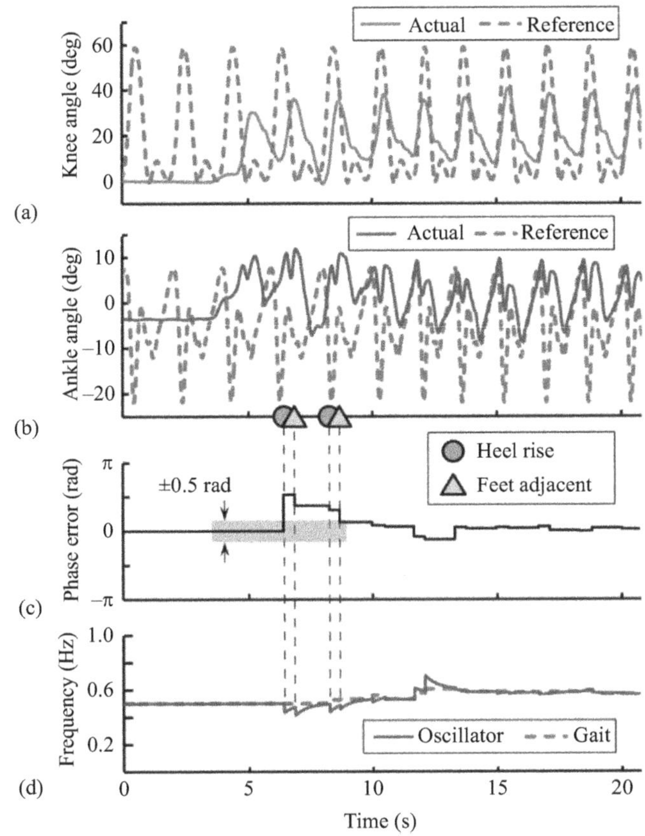

Figure 8.12 Experimental results of SAW using Heel Rise and Feet Adjacent for synchronization, including the actual knee (a) and ankle (b) joint angles, and their corresponding oscillating reference trajectory; (c) phase difference between the adaptive oscillator and the actual gait; (d) frequency of the adaptive oscillator and estimated frequency of the actual gait

8.5.1.3 SAW test

SAW test was conducted to evaluate the reliability of the proposed strategy. Another two different gait events, Heel Rise ($\Phi_1 = 48\%$) and Feet Adjacent ($\Phi_2 = 77\%$), were used for synchronization.

Under this condition, the gait patterns of the knee and ankle joints were seriously altered (Figure 8.12(a) and (b)). The first peak of the knee joint angles was suppressed and the range of knee motion was significantly reduced; the plantar flexion of the ankle during push-off was also limited. However, the gait events could still be detected reliably, as shown in Figure 8.12(c) (circle and triangle). Synchronization was achieved at the third step (Figure 8.12(c) gray shaded bar). The average phase error in the last six steps in this figure is around 0.037 rad. This

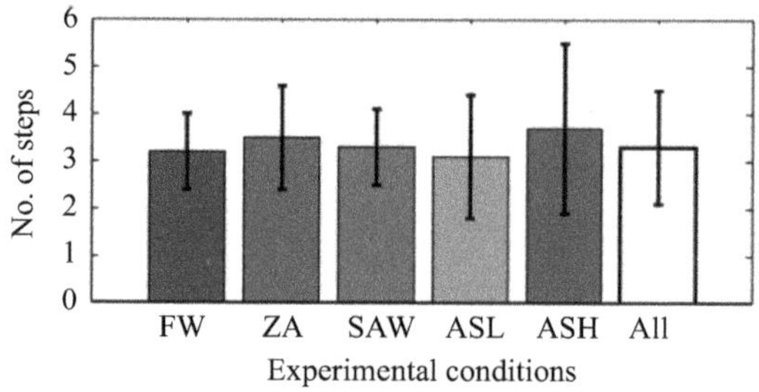

Figure 8.13 Statistical results of all subjects regarding the number of steps needed to reduce the phase error to below 0.5 rad in different experimental conditions

Table 8.2 Relevant variables with different conditions

Variable	FW	ZA	SAW	ASL	ASH
Ankle movement range (deg)	21.9 ± 5.6	20.8 ± 4.0	12.7 ± 1.9	20.3 ± 2.8	21.4 ± 3.5
Knee movement range (deg)	55.4 ± 2.7	52.7 ± 5.2	32.4 ± 3.0	44.1 ± 5.4	49.8 ± 3.2

result indicates the robustness of this synchronization method, since the gait events are the only information used for synchronization which minimizes the influence of the abnormality in the gait pattern. The trajectory thus could be employed in the robot for reference.

8.5.2 Efficiency of the adaptive oscillator

Statistical results of experiments involving 15 subjects are presented in this section. Figure 8.13 and Table 8.2 show the mean and standard deviation of the steps needed to achieve human–robot synchronization ($|\Delta\phi| < 0.5$ rad) among all experiment conditions. From these results, it can be seen that the synchronization was achieved in less than four steps in all experiment conditions. The adaptation process was not elongated when an abnormal gait pattern was involved and repeated measures ANOVA failed to reach significance [$F(4,56) = 0.078$, $P = 0.99$]. This was because the gait event was the only information needed to achieve synchronization and it minimized the influence of the abnormality on the gait pattern. This result demonstrated that the proposed control strategy was efficient and robust in achieving human–robot synchronization.

8.5.3 Evidence of assistance

In this subsection, we first provide results of a representative subject to illustrate the robot performance; then provide statistical results across all subjects.

The results of the ASH test on a subject are shown in Figure 8.14. According to the protocol, the robot worked in zero-assistive mode in the first six steps in

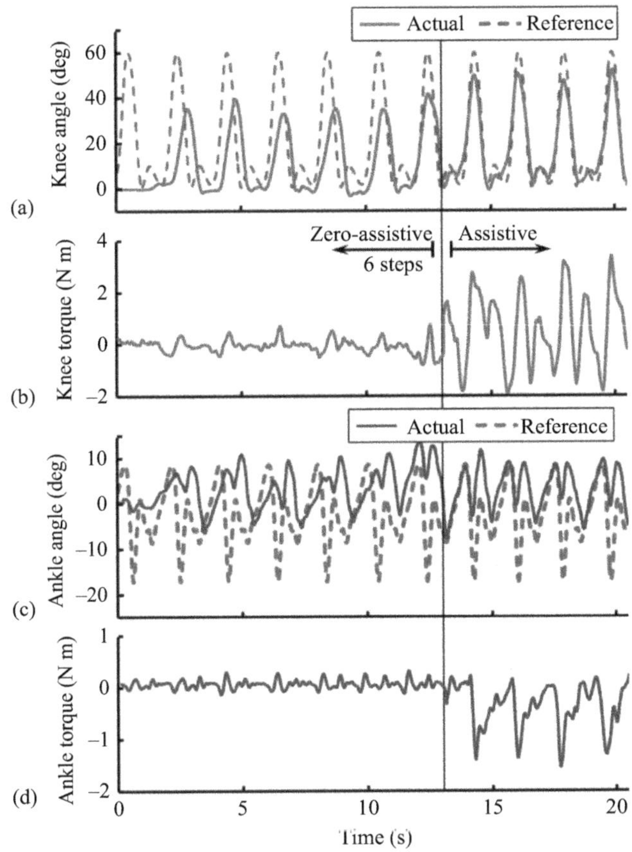

Figure 8.14 Experimental results of ASH, including the actual knee (a) and ankle (c) joint angles, and their corresponding oscillating reference trajectory; robotic assistive torque profiles on the (b) knee and (d) ankle joints

order to achieve human–robot synchronization. In these steps, the gait patterns of the knee and ankle joints were seriously altered as in the SAW condition (Figure 8.14(a) and (c)). The desired assistive torque was zero and the robot tried to minimize the interaction torque (Figure 8.14(b) and (c)). Starting from the seventh step, the robot provided assistive torque during walking. The figure shows that the human–robot synchronization has been achieved and the assistive torque was provided based on the deviation of the actual gait and the reference trajectories. The robotic assistive torque improved the gait pattern and pushed the actual gait pattern closer to the reference.

The averaged angles of the knee and ankle joints in different experimental conditions are shown in Figure 8.15. The movement ranges of the ankle and knee joints are shown in Table 8.2. It can be seen that the kinematics of both ankle and

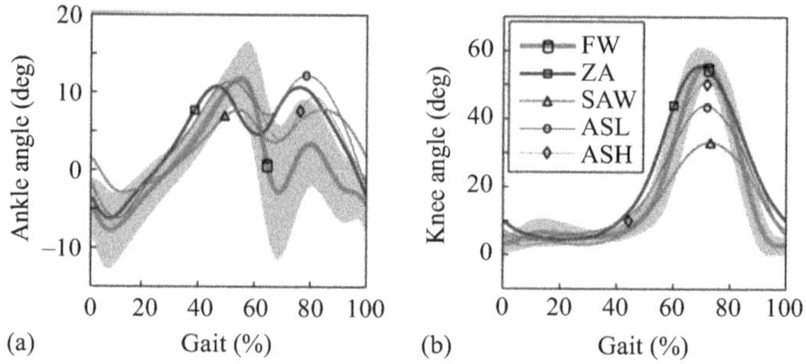

Figure 8.15 Statistical results of all subjects regarding the joint angles of (a) knee and (b) ankle in different experiment conditions. The gray shading is the standard deviation of the joint angles in the FW condition

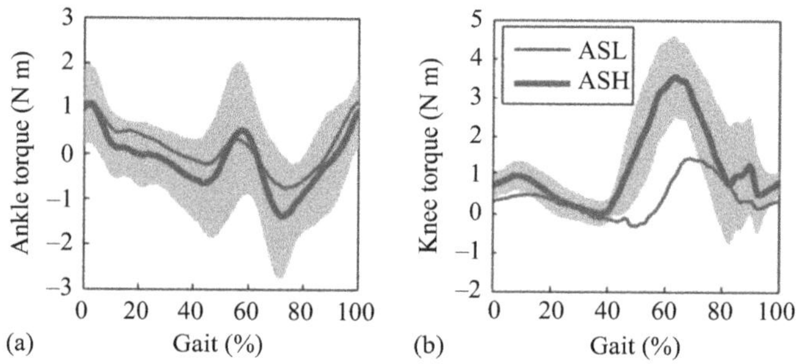

Figure 8.16 Statistical results of all subjects regarding the assistive torque of knee and ankle in ASL and ASH conditions. The gray shading is the standard deviation of the torque in the ASH condition

knee were similar in FW (thick curve with cylinder) and ZA (solid line with square). The ROM for both knee and ankle joints were reduced in the SAW condition.

The gait pattern of the knee and ankle was improved with the assistance of the robot. The ankle angles in ASL and ASH were closer to that in FW, which was the reference trajectory of the robot. The motion range of the knee was extended compared to the SAW scenario. The peak knee angle was about 45° in ASL and 52° in ASH, which were significantly larger than that in the SAW condition. Repeated measures ANOVA reached significance, with a post hoc test establishing a significant difference among the two assisted conditions and the abnormal condition [$F(4,56) > 18.0$, $P < 0.0001$].

The averaged assistive torque profile across all subjects on the knee and ankle provided by the robot in ASL and ASH is shown in Figure 8.16. The standard deviation of the torque profile in ASH is also shown in the figure (gray-shaded bar). It can be seen that assistive torque provided by the robot was based on the deviation of the actual knee and ankle positions from the reference trajectory. The assistive torque in ASH was relatively larger in amplitude than that in ASL, which resulted in a more improved gait pattern. This result indicated that the robot, by employing the proposed control strategy, was able to synchronize with the human gait and provide assistive torque to improve the gait pattern during walking.

8.6 Conclusion

In this chapter, we present the mechanical design and synchronization control of a portable knee–ankle–foot robot aimed for stroke patients to carry out overground gait training. A novel compliant SEA utilizes two springs in series with different stiffness values to achieve back-drivability, large force range and low output impedance for safe human–robot interaction. An adaptive-oscillator-based method using gait events information has been proposed for robot-aided gait training. The principle of the method is through estimating the stride percentage based on the gait-event information. Our study provides a new solution for achieving human–robot synchronization in gait rehabilitation robotics. This method can be applied to a wide range of control strategies, such as identifying the stance and swing phase [33–35]. We will develop advanced control strategies for robot-aided gait training based on the synchronous trajectory in our future work. This robotic system also provides a good platform to investigate the biomechanical effects of human–motor adaptation in the robot-aided walking. This will be further evaluated with clinical trials.

References

[1] Pennycott A., Wyss D., Vallery H., Klamroth-Marganska V, Riener R. 'Towards more effective robotic gait training for stroke rehabilitation: a review'. J. NeuroEng. Rehabil. 2012, vol. 9(65), pp. 1–13.

[2] Dollar A.M., Herr H. 'Lower extremity exoskeletons and active orthoses: challenges and state-of-the-art'. IEEE Trans. Rob. 2008, vol. 24(1), pp. 144–158.

[3] Chen G., Chan C.K., Guo Z., Yu H. 'A review of lower extremity assistive robotic exoskeletons in rehabilitation therapy'. Crit. Rev. Biomed. Eng. 2013, vol. 41(4–5), pp. 343–363.

[4] Riener R., Lunenburger L., Jezernik S., Anderschitz M., Colombo G., Dietz V. 'Patient-cooperative strategies for robot-aided treadmill training: first experimental results'. IEEE Trans. Neural Syst. Rehabil. Eng. 2005, vol. 13(3), pp. 380–394.

[5] Fisher S., Lucas L., Thrasher T.A. 'Robot-assisted gait training for patients with hemiparesis due to stroke'. Top. Stroke Rehabil. 2011, vol. 18(3), pp. 269–276.

[6] Veneman J.F., Kruidhof R., Hekman E.E.G., Ekkelenkamp R., Van Asseldonk E.H.F., van der Kooij H. 'Design and evaluation of the LOPES exoskeleton robot for interactive gait rehabilitation'. IEEE Trans. Neural Syst. Rehabil. Eng. 2007, vol. 15(3), pp. 379–386.

[7] Banala S.K., Kim S.H., Agrawal S.K., Scholz J.P. 'Robot assisted gait training with Active Leg Exoskeleton (ALEX)'. IEEE Trans. Neural Syst. Rehabil. Eng. 2009, vol. 17(1), pp. 2–8.

[8] Hussain S., Xie S.Q., Liu G. 'Robot assisted treadmill training: mechanisms and training strategies'. Med. Eng. Phys. 2011, vol. 33(5), pp. 527–533.

[9] van den Brand R., Heutschi J., Barraud Q., *et al.* 'Restoring voluntary control of locomotion after paralyzing spinal cord injury'. Science 2012, vol. 336(6085), pp. 1182–1185.

[10] Lee S.J., Hidler J. 'Biomechanics of overground vs. treadmill walking in healthy individuals'. J. Appl. Physiol. 2008, vol. 104(3), pp. 747–755.

[11] Blaya J.A., Herr H. 'Adaptive control of a variable-impedance ankle–foot orthosis to assist drop-foot gait'. IEEE Trans. Neural Syst. Rehabil. Eng. 2004, vol. 12(1), pp. 24–31.

[12] Shorter K.A., Kogler G.F., Loth E., Durfee W.K., Hsiao-Wecksler E.T. 'A portable powered ankle–foot orthosis for rehabilitation'. J. Rehabil. Res. Dev. 2011, vol. 48(4), pp. 459–472.

[13] Noel M., Cantin B., Lambert S., Gosselin C.M., Bouyer L.J. 'An electrohydraulic actuated ankle foot orthosis to generate force fields and to test proprioceptive reflexes during human walking'. IEEE Trans. Neural Syst. Rehabil. Eng. 2008, vol. 16(4), pp. 390–399.

[14] Fleischer C., Hommel G. 'A human—exoskeleton interface utilizing electromyography'. IEEE Trans. Rob. 2008, vol. 24(4), pp. 872–882.

[15] Horst R.W. 'A bio-robotic leg orthosis for rehabilitation and mobility enhancement'. IEEE Int Conf on Medicine and Biology Society (EMBC), 2009, pp. 5030–5033.

[16] Shamaei K., Napolitano P.C., Dollar A.M. 'Design and functional evaluation of a quasi-passive compliant stance control knee–ankle–foot orthosis'. IEEE Trans. Neural Syst. Rehabil. Eng. 2014, vol. 22(2), pp. 258–268.

[17] Yu H., Huang S., Chen G., *et al.* 'Design and analysis of a novel compact compliant actuator with variable impedance'. IEEE Int Conf on Robotics and Biomimetics (ROBIO), 2012, pp. 1188–1193.

[18] Cao J., Xie S.Q., Das R., Zhu G.L. 'Control strategies for effective robot assisted gait rehabilitation: the state of art and future prospects'. Med. Eng. Phys. 2014, vol. 36(12), pp. 1555–1566.

[19] Riener R., Lünenburger L., Jezernik S., Anderschitz M., Colombo G., Dietz V. 'Patient-cooperative strategies for robot-aided treadmill training: first experimental results'. IEEE Trans. Neural Syst. Rehabil. Eng. 2005, vol. 13(3), pp. 380–394.

[20] Aoyagi D., Ichinose W.E., Harkema S.J., Reinkensmeyer D.J., Bobrow J.E. 'A robot and control algorithm that can synchronously assist in naturalistic motion during body-weight-supported gait training following neurologic injury'. IEEE Trans. Neural Syst. Rehabil. Eng. 2007, vol. 15(3), pp. 387–400.

[21] Jezernik S., Colombo G., Morari, M. 'Automatic gait-pattern adaptation algorithms for rehabilitation with a 4-DOF robotic orthosis'. IEEE Trans. Rob. Autom. 2004, vol. 20(3), pp. 574–582.

[22] Ronsse R., Vitiello N., Lenzi T., van den Kieboom J., Carrozza M.C., Ijspeert A.J. 'Human–robot synchrony: flexible assistance using adaptive oscillators'. IEEE Trans. Biomed. Eng. 2011, vol. 58(4), pp. 1001–1012.

[23] Petrič T., Gams A., Ijspeert A.J., Žlajpah L., 'On-line frequency adaptation and movement imitation for rhythmic robotic tasks'. Int. J. Rob. Res. 2011, vol. 30(14), pp. 1775–1788.

[24] Lenzi T., Carrozza M.C., Agrawal S.K., 'Powered hip exoskeletons can reduce the user's hip and ankle muscle activations during walking.' IEEE Trans. Neural Syst. Rehabil. Eng. 2013, vol. 21(6), pp. 938–948.

[25] Ronsse R., Lenzi T., Vitiello N., *et al.* 'Oscillator-based assistance of cyclical movements: model-based and model-free approaches'. Med. Biol. Eng. Comput. 2011, vol. 49(10), pp. 1173–1185.

[26] Ermentrout B. 'An adaptive model for synchrony in the firefly Pteroptyx malaccae'. J. Math. Biol. 1991, vol. 29, pp. 571–585.

[27] Yu H., Cruz M., Chen G., *et al.* 'Mechanical design of a portable knee–ankle–foot robot'. IEEE Int Conf on Robotics & Automation (ICRA), 2013, pp. 2183–2188.

[28] Whittle M.W. '*Gait analysis: an introduction*'. Butterworth-Heinemann, 4th ed., Elsevier Ltd, 2007.

[29] Meng X., Yu H., Tham M.P. 'Gait phase detection in able-bodied subjects and dementia patients'. Proc. Int. Conf. Eng. Med. Biol. Soc (EMBS), 2013, pp. 4907–4910.

[30] Chen G., Guo Z., Yu H. 'A novel gait phase-based control strategy for a portable knee–ankle–foot robot'. IEEE Int. Conf. Rehabil. Robot. (ICORR), 2015, pp. 571–576.

[31] Rabiner L., 'A tutorial on hidden Markov models and selected applications in speech recognition'. Proc. IEEE, 1989, pp. 257–286.

[32] Balaban B., Tok F. 'Gait disturbances in patients with stroke'. Pm&R 2014, vol. 6, pp. 635–642.

[33] Blaya J.A., Herr H. 'Adaptive control of a variable-impedance ankle–foot orthosis to assist drop-foot gait'. IEEE Trans. Neural Syst. Rehabil. Eng. 2004, vol. 12, pp. 24–31.

[34] Shamaei K., Cenciarini M., Adams A.A., Gregorczyk K.N., Schiffman J.M., Dollar A.M. 'Design and evaluation of a quasi-passive knee exoskeleton for investigation of motor adaptation in lower extremity joints'. IEEE Trans. Biomed. Eng. 2014, vol. 61(6), pp. 1809–1821.

[35] Bae J., Kong K., Tomizuka M. 'Gait phase-based control for a rotary series elastic actuator assisting the knee joint'. ASME Trans. J. Med. Dev. 2011, 5(3), pp. 031010.

Section 3

Devices

In this section, a variety of different types of exoskeletons are described, with details of the designs, how they were built, and tested with users. The exoskeletons have been created for different types of markets such as spinal cord injuried persons, medical rehabilitation devices, user assistance exoskeletons, military, industrial, and recreation systems. A good description of commercial devices can be found at www.exoskeletonreport.com and www.wearablerobotics.com.

In this section, three chapters are included describing exoskeletons used in the medical markets. The medical, exoskeleton market is the most mature as researchers have focused on this area. In Chapter 9, a medical exoskeleton to aid persons with spinal cord injury is described. The chapter focuses on generating a stable gait to ensure balance. The user's legs are powered at the hip and knee, and crutches are added for maintaining balance. The gait which includes both two legs and two crutches is analyzed as a quadruped gait. When one leg or crutch is not touching the ground in the swing phase, a tripod structure is formulated to ensure balance. Experimental results show a user able to walk with the system. There are many spinal cord injury exoskeletons and readers are encouraged to review Chapter 1 as well and study systems by Ekso Bionics, ReWalk, Indego, Rex Bionics, and Fourier Intelligence.

In Chapter 10, soft wearable robotic systems are described as well as the chronology of robot development from typical, stiff actuation to compliant actuation allowing for human–robot interaction. The compliant actuators allowed worldwide researchers to focus on rehabilitation robotics where systems can assist as needed and guide and improve human function such as reaching and walking. Recently, soft robotic solutions were employed to design systems that can compliantly interact with the human to improve safety and ergonomics and reduce weight. Typically, rigid structures strapped or placed in-parallel with the human limb structure can be uncomfortable, heavy, and limit sideways motion. The authors believe that soft robotics is a natural progression in the research line from stiff actuation, to series and parallel elastic actuators, to compliant exosuits to improve safe human–robot interaction. Examples are described in the chapter: a soft-robotic arm exoskeleton is described in detail along with the Bowden-type actuators. A soft-glove system is used to aid picking up objects and a unique supernumerary finger is added to a user's arm to aid in grasping objects improving dexterity and aiding in activities of daily living of stroke survivors.

In Chapter 11, lower limb devices are described to aid in gait training and assist users with reduced leg muscle strength. One exoskeleton uses a leg mechanism with a foot plate that helps to lift the foot at the correct time. A second exoskeleton uses pneumatic muscles to assist the whole body. The system assists the legs and arms to enhance walking by swinging the arms and legs in a synchronized rhythm. Last, a lightweight ankle device, RE-Gait was designed to assist push-off. This device has been commercialized in 2016.

As the field is growing rapidly, we foresee new devices also emerging in the industrial area. Commercialized devices are already assisting the lower back when bending over and reaching in various pick-and-place tasks. Passive exoskeletons allow the user to hold heavy grinders and other objects by transferring the load to the ground. Typical devices use some type of gravity balancing mechanism.

Chapter 9

Real-time gait planning for a lower limb exoskeleton robot

Xinyu Wu[1], Can Wang[1], Yue Ma[1], and Duxin Liu[1]

Abstract

This chapter presents a medical robot exoskeleton developed to assist spinal cord injury patients to walk independently, focusing on real-time gait planning. The exoskeleton robot is built with actuators at the hip and knee joints, and a pair of crutches to help paraplegic patients maintain balance during walking. A real-time gait planning strategy is developed to allow the users to walk stably in a natural manner. This gait planning treats the human–machine coupling system as a quadruped robot and adjusts and corrects gait before each step according to real-time responses of multiple sensors. Taking into account of the kinematic model of people wielding crutches, this gait planning strategy provides a larger stability margin for the system. Results of experiments are presented to illustrate the effectiveness of the proposed gait planning method.

Keywords: exoskeleton; spinal cord injury; real-time gait planning

9.1 Introduction

With the rapid development of robot technologies [1], many medical robots were developed for improving people's quality of life. Of them, lower extremity exoskeleton robots are of great significance for elderly people and spinal cord injury (SCI) patients to improve their walking ability. The elderly or patients with limited mobility can be restored lower limb motor ability and daily activities and improved quality of life dramatically.

The statistic data of patients with SCIs have showed an increasing trend in the last decade. Most SCI patients are suffering from walking impairment which seriously affected their quality of life. Furthermore, as they spend much time sitting in wheelchairs, their lower limbs are short of exercise and have trouble with blood

[1]Center for Intelligent and Biomimetic Systems, Shenzhen Institutes of Advanced Technology (SIAT), Chinese Academy of Sciences, China

circulation [2]. To solve such problem, exoskeleton robots have been introduced to provide an effective method for lower limb rehabilitation.

For lower limb exoskeletons, gait planning plays a key role, as it directly affects the control strategy and performances of robots. Most existing methods of gait planning utilize the theories of Biped robot based on Zero-Moment Point (ZMP) [3], fuzzy control [4] as well as data sampling from healthy people [5]. For example, the Atlas robot of Boston Dynamics adopts a parametric pendulums swinging leg model [6]. Stance and swing phase transitions between tracks smooth Gaussian basis function filters. In the Hybrid Assistive Limb (HAL) system, support and swing phases are controlled by the movement of hip and knee joints [7]. Gaits of healthy people were collected in advance. By comparing the stored and real-time sensor data, leg movement of different phases can be distinguished. Mina [8] is a new version of Institute for Human and Machine Cognition (IHMC) [9], which allows patients walk with preset cycles through predefined gait. ReWalk lets wearer control the exoskeleton to follow a certain gait. More than 12 patients have been tested and it is by far the most successful walking exoskeleton [10]. eLegs is controlled via a walking stick. The body inclination triggers the robot walking gait [1].

While collecting gaits from normal people is convenient, the gait does not well fit for exoskeletons, due to the fact that exoskeleton robot is an underactuated human–machine coupling system which not only contain the human being but also include the exoskeleton. Moreover, in our case, the exoskeleton needs crutches to assist walking rather than a system similar to a biped robot. Due to restriction of mechanical design, exoskeleton system needs a static gait to assist patient walking. Crutches have a significant impact on the system's stability [11]. The proposed method models this system as a quadruped robot. Such system can improve its gait stability through predetermining landing positions of robot feet.

9.2 SIAT lower limb exoskeleton robot

A mobile medical exoskeleton named SIAT exoskeleton has been developed at Shenzhen Institute of Advanced Technology, Shenzhen, China. The exoskeleton can allow SCI patients to walk independently.

9.2.1 System and structure

Figure 9.1 shows the third-generation exoskeleton robot developed. Each hip and knee joint of the robot is equipped with a Direct Current (DC) servo motor. A backpack containing rechargeable batteries and a central control unit is connected to the main body of the robot. The knee and hip joints rotate in the sagittal plane. Ankle joint of the robot has only one degree of freedom to prevent ankle sprains. A spring is attached to the ankle joint to ensure that the patient can lift toes without rubbing with the ground. The degrees of freedom are much limited for this exoskeleton robot. Thus, the design minimized complexity and total weight while still ensuring necessary flexibility and the function of walking and sitting down/standing up. Waist belt and bandages on the legs are used to attach the user and the robot. A pair of forearm

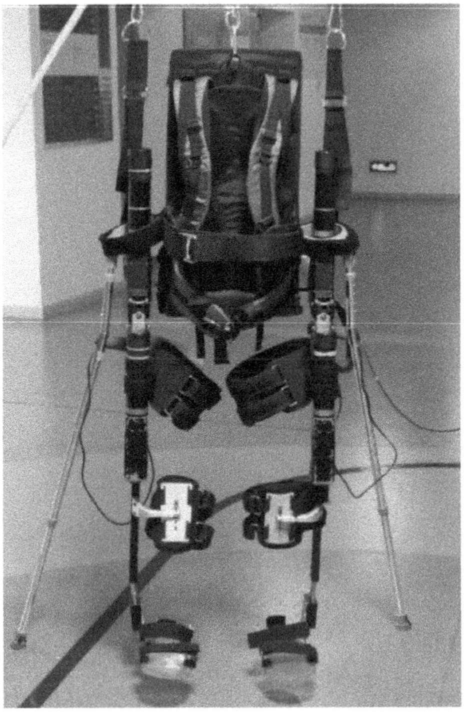

Figure 9.1 The SIAT exoskeleton robot

crutches is included to maintain balance, as well as giving orders to the robot through Bluctooth.

All active joints of robot are driven by DC servo motors. The controller provides voltage, current, speed, and position loop, the servo motor walking in position–velocity–time modes. Hall angle sensor (Figure 9.2) is installed on each joint. The sensor data could compensate and correct the position of the encoder and could also obtain real-time joint angle value. A gyroscope is mounted on backpack for falling detection. Cane is equipped with buttons, the wearer can control the robot key combination, stand up and sit down.

The robot's motion control system adopts a distributed control structure, as shown in Figure 9.3. It is mainly composed of main control computer and Controller Area Network (CAN) bus controller such as joint drive module. Each driven joint servo control system with CAN bus interface circuit is connected to the CAN bus interface card. This kind of robot control system provides a good scalability, convenient removability, and well maintainability.

The distributed system requires little computational resources of the master computer and each controller have light burden on computation. The master computer only need to concentrate on trajectory planning and the distributed controller takes care of trajectory tracking. Each controller is based on control command of Master computer to complete motion controls such as position, velocity, current

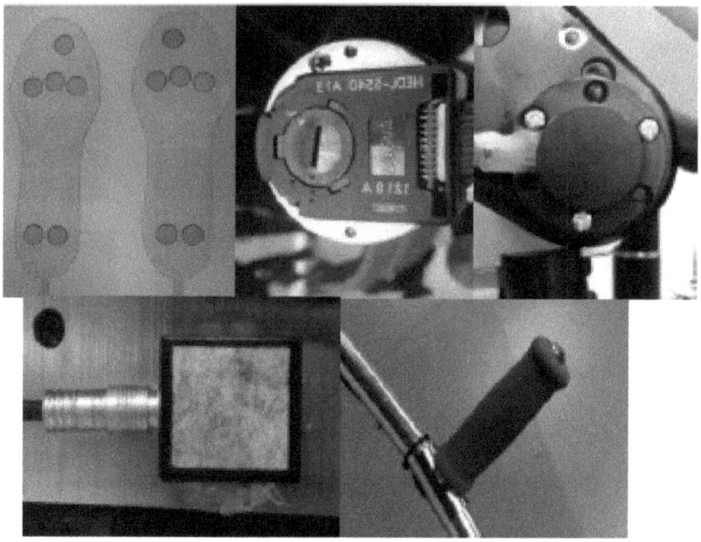

Figure 9.2 Sensors assembled on the exoskeleton robot

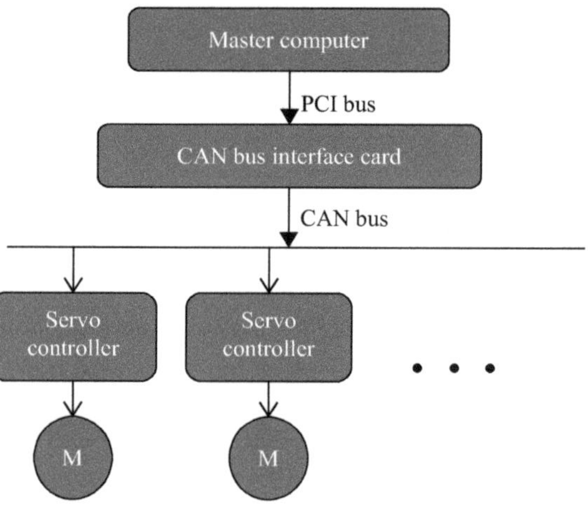

Figure 9.3 Robot control structure based on CAN bus

feedback, and output Pulse Width Modulation (PWM)-driven signals to make motor running to the specified location.

The servo control system at exoskeleton joints is showed in Figure 9.4. Each single joint servo control system uses a closed-loop control and real-time feedback signal for the current speed adjustment. DC motor equipped with incremental encoder and high-precision hall can get motor error between actual speed and commanding speed adjusted by the controller according to the error value.

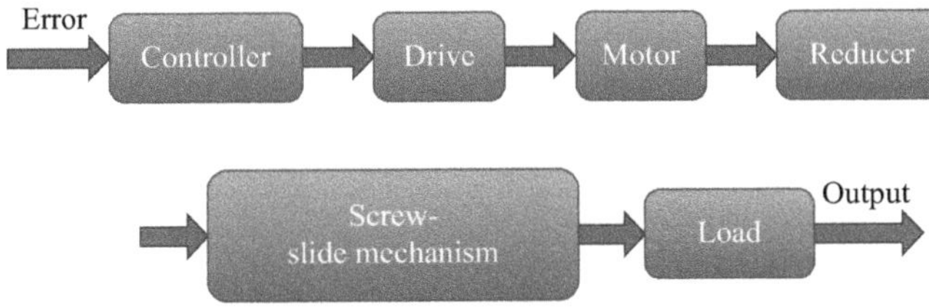

Figure 9.4 Single joint servo control system

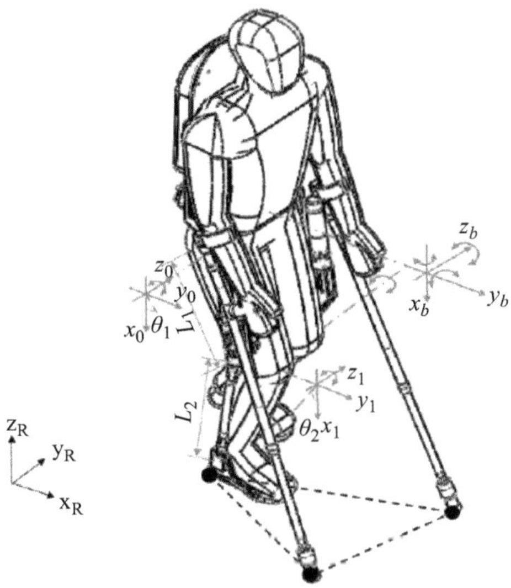

Figure 9.5 The kinematic model

9.2.2 Kinematics modeling

When people walk, walking is performed by hip and knee flexion/extension, and the legs usually swing in the sagittal plane. We define base coordinate system in the waist. As shown in Figure 9.5, set up a coordinate system for each connecting rod.

As shown in Figure 9.6, the common perpendicular of two axes is a_i, the axis Z_i along the motion, the axis X_i along the common normal line of Z_i and Z_{i-1}. The relative position between connecting links i and $i-1$, and the i coordinate system relative to the base coordinate system posture ${}_i^0T$ can be calculated through the following formula:

$$ {}_i^0T = A_1 \cdot A_2 \ldots A_i = \prod_{j=1}^{i} A_i = \begin{bmatrix} n_i & o_i & a_i & p_i \\ 0 & 0 & 0 & 1 \end{bmatrix} \tag{9.1} $$

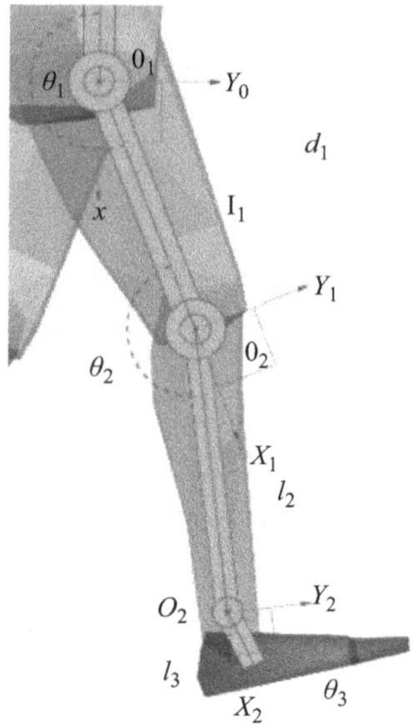

Figure 9.6 Legs structure of the robot

Table 9.1 D–H *parameters and joint variables*

Number	a_i	a_i	d_i	θ_i	Parameter
1.	0	l_1	0	$\theta_1 - \pi$	θ_1
2.	0	l_2	0	$\theta_2 + \pi$	θ_2
3.	0	l_3	0	θ_3	θ_3

where transformation matrix A_i is calculated through *D–H* parameters given in Table 9.1 with

$$A_i = R(z, \theta_i)T(0, 0, d_i)T(a_i, 0, 0)R(x, a_i). \tag{9.2}$$

Each leg can be deemed as a kinematic chain, formed by the range of joints and rod, start point is the origin of based coordinate systems on the waist, end in the foot. Hip and knee are driven by servo motors, let the ends joints move toward the target position.

In order to analysis the kinematic model more convenient, we deem the thighs and shank as a cylinders with uniform quality and constitutes a connecting rod with robot. L_1, L_2, and L_3 representing the length of thigh, shank, and foot, respectively. Ankle rotation angle of hip, knee, and ankle is θ_1, θ_2, and θ_3, respectively. At first, according to the theory of *D–H* (Denavit–Hartenberg) to establish coordinates $O_0X_0Y_0Z_0$, $O_1X_1Y_1Z_1$, and $O_2X_2Y_2Z_2$ for this kinematic chain-link.

Using *D–H* parameters in formula (9.1), we can obtain the following formulas:

$$\begin{aligned}
A_1 &= \begin{bmatrix} -c_1 & s_1 & 0 & -l_1c_1 \\ -s_1 & -c_1 & 0 & 0 \\ 0 & 0 & 1 & 0 \\ 0 & 0 & 0 & 1 \end{bmatrix} \\
A_2 &= \begin{bmatrix} -c_2 & s_2 & 0 & -l_2c_2 \\ -s_2 & -c_2 & 0 & -l_2s_2 \\ 0 & 0 & 1 & 0 \\ 0 & 0 & 0 & 1 \end{bmatrix} \\
A_3 &= \begin{bmatrix} c_3 & -s_3 & 0 & -l_3c_3 \\ s_3 & c_3 & 0 & -l_3s_2 \\ 0 & 0 & 1 & 0 \\ 0 & 0 & 0 & 1 \end{bmatrix} \\
{}_3^0T = A_1A_2A_3 &= \begin{bmatrix} c_1c_2c_3 & -s_1s_2s_3 & 0 & l_3c_1c_2c_3 + l_2c_1c_2 - l_1c_1 \\ s_1s_2s_3 & c_1c_2c_3 & 0 & l_3s_1s_2s_3 + l_2s_1s_2 - l_1s_1 \\ 0 & 0 & 1 & 0 \\ 0 & 0 & 0 & 1 \end{bmatrix}
\end{aligned} \tag{9.3}$$

where $c_i = \cos\theta_i$ and $s_i = \sin\theta_i$.

After getting T_i, the Jacobian matrix can be obtained readily by differential and integral manipulations. Through the analysis of kinematics and inverse kinematics of the robot, we can calculate the robot leg joint angles values as needed.

9.3 Crutches-walking gait analysis

Gait is the coordination of legs. Through the clinical experiment, we found that the exoskeleton should ensure the sequence of the phases in the whole stage of each gait cycle. In the SIAT designed exoskeleton, a control button embedded on crutches is used to send control command to the processor via Bluetooth. After a period of training, it is proved that patients will get used to this kind of control method.

In this exoskeleton, we use the classic four-legged walking gait. The legs and crutches are put forward sequentially. Every step the system will ensure a large enough margin for stability, which has mainly two factors: (1) a triangle determined by robot's feet and the supporting crutch and (2) the minimal distance from the triangle's edge to center of pressure (COP). To ensure walking stability, at least three supporting points should exist at any given time. At the same time, with the consideration of the effects in different terrain, path planning strategies of quadruped robots [12,13] could be used in this model easily. Imitating the walking sequence of a quadruped such as turtles, lizards, and quadruped robot [14], the moving sequence is right crutch, left leg, left crutch, and then right leg. Figure 9.7 indicates the four phases of the proposed gait.

It is clear that a reasonable support triangle is helpful to enhance the disturbance rejection ability. When the area of support triangle equals to zero, the robot will come into the critical stable state and supported by only two

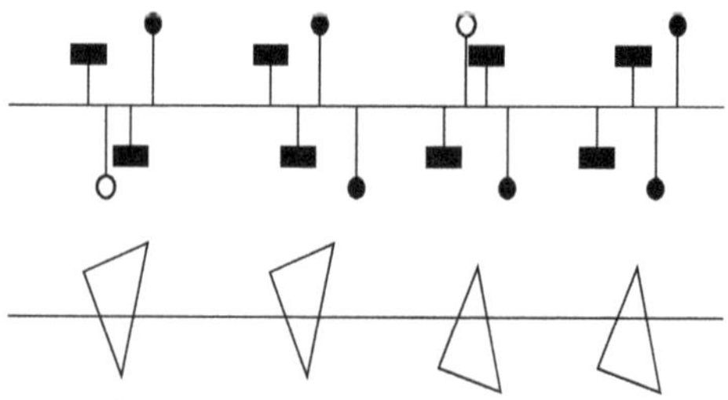

Figure 9.7 Four-legged gait and supporting triangle in each phase. The system moves the right crutch forward at first, followed by the robot's left leg, then the left crutch, finally the right leg. This pattern will repeat as a gait cycle during walking in each phase

points. This state of the system is unstable and prone to fall as a result of any disturbance from outside. So a reasonable gait of crutches and legs is very critical for walking stability.

9.4 Real-time gait planning

In order to get the phases of gait used in exoskeleton, we need to analyze the normal walking gait. A complete gait cycle is usually divided into four phases, right double support phase, right single support stance, left double support phase, and left single support phase. Each leg foot from heel to toe off the ground, represents 60% of a cycle, which contains two double support phase and a single support phase. Single leg support phase possesses 40% of a gait cycle. Single support phase is also called swing phase, the swinging legs leave the ground and moving forward, the main responsibility of body center of gravity moves forward. Bipedal stance also known as stance phase, both feet contact with the ground, provide support to the body.

It is seen from the above analysis that normal walking must complete three processes: (1) the legs support the body weight, (2) support user's body and lift user's weight through one leg, (3) the swing leg move forward, support leg move back, letting the center of gravity move forward. During a gait cycle, these important parts operate alternatively, allow the body to move. Normal people cannot walk forward if missing any of the parts (Figure 9.8).

It is needed to obtain a normal walking gait and joint torque information based on the gait analysis. Looking up to a normal gait through Clinical Gait Analysis (CGA) gait database, with these normal human gait simulation data, using the model given on the OpenSim we can simulate normal walking gait. Torque, joint information and joint angles values can be obtained by simulation of the normal walk. The simulation is shown in Figure 9.9.

Through repeated experiments, we found several useful features of human walking. It can be observed that body weight moves in the sagittal plane in

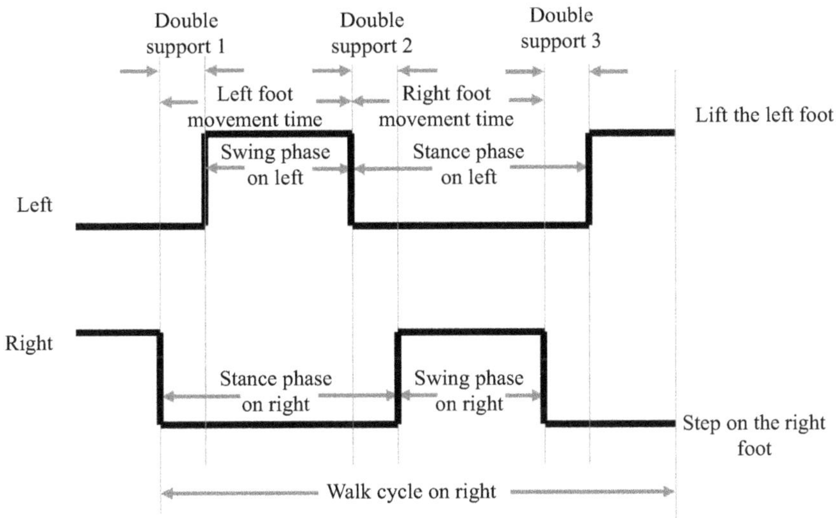

Figure 9.8 Four phases of a gait cycle

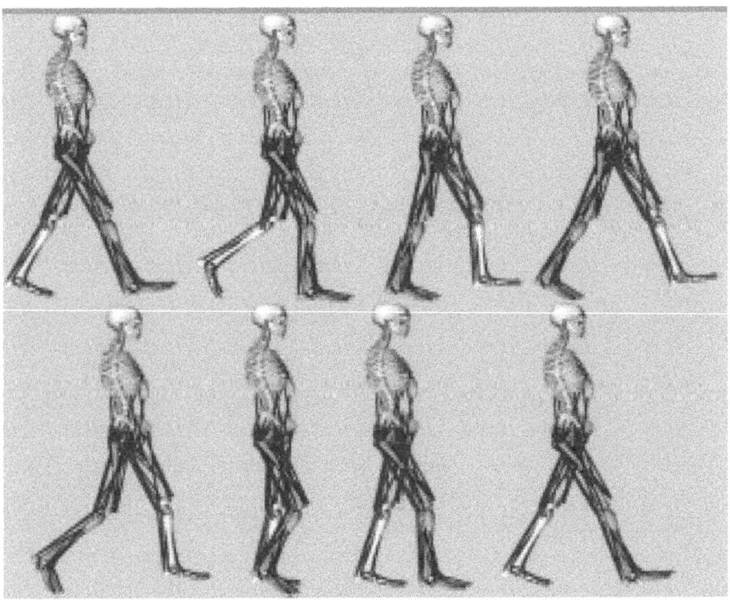

Figure 9.9 Gait simulation in OpenSim

repetitive motion, while the center of gravity trajectory can be approximated sine waves, with peaks appearing in the support period and lowest points in bipedal support. Through the study with a 1.75-m sport model, the lift of the center of gravity is found approximately as 4.6 cm. When viewed over from a bipedal stance to single support phase, the pelvis will make the weight move to the supporting leg. The projection of the center of gravity in the horizontal plane is approximate to sine wave, with an amplitude of about 5 cm.

Through the curves of hip, knee, and ankle in a cycle, as can be seen from Figure 9.10, the swinging leg knee keep bend after swing leg's heel landing on the ground. This way can reduce the Centre of gravity's shifting magnitude. During supports phase, legs and hips will rotate forward, knees changing straight, let the gravity lift, ankle rotation letting the weight move forward. The hip of swing leg will rotate reverse, driving leg moves forward. During this phase, the swinging legs move straight after the leg move forward.

Double support phase, the support leg replaces alternatively. Support time is short, but it directly determines the stability of gait. Proper foot support can ensure that alternated between body weight on two legs, time is moving away before your body weight on the leg, and walking instability. Torque value of three rotating joints in a gait cycle can obtain from Figure 9.11.

9.4.1 Gait planning strategy

Through the analysis of normal gait, we learned that a gait cycle consists of four stages, and the roles of each phase are clearly understandable. But for the exoskeleton, the normal gait cannot be used directly. Each SIAT exoskeleton leg has only two motors, which drive hip and knee joints, while ankle joints are passive. So the system cannot simply adopt the normal gait to walk. However, by rational using of crutches, the underactuated system could assist people walking.

Since the ankle is free, a user needs to walk with crutches to support their body to maintain balance. When the body is slightly tilted forward, crutches and supporting legs constitute a support polygon. If the projection of weight falls within the supporting polygon the walking is stable, as shown in Figure 9.12.

The patients are expected to move each step with their crutches to keep balance in daily life. Except for stability, a more natural gait is also expected. The crutches and legs play a crucial role for supporting the body weight as well as providing necessary flexibility during walking. So we need a proper criterion to evaluate the body's stability. As for our static gait, only the forces from the ground are taken into consideration, such system can be seen as a static equilibrium with inertia ignored.

We define the coordinate system, which is shown in Figure 9.13. The *fB* is the world coordinate system, b1 and b2 are located in the horizontal plane. The fN is the reference frame, $\vec{n}_1$ and $\vec{n}_2$ are located on the supporting plane. Point G is the center of gravity of system, whose projection on the supporting surface is G'. The angle between $\vec{n}_3$ and b3 is the ground elevation. $P1$, $P2$, and $P3$ are the support points on the ground, which consist by legs or crutches. The support forces produced by the ground are $\vec{R}_1$, $\vec{R}_2$, and $\vec{R}_3$. The resultant force and the resultant moment are generated by the external environment, which is Fg and Mg. The total

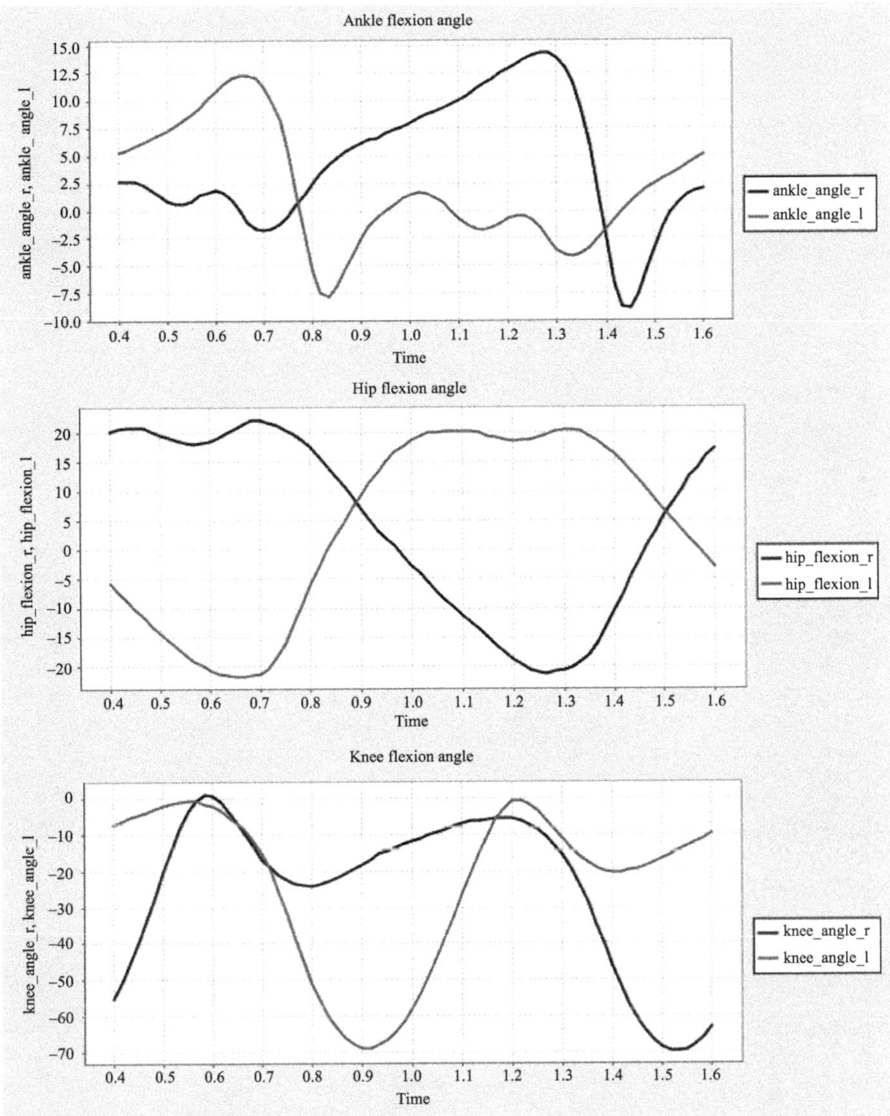

Figure 9.10 Joint angle value in a gait cycle

mass of the system is m, and we assume the COP is C. The COP can be determined by the following equation:

$$\vec{r}^{\,\mathrm{OC}} = \frac{1}{\left(\vec{n}_3 \cdot \sum_i \vec{R}_i\right)} \vec{n}_3 \times \left(\sum_i \vec{r}_i^{\,\mathrm{OP}} \times \vec{R}_i\right) \tag{9.4}$$

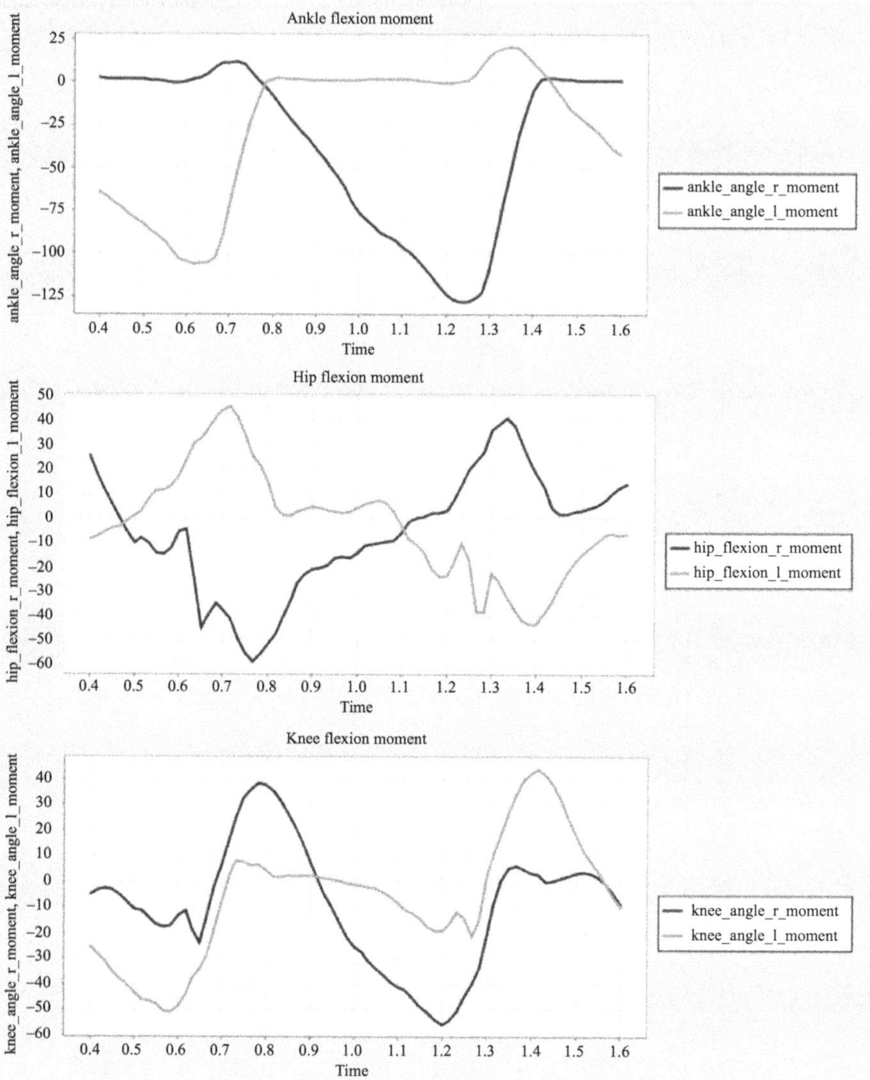

Figure 9.11 Joint angle value in a gait cycle

We define the stability margin is the minimal distance from the triangle's edge to COP.

The COP is determined by the state of this system. Equation (9.4) can be obtained from [15]. When $\vec{M}_G$ and $\vec{F}_G$ is zero, the COP coincides with the projection of the center of gravity (COG) on the supporting surface.

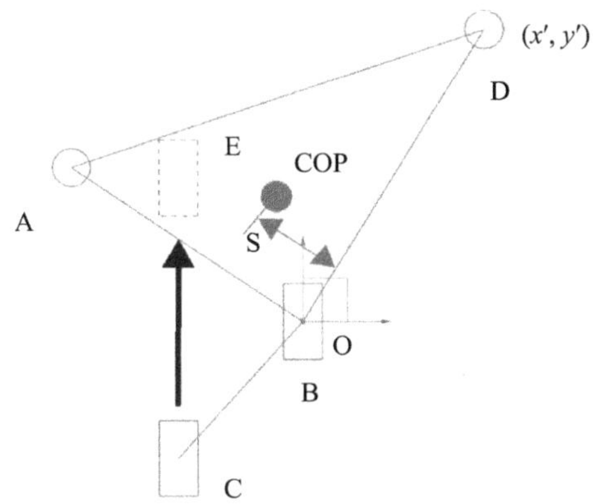

Figure 9.12 Stable polygon is consisted by feet and crutches and G is the projection of body weight

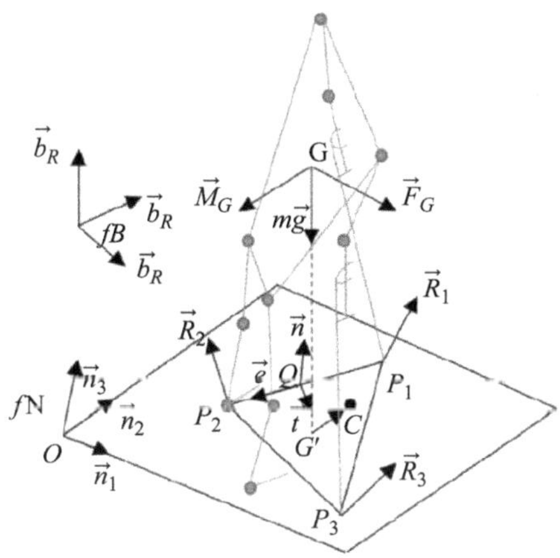

Figure 9.13 Coordinates and static forces acting on system

$$\lambda > \frac{\vec{n}_3 \times \left(\vec{M}_G + \vec{r}^{G'G} \times \vec{F}_G\right)}{\vec{n}_3 \cdot \left(\vec{F}_G + m\vec{g}\right)} = \lambda_{\min} \tag{9.5}$$

where λ is the minimum distance between the support edge $P_iP_j(i,\ j = 1,\ 2,\ 3;\ i \neq j)$ to COG. The stability margin is $S = \lambda - \lambda_{\min}$. This definition of stability margin makes judgment of stability become more convenience.

It is clear that stability margin stands for the disturbance rejection ability. When the stability margin equals to zero, the robot is in a critical stability state and supported by only two points. This state of the system is unstable and easy to fall as a result of any disturbance from outside. So finding a reasonable support-point of crutches and legs is very significant for stability of the gait. The software needs to find a position to ensure that the stability margin is sufficient.

The most-weighted component of the exoskeleton is the leg structure. A robot needs swing the leg when moving. All the parts of the system determine the COG. So the center of gravity is changing when the system is walking. In order to minimize the complexity of software, we ignore the effect from movements of the arm and deem the COG of each leg for they are light. Assuming the center of mass coincides with geometric center for each parts, we can obtain the following equation:

$$\vec{C} = \frac{\sum_{j=1}^{N} \vec{m}_j \left(p_j + R_j \vec{c}_j\right)}{m} \tag{9.6}$$

where the $\vec{c}$ is the relative coordinate of COG, $\vec{m}_j$ is the mass of each part, and m is the total mass of the system. (p_j, R_j) is the position and posture for each part. As for the system, we can get the COG via the sensors and calculate $\vec{c}$ in real time, which is significant for the calculate stability margin of robot system.

9.4.2 Joint servo system

This chapter should undertake external bone single robot joint adaptive control, dynamic adjustment of a proportional-integral-derivative (PID) control parameters according to the system response, so get a single joint servo system control model is necessary. Servo motor directly connected to screw shaft, screw rotates to drive the slider movement, the slider drive the connecting rod drive humanoid robot legs do movement.

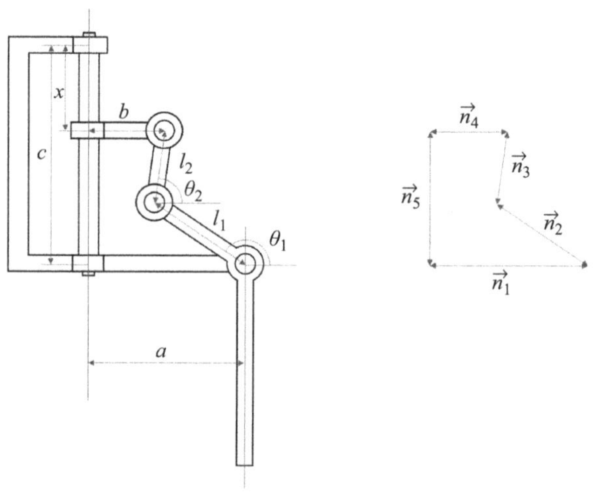

Figure 9.14 Single DC motor servo system

As shown in Figure 9.14, the loop-closure equation of the joint actuation gives

$$\begin{aligned}
&\sum_{i=1}^{5} \vec{n}_i = 0 \\
&\Rightarrow \begin{cases} a + l_1 \cos\theta_1 + l_2 \cos\theta_2 - b = 0 \\ x + l_1 \sin\theta_1 + l_2 \sin\theta_2 - c = 0 \end{cases} \\
&\Rightarrow l_2^2 = (b - l_1 \cos\theta_1 - a)^2 + (c - l_1 \sin\theta_1 - x)^2 \\
&\Rightarrow x = -\sqrt{l_2^2 - (b - l_1 \cos\theta_1 - a)^2} + c - l_1 \sin\theta_1
\end{aligned} \tag{9.7}$$

Differentiating both sides of the above equation yields:

$$\dot{x} = \left[\frac{l_1 \sin\theta_1 (b - l_1 \cos\theta_1 - a)}{\sqrt{l_2^2 - (b - l_1 \cos\theta_1 - a)^2}} - l_1 \cos\theta_1 \right] \dot{\theta}_1 \tag{9.8}$$

Reducer have the following relationship with screw lead s:

$$\dot{x} = \frac{s}{n \cdot i} \dot{\theta}_m \tag{9.9}$$

According to conservation of energy, we have

$$\frac{\pi M n}{30} \cdot \eta = T \cdot \dot{\theta}_m \tag{9.10}$$

The motor output torque is calculated with:

$$\begin{cases} M = \dfrac{30 i \cdot D(\theta_1)}{\pi \eta s} \cdot T \\ D(\theta_1) = \dfrac{l_1 \sin\theta_1 (b - l_1 \cos\theta_1 - a)}{\sqrt{l_2^2 - (b - l_1 \cos\theta_1 - a)^2}} - l_1 \cos\theta_1 \end{cases} \tag{9.11}$$

where T is joint output torque, M is the motor output torque, η is the kinematic chain comprehensive efficiency, s stands for screw lead, and i is reduction ratio for the reducer.

The moment of inertia of joints is the sum of moments of inertia of all bodies in a single joint structure, mainly including the slider, screw, shaft coupling, connecting rod, and turning block k. The system total moment of inertia is found as

$$\begin{aligned} J_{\text{load}} &= J_h + J_s + J_l + J_1 + J_K \\ &= M_h \left(\frac{t}{\pi} \right)^2 + \frac{\pi \rho_s d_s^4 l_s}{32} + \frac{\pi \rho_l d_l^4 l_l}{32} + \frac{1}{3} m_1 l_1^2 \end{aligned} \tag{9.12}$$

Among them, t for screw lead, M for the slider and the traction part weight, ρ_s and ρ_l, respectively, for the screw and the density of coupling, d_s and d_l are screw and the diameter of the shaft coupling, l_s and l_1 for screw length, m_1 and m_k rod l_1 and the quality of the block k, respectively, and the l_1 and l_k rod l_1 and the length of the block k, respectively.

The motor torque is calculated by

$$T_{\text{motor}} = (J_{\text{motor}} + J_{\text{load}}) \cdot \frac{d\omega_n}{dt} + T_{\text{load}} \tag{9.13}$$

On the other hand, the voltage equation is as follows:

$$U_m = I_m R_m + L_m \frac{dI_m}{dt} + \omega_n K_u \tag{9.14}$$

where U_m is motor armature voltage, I_m is armature current, L_m is armature inductance, and R_m is motor electronic resistance. K_u is the counter potential coefficient of motor.

The speed and current double closed-loop servo system is defined as shown in Figure 9.15. In order to facilitate speed controller, we can put the motor model and current closed loop of unification as a controlled object. In Figure 9.15, ω^* stands for a given motor speed, ω for the actual output of the motor rotor speed, and α is current feedback amplification coefficient.

Ignoring the counter electromotive force produced by the motor armature, the current loop of the closed-loop transfer function is

$$G_l = \frac{\left(\frac{1}{L_s s + R_s}\right)\left(K_{pl} + \frac{1}{T_{il} s}\right)}{1 + \alpha\left(\frac{1}{L_s s + R_s}\right)\left(K_{pl} + \frac{1}{T_{il} s}\right)} \tag{9.15}$$

As in the current controller the time constant is very small, we can put the current controller as proportion link. The current loop of the closed loop transfer function can be approximated into first-order inertia link, and define T_n as the time constant, for K_n proportion coefficient. We can thus simplify it as

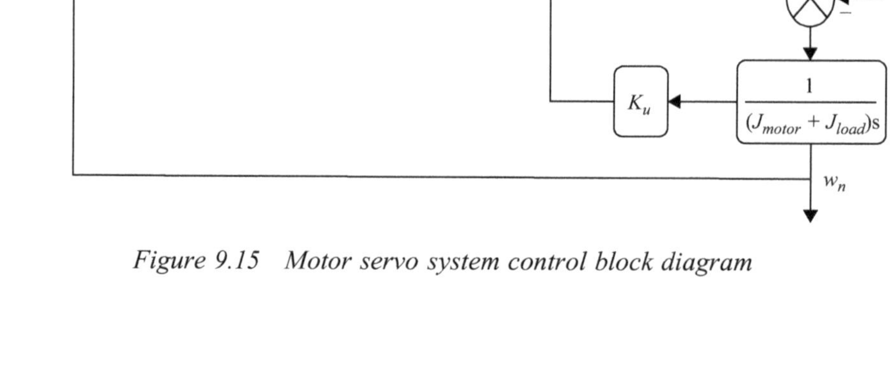

Figure 9.15 Motor servo system control block diagram

$$G_l(s) = \frac{K_{pl}}{L_s s + \left(\alpha K_{pl} + R_s\right)} = \frac{K_n}{T_n s + 1} \tag{9.16}$$

where

$$K_n = \frac{K_{pl}}{\alpha K_{pl} + R_s}, \ T_n = \frac{L_s}{\alpha K_{pl} + R_s}$$

The motor system's transfer function finally becomes

$$G_v(s) = \frac{K_n K_e}{s(T_n s + 1)(J_{\text{motor}} + J_{\text{load}})} - \frac{T_l}{s(J_{\text{motor}} + J_{\text{load}})} \tag{9.17}$$

The traditional PID control is matured with high reliability, simple algorithm, and easy parameter setting. But it relies on a known system model, and parameter setting is limited to a certain range. As an alternative, fuzzy control method can be considered for real-time control, which provide certain adjustments to the PID parameters and enables the controller applicable to a large load variation range, and thus to maintain good control performance and robustness.

The architecture diagram of fuzzy controller is shown in Figure 9.16, where $y^*(k)$ is k point reference speed, $y(k)$ is k times the actual speed. Adaptive controller input from the real-time operational status under system status in real time to calculate output values and adjusting k_p, k_i, and k_d values (Figure 9.17).

9.4.3 Control software

The software of this gait planner mainly focuses on finding the new support position of robot feet and adjust the gesture to ensure stability of the system. The processer sends angle values to actuators in real-time to control walking. At first, when the crutch is on the ground and the button on the crutch is pressed, the "start" command is sent to the processer by Bluetooth. Then the processor reads the relative position of feet and crutches (measured by the Large Area Tracking Untethered System). As a symmetric gait is adopted, we need to describe only half of one full gait cycle. The software will find the landing position E for robot leg based on relative distance between D and B (Figure 9.7). The procedures of the software processing are shown in Figures 9.18 and 9.19.

The software needs to calculate two times during a complete gait cycle. Each phase changes regularly. This software uses a finite state automation model to record

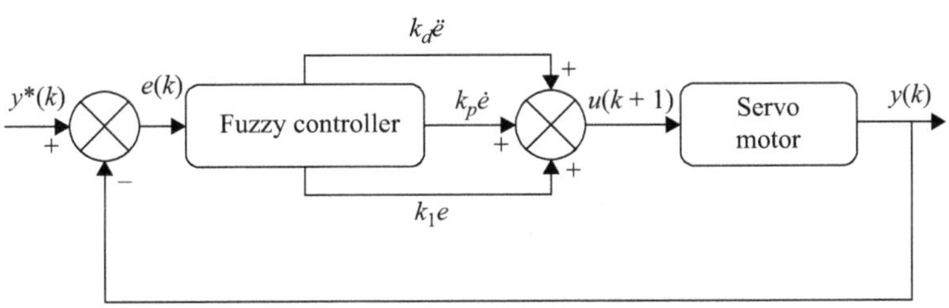

Figure 9.16 Control system based on fuzzy controller

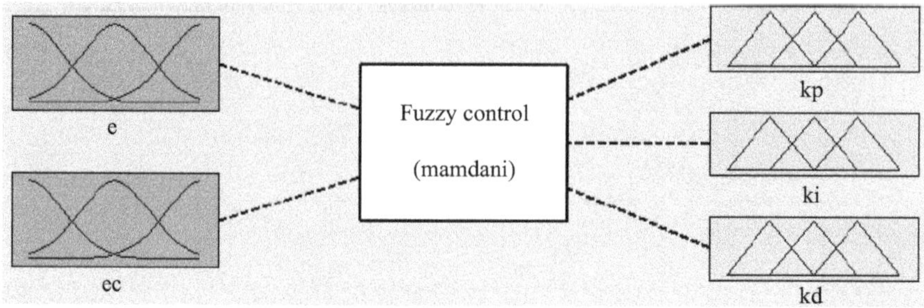

Figure 9.17 Fuzzy controller

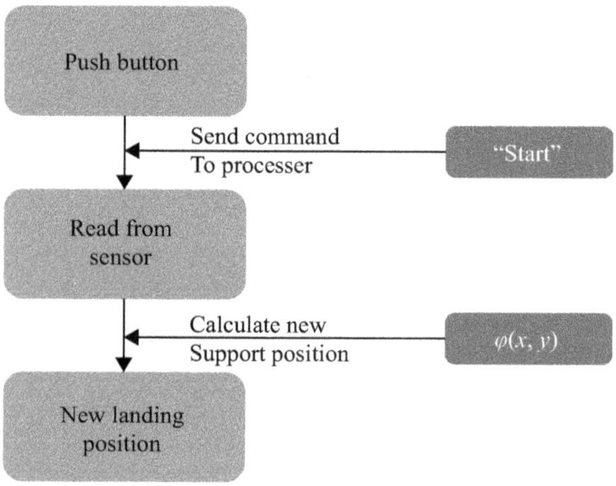

Figure 9.18 Procedures of getting the landing position

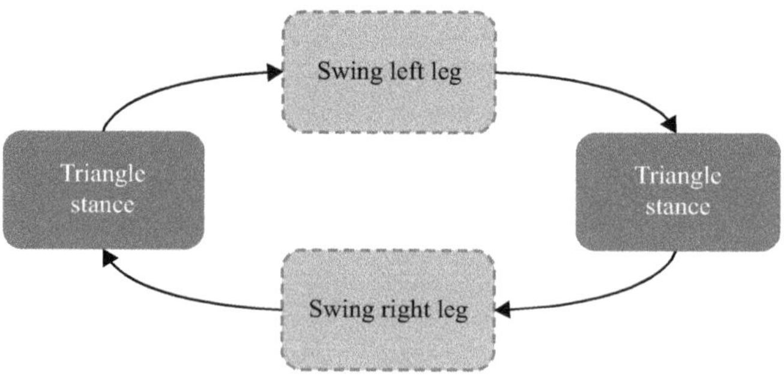

Figure 9.19 The walking cycle consists of the swing and stance phases

and determine current phase based on the sensor readings. In terms of the first triangle stance phase, the initial state of the system was mainly supported by two feet and left crutch, and the COP is contained in this triangle. Basing on this sensor information, the software can determine current phase is first triangle stance phase.

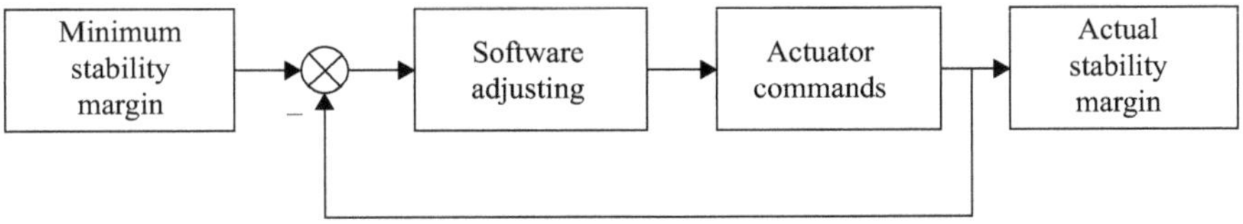

Figure 9.20 Control block diagram of adjusting the stability margin. The controller will adjust the gesture of the robot in the swing phase to ensure the system have enough stability margin to reject disturbance from outside

After the right crutch touches and supports on the ground and the button is pressed, the system enters the phase swing left leg, meanwhile the body moves forward.

After the landing position is determined, the processor need to send angle values to actuators to drive robot move forwards. The movement of body is very important for maintaining walking stability. Stability can be evaluated by the stability margin we defined before. The software will mainly generate the initial trajectories of the COG and each joint, which are determined by the landing position we calculated before, and control the trace of robot's COG in real time. In order to enhance the ability of disturbance rejection for this system, gait controller will adjust the step length in swing phase based on sensor readings, as shown in Figure 9.20. Feet also need moving smoothly, using a parabolic to represent the trace of foot, and keeping it raised to a certain level to avoid obstacle or ground friction.

9.5 Experiments and discussion

To prove that gait can be used to improve the stability of walking and rectify the gait in real-time, the gait planner has been testified on the SIAT exoskeleton at Shenzhen Institutes of Advanced Technology.

Two paraplegic patients involved in these experiments with this robot (Figure 9.21 shows a test with one patient). Each patient participated with walking experiments many times, and familiar with the way that control this robot start walking via the button assembled on the crutch. The walking speed is setting the same for each experiment. The experiment environment we select is a flat and level ground. The knee and hip joints are actuated in the sagittal plane, the reasonable landing position for each phase can be presented by the step length. So we need determine the reasonable step length. Moreover, for this static crutches walking gait, the gait stability largely depends on the stability margin during triangle stance phase. However, for this human–machine system, it is difficult to define a suitable step length which will make this gait become more coordinated and natural. But a more comfortable step length can be achieved based on patients own evaluation and reference the stability margin read from sensors. The stability and stability margin can be measured by sensor readings.

At first, we mark nine fixed points on the ground, which represent points of crutch support (Table 9.2). Then a patient uses their right crutch supports on the

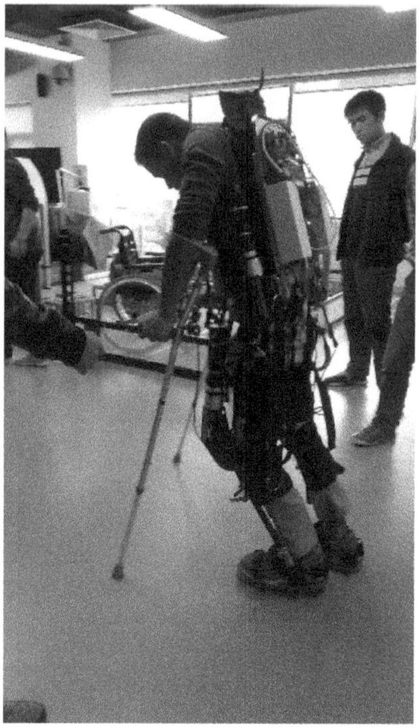

Figure 9.21 Test with paraplegia patient

Table 9.2 Average step length for left leg. The experiment data of a group is collected from a patient of 1.72 m tall, and another in B group from a patient of 1.65 m tall. The right crutch support points' coordinate relative to right foot listed in this table is the relative position between right crutch and robot left leg

Number	Right crutch coordinates	Step length (mm) group A	Step length (mm) group B
1.	(325, 330)	262	288
2.	(340, 330)	318	346
3.	(355, 330)	369	383
4.	(325, 345)	222	229
5.	(340, 345)	258	263
6.	(355, 345)	281	296
7.	(325, 360)	143	182
8.	(340, 360)	190	208
9.	(355, 360)	231	242

selected point, in the condition that patients use their right crutch support on the same position in each gait cycle. We change gradually the step size to determine the step length. Each patient repeated walking five times to get average results. Finally, we record the reasonable step lengths of left foot during walking.

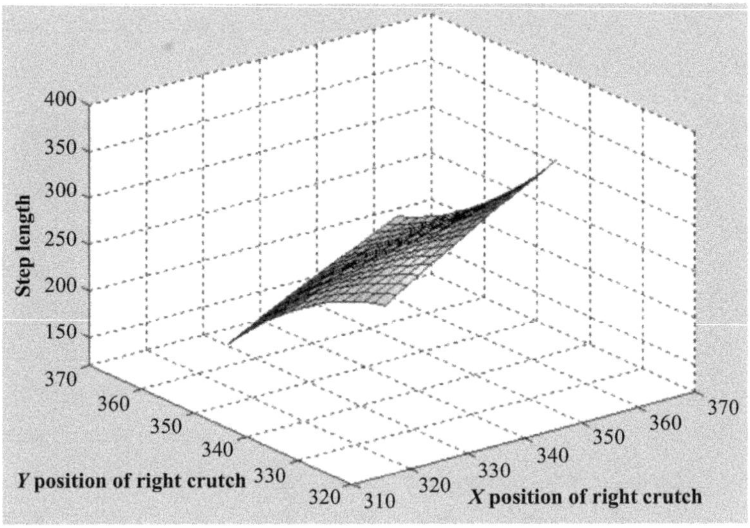

(a)

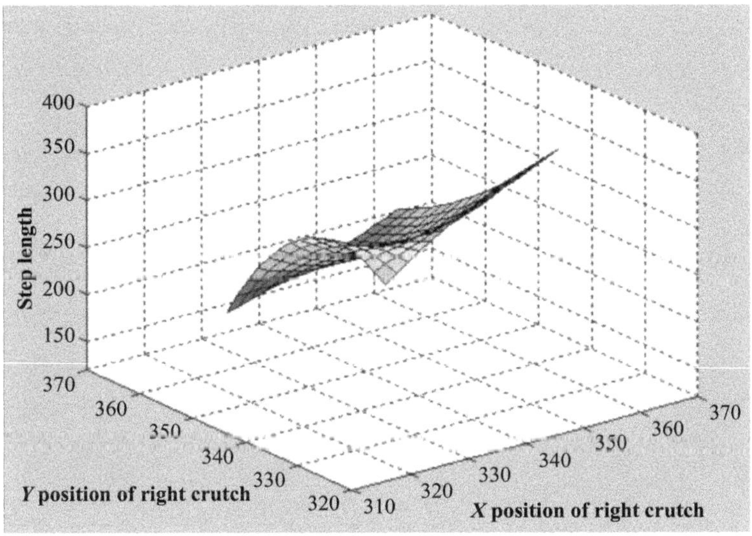

(b)

Figure 9.22 Fitting experiment data. Surface A is generated from data with the person of 165 cm tall, and B from the person of 172 mm tall

We mainly collected the step length of left foot as the change of support point of right crutch. And the other side is similar. In order to predetermine the step length after the crutches support on the ground, we use a surface to fit the date. As shown in Figure 9.22, we can predetermine the step length from this surface. The boundary of this surface is determined by the actual conditions. The maximum step length is 421 mm, which is mainly restricted by the mechanical structure.

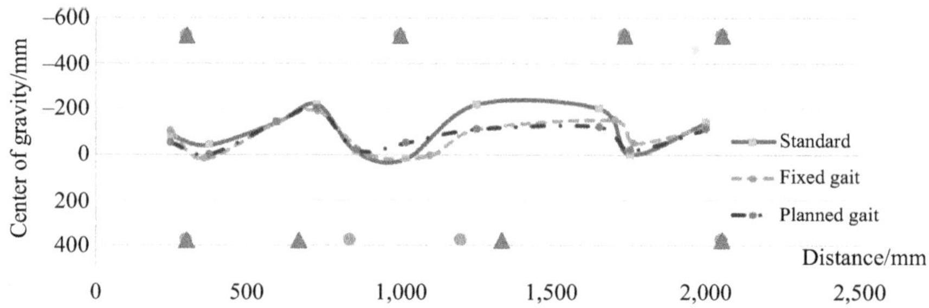

Figure 9.23 Comparison of two methods

Figure 9.24 COG variation for a whole gait cycle

And the minimum step length we defined is 120 mm. These boundaries will also ensure this system walking natural.

In the second experiment, we compared the two methods of variable step length gait and fixed step length gait. The trajectory of the weight can be measured through sensor readings. As shown in Figure 9.23, these curves represent the trajectories projected on the ground. The dashed curve represents the ideal gait, which gait the robot feet and crutches support on the given position (little triangle icon is the landing position for crutches). We expect that the patient can achieve this perfect gait by adequate training. However, usually people cannot support crutches on the ideal point we given previous, they often landing crutches on a random position such as the little round icon shown in Figure 9.23. The solidline curve represents the person use fixed step length gait, and the dash dot curve represents the one use variable step length gait. It is easy to known from this curve, the solidline curve indicates that this system need to do more walk to adjust their weight keep their stability. Especially during patient training. So we can achieve that the variable step length gait is smoother and labor saving that the fixed step length gait.

The data obtained by a paraplegic patient who wearing this robot through this real-time gait we proposed, and average walking speed is 0.41 m/s as shown in Figures 9.24–9.26. The gait in SIAT robot is not symmetric, so the gait data we

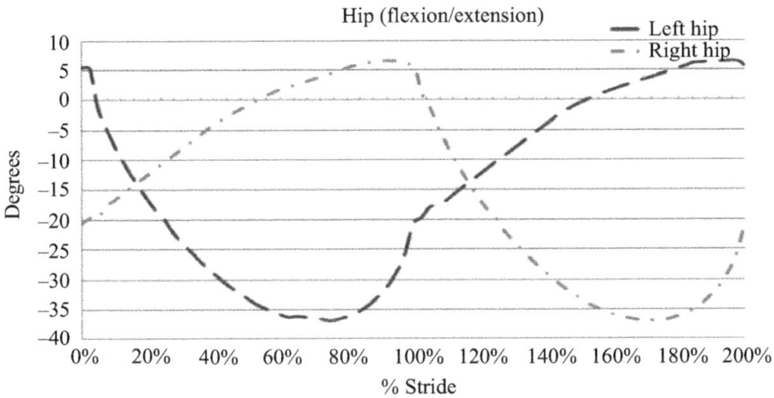

Figure 9.25 Hip flexion/extensions for a whole gait cycle

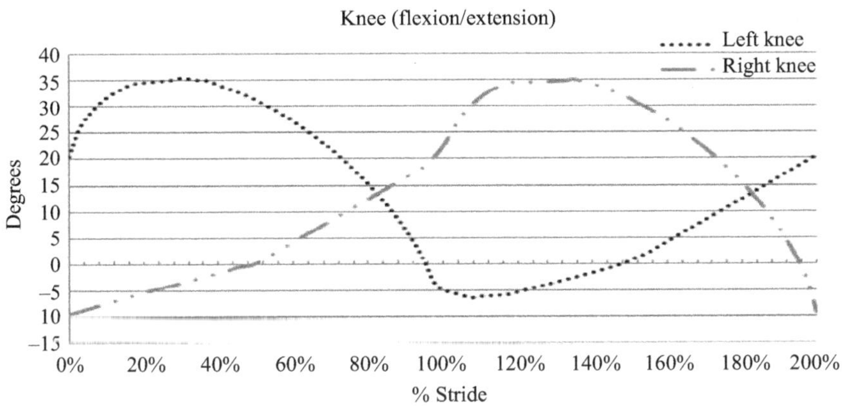

Figure 9.26 Knee flexion/extensions for a whole gait cycle

chose is not symmetric too. Graphic data were generated from a whole gait cycle. Turning points on the curve show that the controller adjusts the trunk pose to ensure sufficient stability margin.

Even if this gait is not asymmetry, but the movement of weight and rotation of joints hip and knee is regularly. And the time of triangle stance phase largely determine the walk speed. For this exoskeleton controlled by the button on the crutch, so the walking speed also depends on using proficiency. After a period of training, walking speed can achieve to 0.45 m/s.

9.6 Conclusions

The real-time crutches-walking gait presented in this chapter can make the exoskeleton walk in a coordinated and natural manner. This gait offers more stabilize margin for the system, so this gait can correct the external disturbances. The gait planner has been testing in clinical trials and has helped several paraplegic patients walk again. The walking speed is much slower than normal people, due to the reason coming from the mechanic design and proficiency of patients using and so on. This method by changing the step length adaptively will help patients learn to use this medical exoskeleton faster, and they will not fall even in the case of unreasonable using of crutches in some extents.

References

[1] K. A. Strausser and H. Kazerooni, "The development and testing of a human machine interface for a mobile medical exoskeleton." pp. 4911–4916.

[2] M. Regan, R. Teasell, D. Wolfe, D. Keast, W. Mortenson, and J. Aubut, "A systematic review of therapeutic interventions for pressure ulcers after spinal cord injury," Arch. Phys. Med. Rehabil., vol. 90, pp. 213–231, February 2009.

[3] K. H. Low, X. Liu, C. H. Goh, and H. Yu, "Locomotive control of a wearable lower exoskeleton for walking enhancement," J. Vib. Control, vol. 12, pp. 1311–1336, December 2006.

[4] J. G. Juang, "Fuzzy neural network approaches for robotic gait synthesis," IEEE Trans. Syst. Man Cybern. B Cybern., vol. 30, pp. 594–601, August 2000.

[5] I. H. Jang, J.-Y. Jung, D. Y. Lee, D. W. Lee, and H. S. Park, "Crutch gait pattern for robotic orthoses by the use of feature extraction," Artif. Life Rob., vol. 16, pp. 262–265, September 2011.

[6] D. Sanz Merodio, M. Cestari Soto, J.C. Arevalo, and E. García Armada, "Control motion approach of a lower limb orthosis to reduce energy consumption," Int. J. Adv. Robot. Syst. vol. 9, pp. 1–8, 2012.

[7] K. Suzuki, G. Mito, H. Kawamoto, Y. Hasegawa, and Y. Sankai, "Intention based walking support for paraplegia patients with robot suit HAL," Adv. Robot. vol. 21, pp. 1441–1469, 2007.

[8] P. D. Neuhaus, J. H. Noorden, T. J. Craig, T. Torres, J. Kirschbaum, J. E. Pratt, and Ieee, "Design and Evaluation of Mina a Robotic Orthosis for Paraplegics," 2011 Ieee International Conference on Rehabilitation Robotics, International Conference on Rehabilitation Robotics ICORR, New York: Ieee, 2011.

[9] H. K. Kwa, J. H. Noorden, M. Missel, T. Craig, J. E. Pratt, P. D. Neuhaus, and Ieee, "Development of the IHMC mobility assist exoskeleton," Icra: 2009 Ieee International Conference on Robotics and Automation, Vols 1–7, IEEE International Conference on Robotics and Automation-ICRA, pp. 1349–1355, New York: Ieee, 2009.

[10] T. Ikehara, E. Tanaka, K. Nagamura, *et al.*, "Development of closed fitting-type walking assistance device for legs with self-contained control system," J. Robot. Mechatron. vol. 22, pp. 380, 2010.

[11] S. Li, C. W. Armstrong, and D. Cipriani, "Three-point gait crutch walking: variability in ground reaction force during weight bearing," Arch. Phys. Med. Rehabil., vol. 81, no. 1, pp. 86–92, 1989.
[12] S. Bai, and K. H.Low, "Terrain evaluation and its application to path planning for walking machines," Adv. Robot., vol. 15, no. 7, pp. 729–748, 2001.
[13] S. Bai, K. H. Low, "Path generation of walking machines in 3D Terrain", Proc. IEEE Conference on Robotics and Automation (ICRA 2002), pp. 2216–2221, Washington, 2002.
[14] J. P. Schmiedeler, "The mechanics of and robotic design for quadrupedal galloping," Ph. D Dissertation, The Ohio State University, 2001.
[15] Wang, Pengfei, and L. Sun. "The stability analysis for quadruped bionic robot." Ieee/rsj International Conference on Intelligent Robots and Systems IEEE, 2007:5238–5242, 2007.

Chapter 10
Soft wearable assistive robotics: exosuits and supernumerary limbs

Lorenzo Masia[1], *Irfan Hussain*[2,3], *Michele Xiloyannis*[1], *Claudio Pacchierotti*[4], *Leonardo Cappello*[5], *Monica Malvezzi*[2], *Giovanni Spagnoletti*[2], *Chris Wilson Antuvan*[1], *Dinh Binh Khanh*[1], *Maria Pozzi*[2], *and Domenico Prattichizzo*[2,6]

Abstract

The intrinsic soft nature of compliant supernumerary limbs and exosuits makes them appealing candidates for assisting human movements, with potential applications in healthcare, human augmentation and logistics. In the following chapter, we describe the technology used in exosuits and supernumerary limbs for assistance of activities of daily living, with emphasis on aiding grasping and flexion/extension of the elbow joint. We discuss the mechanical design principles of such devices, detail the control paradigms that can be used for intention-detection and present the design and evaluation of cutaneous interfaces used for force feedback rendering. Tests on healthy and impaired subjects highlight that exosuits and supernumerary limbs are potential cost-effective and intrinsically safe solutions for increasing the capabilities of healthy subjects and improving the quality of life of subjects suffering from motor disorders.

Keywords: Assistive devices; wearable robots; supernumerary limbs; rehabilitation robotics; medical robots; soft robotics

10.1 Introduction

The introduction of compliant elements in robotics revolutionized the path to the design of applications with human in the loop; series elastic actuation [1] was the

[1]Robotics Research Center, Nanyang Technological University, Singapore
[2]Department of Information Engineering and Mathematics, University of Siena, Siena, Italy
[3]Khalifa University Robotics Institute, Khalifa University of Science Technology and Research, Abu Dhabi, United Arab Emirates
[4]CNRS, Univ Rennes, Inria, IRISA, France
[5]The Biorobotics Institute, Scuola Superiore Sant'Anna, Pisa, Italy
[6]Department of Advanced Robotics, Istituto Italiano di Tecnologia, Genoa, Italy

ignition to this new approach which has been widely accepted by a large part of the robotic community that considered the use of compliant elements as the natural solution to solve the problem of human–robot interaction.

Haptics, robot-aided rehabilitation and assistive technology embraced the new idea of designing devices which included compliance among the joints and at the end effector, with the final purpose to optimize the human interaction by reducing the contact force and filtering fast and abrupt dynamics. Literature reports a multitude of devices employing series elastic [2], pneumatic [3], and variable impedance actuators [4] which found their natural application in rehabilitation [5], telemanipulation [6], and in general for all those tasks which can benefit from force limitation and fine regulation of the interaction at the end effector.

Compliant actuation opened a new scenario, alternative to the classic robotics precept "stiffer is better"; a new control approach was adopted with the purpose of transforming the force problem into a position problem where the deformation of the compliant element was the main indicator of the interaction forces between the human and robotic device. Rehabilitation robotics and assistive technology were the applications which benefited most from the new gentle approach. Worldwide scenario comprises two different approaches which are mainly driven by the residual capacity of motion from the final users.

Technology for rehabilitation and assistance has principally targeted stroke as the leading cause of permanent disability worldwide and its efficacy is strictly linked to the mutual interaction between the patient and the devices.

Despite the large diffusion there are still several drawbacks which are far from solved, especially in assistive technology. The aforementioned improvements in actuation and control solutions did not prevent the majority of devices from employing rigid components in parallel with the human biomechanics, with a resulting ergonomics and usability which are still far from optimal. Furthermore, the weight of the current devices requires unacceptable metabolic costs from the wearer, with a consequence of movement restriction imposed on the user's joints which may result in misalignment and parasitic torques on the articulations.

Soft wearable exoskeletons (or exosuit) and supernumerary robotics are the successive step to the introduction of compliant actuation, where not only the actuators but also the structure of the device itself is designed to be compliant. This new vision has been gradually arousing the interest of the robotic community, proving to be a viable complementary solution to rigid robotics. While not being suitable for the applications requiring large forces and torques where rigid devices still show higher performance, exosuits' and supernumerary limbs' intrinsic compliance, portability, and low-power consumption make them ideal for partly augmenting the muscular strength or providing additional support in activities of daily living (ADL) such as walking [7–9], hand grasping [10], and stabilization tasks [11].

Exosuits overcome the limitations introduced by the conventional exoskeletons, where the lack of mechanical compliance in their kinematic structure represents the main factor limiting a wider diffusion of such technology.

As an alternative to the stiff links of conventional exoskeletons, the exosuit design comprises fabrics and meta-materials to connect the human limb to the actuation stage, while the support is demanded to the human musculoskeletal system: this solution results particularly suitable in those applications where the human biomechanics is able to provide a supporting structure while actuation and transmission can deliver torque and force across the human articulations [12].

There are not many examples of exosuits in the literature, and the majority is focused on lower limb. Due to the cyclic nature of walking, the great control challenge is triggering the assistance at the right phase of walking, and delivering it to the correct anatomical joints which are usually restricted in the sagittal plane of motion: examples of exosuit used in delivering assistance to both healthy and impaired subjects are from Panizzolo and Rossi *et al.* [8,13] where a Bowden cable-driven exosuit for ankle and hip assistance has been used on stroke subjects showing tangible benefits in improvement of metabolic costs and walking efficiency.

One of the most significant examples of an upper limb exosuit, which employed a fine model-based controller implementation, was from Ueda *et al.* and Ding *et al.* [14,15]; consisting in a wearable fully compliant exosuit at the shoulder/elbow joints. The device was driven by flexible pneumatic actuators with ends anchored to rigid plastic frames, and equipped with force and EMG transducers to apply force feedback control. The novelty of the design was in the integrated human–exoskeleton model used to compute the interaction between the human muscle forces and torques generated by the exosuit.

Other examples of wearable textile-garment based exosuits for hand rehabilitation come from Lee *et al.* [16] and In *et al.* [17], where and exotendon device and exo-glove, respectively, used cable actuation to promote the recovery of fingers coordination and restoring functional hand movement after stroke or spinal cord injury. The two devices used different approaches: the exotendon has been designed to be a rehabilitation device for clinical therapy, where the geometry and the disposition of the tendons, driven by seven motors, aimed at finely replicating different hand gestures (i.e., grasping, lateral grip, pinching, etc.) with a passive motion paradigm; the exo-glove relied on a more compact design, employing an underactuated mechanism by Jeong *et al.* and focusing its strength on the portability and versatility, where assistance was triggered by wrist motion detection.

Supernumerary limbs contrarily to the exosuits are not mounted in parallel respect to the human articulations but they are additional limbs. The main function of such architectures is not intended to provide additional torque at the level of human joints but mostly to replace the lost functionality, improving the dexterity of the users.

Supernumerary robotic limbs (SRLs) were initially conceived for industrial applications; attached to the human waist to support the body, powered joints were attached to the human joints and are constrained to move together with the human limb. The SRLs are designed to take a set of postures to maximize the load bearing efficiency. The SRLs can bear a large load with small power consumption and provide the users with additional dexterity which can facilitate a specific set of tasks.

Supernumerary limbs have been also been recently employed on impaired subjects: more specifically when the level of neurological impairment is severe and the patient's residual capacity of motion is too low, supernumerary robotics represents a complementary option to regular exoskeletons and exosuits. For example, patients affected by hypertonia or muscular atrophy due to a prolonged limb immobilization, will be unable to use an exoskeletal structure, because of the lack of biosignals useful to drive/trigger the exosuits, and their compromised biomechanics which avoid the exoskeleton to properly works (i.e., joint misalignment and residual muscular cocontraction).

For the above-mentioned reasons, supernumerary limbs represents a viable option to provide severely affected patients with an additional dexterity, which is able to partly restore the functionality of a limb and allow them to perform basic ADL which before would have been almost impossible.

The following sections will describe the technology used in exosuits and supernumerary limb applications. Examples and previous work from the authors will provide an overview of the potentials and limitations of such technology, highlighting their complementarity and possible future trends.

10.2 Exosuits

A wealth of devices has been engineered to assist the upper-limbs in physical therapy [18–27] mostly consisting of load-bearing exoskeletons made of rigid links that operate in parallel to the human skeleton.

One of the most common limitations of these exoskeletons is given by the kinematic constrains imposed on the wearer's joints by the rigid frame. Misalignment between the robot's joints and the biological ones results in hyperstaticity [28], i.e., the application of uncontrolled interaction forces, which upsets the natural kinematics of human movements. Various methods have been proposed to avoid hyperstaticity, such as adding passive DOFs [29], self-aligning mechanisms [30], or remote centers of rotation [31]. These solutions come at the cost of increasing the size and mass of the device, which makes it intrinsically unsafe and thus unsuitable for unsupervised, at-home use.

A recent and promising paradigm consists of delivering forces to the human skeletal system by means of soft, clothing-like frames powered either by pressurizable elastomeric actuators [3,32–35] or by Bowden cables moved by proximally located motors [12,17,36,37].

The use of clothing-like frames, known as *exosuits*, for transmitting forces to the human body represents an appealing solution for human motion assistance. Their intrinsic compliance, low profile and quasi-negligible inertia make them likely candidates for use on a daily basis. The absence of a rigid structure, moreover, avoids the joint-misalignment problem and makes the device completely transparent to human kinematics.

The downside of exosuits is their inability to apply high forces: being there no external rigid frame, loads are born by the wearer's joints. Nevertheless Quinlivan *et al.* have demonstrated that applying small forces with the correct timing during the walking cycle can reduce the metabolic cost of walking [9] up to 22.83% with no reported damage on the user's joints and In *et al.*, have experimented with a soft

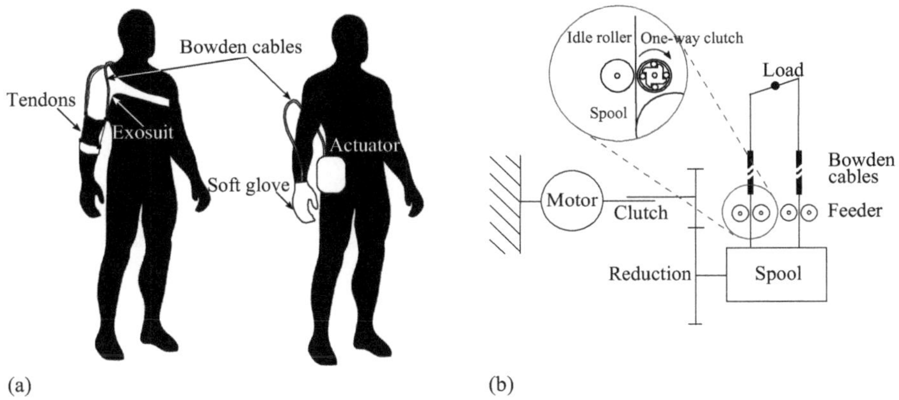

Figure 10.1 Components of the exosuits (a) and schematics of the actuation units (b). (a) Both devices comprise a wearable component (exosuit or glove) and a proximally located actuation unit that transmits power to the joints via Bowden cables. (b) DC brushless motors drive a spool around which the artificial tendons are wrapped; an electromechanical clutch allows to couple and decouple the motor from the transmission, allowing the user to move freely on demand. A feeder mechanism, thoroughly described by In et al. *[38], keeps the cables in tension around the spool*

glove with a bio-inspired tendon routing on a tetraplegic patient for restoring up to a 50-N hand grasp [17].

10.2.1 Design and actuation

The elbow and the hand exosuits comprise an actuation stage and a wearable component. The actuation stages are located proximally (e.g., in a belt around the waist) and transmit power to the suits via Bowden cables (Figure 10.1(a)).

Both tendon-driving units follow the same working principle, schematically shown in Figure 10.1(b): a brushless DC motor drives a spool around which one or more pairs of artificial tendons are wrapped in an antagonistic fashion so that rotation of the motor in one direction causes retraction of the agonist cable and releases its antagonist. An electromagnetic clutch is placed in between the motor shaft and the spool, coupling and uncoupling the motor and the end-effector when engaged and disengaged, respectively. This allows the controller to switch from a driving mode, where the exosuit is assisting its wearer, to a *transparent* mode, where the suit acts as a garment and allows free and unconstrained movements.

It is important to guarantee that the tendons don't slack around the spool. Pretensioning, a strategy commonly used in tendon-driven robots [39], is not a feasible solution due to the stress that a continuous force would introduce on human joints: rather, we adopt a feeder mechanism, thoroughly described in [38], that confines the slack outside of the actuation unit.

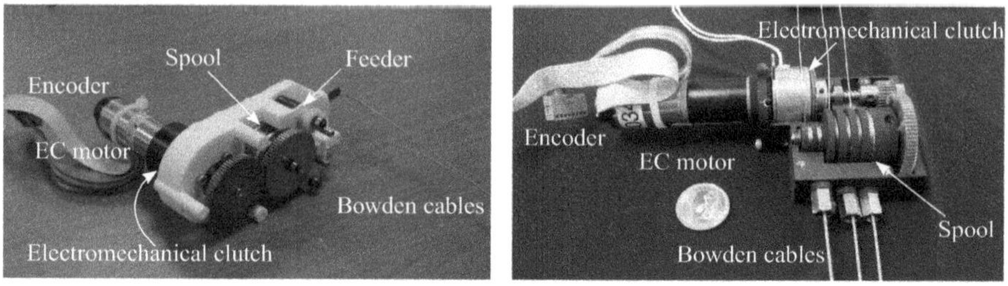

Figure 10.2 Tendon-driving units for the elbow sleeve (a) and the glove (b). Both actuators follow the design schematized in Figure 10.1(b), enclosed in a 3D printed plastic casing. The elbow and hand units weight, respectively, 750 and 420 g

The feeder mechanism comprises two idle rollers and two one-way clutches. The tendons pass between the rollers and the clutches. The one-way clutches' shaft is coupled to the spool's shaft with a spur gear. When the cable is released, the one-way clutches are engaged, thus pulling the cables out and keeping the tension around the spool. When the cable is retracted, the one-way clutches are idle and the feeder mechanism does not interfere with the rewrapping of the cable. In order to increase adhesion, a lining of urethane coating was added on the metallic surface of the one-way clutches.

The elbow actuator (shown in Figure 10.2(a)) comprises the following components: a brushless motor (Maxon EC-max, Ø 22 mm, 50 W) coupled to a gearhead (reduction of 33:1) and whose position is sensed by a quadrature encoder (Maxon Encoder MR, 512 CP), a spool around which two cables are coiled in opposite directions, a feeder mechanism and an electromagnetic clutch (Inertia Dynamics, SO11).

The two tendons, made of superelastic NiTi wire, were routed from the actuator unit on the backpack to the elbow joint through an outer cable housing (Nokon, Sava Industries). The whole mechanism is enclosed in a 3D printed case in ABS plastic. The fundamental components of the actuator driving the cables of the soft glove for grasp assistance are shown in Figure 10.2(b). Schematically, this unit is exactly like the one for the elbow, modeled in Figure 10.1(b), with the only difference that it drives three pairs of antagonistic tendons instead of one. The device consists of a DC brushless motor (Maxon EC-max, Ø 22 mm, 25 W) equipped with a rotary encoder (Maxon Encoder MR, 512 CPT) and a planetary gearhead with a reduction of 23:1. A further 3:1 reduction between the motor and the spool shaft ensures that the electromechanical clutch (Inertia Dynamics, SO11, $\tau_{max} = 0.68$ N m) can withstand higher locking torques.

The exosuit for the elbow (shown in Figure 10.3) was designed by modifying a commercially available passive orthosis (MASTER-03, Reh4mat). The substrate of the suit, having the function of adhering to the body of the user and keeping it in place, is made of a three-layered fabric: an external layer used to attach hard

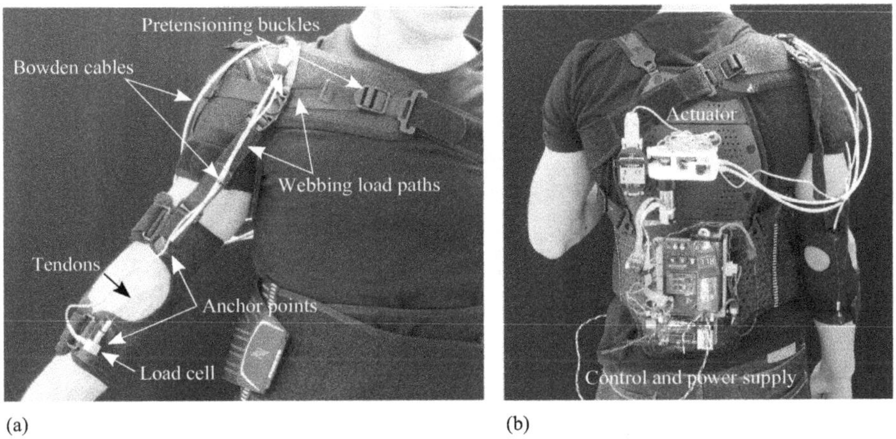

Figure 10.3 Anterior and posterior views of the exosuit for assisting elbow flexion/extension motions. (a) The artificial tendons are routed in a pair of Bowden sheaths from the actuator to the elbow joint; anchor points, on both sides of the joint, attach the tendons to the body, behaving like ligaments, and can be tightened with buckles and a Velcro strap. Loads are directed along paths made of unstretchable fabric that can be pretensioned around the body of the wearer upon donning. (b) The actuator, the electronics, and the power supply are carried in a backpack-fashion and weighs, overall, 2.1 kg

components (buckles and webbing strips), an intermediate ethylene-vinyl acetate foam cushions high loads and avoids high pressures and an internal 3D polyamide structure provides high air permeability and moisture absorption. Additionally, the arm bands are lined with a silicone pattern at the interface with the skin to prevent slipping. Load paths, i.e., the directions along which forces are transmitted through the fabric to the body, need to be as stiff as possible to maximize force transmission efficiency. They are thus made of webbing, i.e., nylon fibers woven in a flat strip, which is virtually inextensible and able to support high loads. To route the tendons along the load paths we sewed 3D printed components on the webbing network on both sides of each joint. These serve as artificial ligaments, anchoring the tendons to the body.

Finally, pretensioning the suit against the body, fundamental to avoid slipping and increase transmission efficiency, is achieved via buckles and Velcro straps around the arm and forearm and along the load-paths (Figure 10.3). In Figure 10.3(b), the subject is also wearing a harness designed to carry the actuation unit on his torso. The harness, that can be tightened through a set of buckles, loads the weight of the device ($\approx$2 kg including actuation, electronics, and power supply) on the wearer's shoulders.

The soft glove for grasping assistance (shown in Figure 10.4) is designed following the same principles. An elastic Lycra layer forms the substrate of the glove, ensuring a snug fit and keeping the anchor points in place while a neoprene layer

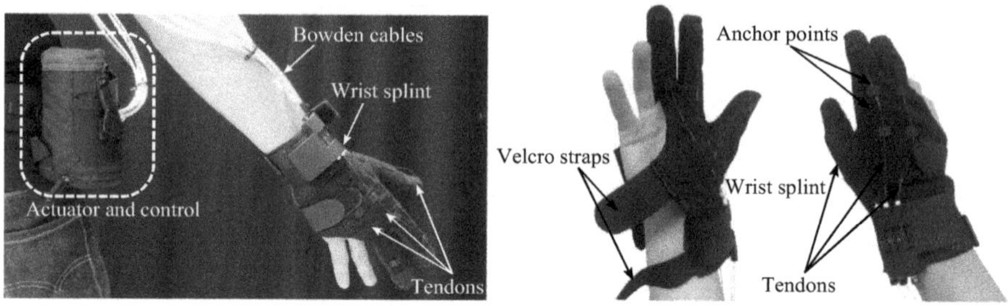

Figure 10.4 Soft glove for grasping assistance; dorsal and palmar view. The glove combines three different fabrics and rigid anchor points to be both comfortable and functional. A substrate in elastic fabric guarantees a snug fit, thus avoiding slipping of the anchor points during usage. A layer of neoprene, under the anchor points, avoids the application of high pressures on the wearer's skin. A pair of Velcro straps facilitates donning and doffing

Table 10.1 Weight and cost of the exosuit and the glove

	Elbow [kg]	**Hand [kg]**	**Elbow [US $]**	**Hand [US $]**
Actuation	0.750	0.420	970	662
Power supply	0.695	0.242	44	23
Electronics	0.270	0.270	222	222
Suit	0.397	0.205	123	97
Total	2.112	1.137	1,359	1,004

ensures comfort where the major forces are applied by the tendons. The anchor points, driving the tendons along the phalanxes and between the joints, are 3D printed in ABS and sewn on the fabric.

A wrist brace and the fingertip fittings are essential for an effective transmission of forces to the body since they are the only points where forces are applied normally to the skeletal structure. Specifically, the wrist brace loads the protruding trapezium and pisiform bones on the wrist and the fingertip fittings act on the distal phalanx of each finger. A pair of Velcro straps facilitates donning and doffing of the glove.

Table 10.1 summarizes the overall weight and materials cost of the exosuit and the glove.

10.2.2 Control

While the use of flexible materials for transmitting forces to the wearer presents many advantages, it also poses unquestionable control challenges: deformation of stretchable materials, friction in the Bowden cables and the viscoelastic properties of human soft tissues make a simple feedback control inadequate for achieving a

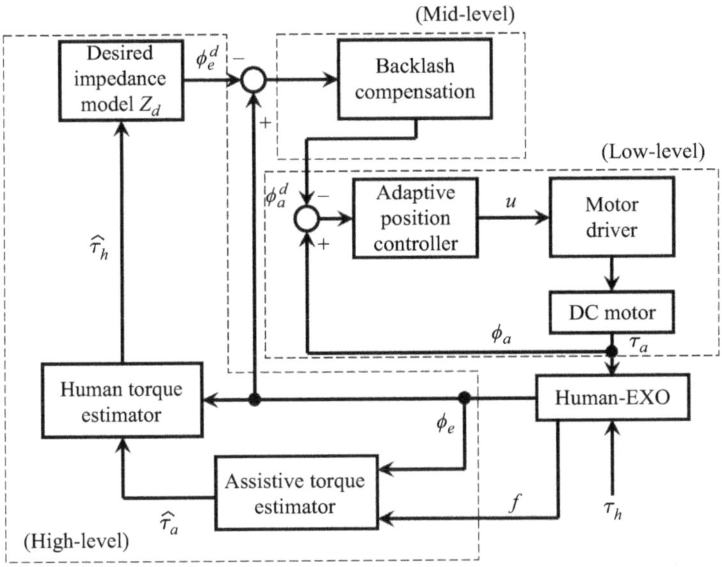

Figure 10.5 Schematics of the controller's architecture. The controller comprises three layers: a high-level one to decode the subject's intention and modulate the level of assistance; a middle-layer to compensate for backlash in the Bowden cable and a low-level controller to compensate for static and dynamic friction in the transmission

reasonable tracking accuracy. Moreover, understanding the intentions of the wearer is a key but challenging task.

We propose to tackle these issues by implementing a layered control paradigm which intuitively operates the exosuit. The proposed controller, shown in Figure 10.5, is referred to be *hierarchical* because it consists of three separate layers in cascade. We aim at considering all the aspects ranging from human intention detection for assistance evaluation (high layer), to adaptive compensation of unwanted effects arising from the presence of nonlinear behaviors in the exosuit (low layer). We test the proposed hierarchical control architecture in a trajectory following task and quantify its level of assistance on healthy subjects by monitoring the electromyographic (EMG) activity of the muscles involved.

10.2.2.1 High-level controller

The high-level controller deals with the task of understanding the intention of the user, converting it into an estimated torque at the joint and successively into a joint position using a dynamic model of the human arm. To do so, we used a mathematical model for Bowden-cable transmission that provides an acceptable estimation of the torque from the cable tensions [40] measured by load cells shown in Figure 10.3. Using simple geometrical considerations, we can derive the *extension functions* $h_i(\phi_e)$ mapping the cables (flexor and extensor) displacement to the joint angle ϕ_e. The flexor cable has an extension function $h_1(\phi_e)$ defined as (Figure 10.5)

$$h_1(\phi_e) = 2\sqrt{a^2 + b^2}\cos\left(\tan^{-1}\left(\frac{a}{b}\right) + \frac{\phi_e}{2}\right) - 2b \tag{10.1}$$

while the extensor cable is described by

$$h_2(\phi_e) = R\phi_e \tag{10.2}$$

where a is half of the width of the arm, b is the distance from the joint center of rotation to the anchor points (rigid braces), R is the radius of the elbow joint, and ϕ_e is the elbow joint angle. From the two extension functions $h_i(\phi_e)$, we can compute the relationship between the cable tensions f_i and the torque $\widehat{\tau_a}$ delivered to the elbow joint. By defining the matrix J as

$$J(\phi_e) = \frac{\partial h^T}{\partial \phi_e}(\phi_e) \tag{10.3}$$

where h represents the vector of cable extensions, the estimated assistive torque generated by Bowden-cable transmission $\widehat{\tau_a}$ is expressed by the following equation:

$$\widehat{\tau_a} = J(\phi_e)f \tag{10.4}$$

where f represents the vector of cable tensions.

The human arm dynamics, used for admittance controller, can be obtained by the Lagrangian formulation [41]:

$$\tau = \tau_h + \tau_a = \frac{2}{3}ml^2\ddot{\phi}_e + b_e\dot{\phi}_e + mgl_c\sin\phi_e \tag{10.5}$$

where τ is the resulting torque of the human action τ_h and the assistive torque from the exosuit τ_a, while ϕ_e, $\dot{\phi}_e$ and $\ddot{\phi}_e$ denote the measured elbow angular position, velocity, and acceleration, respectively; b_e is the viscous damping constant; and $g = 9.81$ m/s^2 represents the gravity constant. The torque deriving from the human muscles τ_h can be obtained from the inverse dynamic model expressed by (10.5), i.e.,

$$\widehat{\tau_h} = \frac{2}{3}ml^2\ddot{\phi}_e + b_e\dot{\phi}_e + mgl_c\sin\phi_e - \widehat{\tau_a} \tag{10.6}$$

where $\widehat{\tau_a}$ is the estimated assistive torque obtained from (10.4).

The parameters of (10.4) can be opportunely tuned on each individual subject following a set of simple known rules [42]: (1) The forearm mass m as 2.2% of the total body weight M, i.e., $m = 0.022M$[kg]; (2) the length l_c from the elbow joint to the center of gravity as 68.2% of the total forearm length l, i.e., $l_c = 0.682l$[m]; (3) the human elbow viscous damping coefficient b_e varies in a range of $0.4 - 2.5$ N m/(rad/s).

The exosuit, which was used in the present work, has been conceived as a device to provide assistance to impaired people, in particular to stroke subjects who do not preserve enough residual voluntary capacity of motion to lift their elbow. Hence after the human torque $\widehat{\tau_h}$ has been estimated by the aforementioned model.

A successive block in the controller must generate a reference trajectory ϕ_e^d which the actuation must deliver to the human joint. In order to provide a smooth intervention of the actuation on the user an admittance controller has been used: the admittance controller has the role of changing the overall dynamics, increasing the transparency of the exosuit itself and decreasing the mechanical impedance of the human biomechanics. The result is an assistance which is gradually modulated and depends on the mutual interaction between the device and the wearer. The admittance control block can be defined as follows:

$$I_e^d \ddot{\phi}_e^d + b_e^d \dot{\phi}_e^d + K_e^d \sin\phi_e^d = \hat{\tau}_h \tag{10.7}$$

where I_e^d, b_e^d, and K_e^d denote the desired inertial, viscous damping, and gravitational torque acting at the elbow joint, respectively, while $\hat{\tau}_h$ represents the estimated human torque obtained from (10.6). The gain of the admittance controller (I_e^d, b_e^d, and K_e^d) were chosen according to the following equation:

$$\begin{cases} I_e^d = \dfrac{2}{3} ml^2 \\ b_e^d = b_e \\ K_e^d = \alpha mgl_c \\ \alpha = \alpha_0 \tanh\left(\dfrac{\dot{\phi}_e}{\varepsilon_\alpha}\right) + \alpha_1 \end{cases} \tag{10.8}$$

where α_0 and α_1 are two constants experimentally chosen to bound the intervention of the assistance and ε_α denotes the sensitivity coefficient of the hyperbolic function tanh(.). The factor α increases with the measured joint velocity $\dot{\phi}_e$, meaning that the level of assistance is strictly dependent on the user's residual motion capacity: a high motion speed from the user (i.e., high motor ability) corresponds to a low assistive torque provided by the exosuit and vice versa, and the device is therefore able to gradually tune the assistive torque based on the a real-time estimation of the subject's capacity of motion.

10.2.2.2 Mid-level controller: adaptive backlash compensation

The purpose of the mid-level control layer is to compensate for the nonlinear backlash phenomenon typical of the Bowden-cable transmission. The desired angular motion ϕ_e^d from the admittance control block must be converted into a motion ϕ_a^d which is successively sent to the servomotor and delivered at the elbow joint. The mid-level controller is implemented with the purpose to specifically compensate for backlash phenomena. The relationship between the desired motion ϕ_e^d and the motion the actuation unit ϕ_a^d is defined adopting the Bouc–Wen hysteresis model [43], such that the desired elbow motion is a function of the actuator rotation and a term representing the backlash uncertainties:

$$\phi_e^d = \alpha_\phi \phi_a^d + D \rightarrow \beta\phi_e^d = \phi_a^d + \beta D \tag{10.9}$$

where α_ϕ represents the positive ratio of ϕ_e^d to ϕ_a^d; β is the inverse of α_ϕ; and D represents the model uncertainties due to the Bowden sheath's configuration variations during operation.

The adaptive controller design uses a reference motion ϕ_e^r and sliding surface s as in:

$$\begin{cases} e = \phi_e - \phi_e^d \\ s = \lambda e + \int_0^t e d\tau \rightarrow \dot{s} = \lambda \dot{e} + e \\ \phi_e^r = \phi_e^d - \lambda \dot{e} \end{cases} \tag{10.10}$$

where e represents the tracking error between the desired elbow joint motion ϕ_e^d and the measured one ϕ_e; and λ is an arbitrarily positive constant. Since the elbow joint is supposed to follow a given trajectory ϕ_e^d, the desired actuator state ϕ_a^d can be chosen as:

$$\phi_a^d = \widehat{\beta}\left(\phi_e^r - \widehat{D_m}\tanh\left(\frac{s}{\varepsilon}\right)\right) - \kappa s \tag{10.11}$$

where $\widehat{\beta}$ and $\widehat{D_m}$ are the estimated value of β and D_m, respectively (such notation will be used for all variables from now on); and κ, ε are positive constants.

Replacing ϕ_a^d from (10.11) to (10.9) leads to the dynamics of the sliding surface s as

$$\beta \dot{s} + \kappa s = \widetilde{\beta}\overline{\phi} - \beta\widetilde{D_m}\tanh\left(\frac{s}{\varepsilon}\right) - \beta\left(D_m \tanh\left(\frac{s}{\varepsilon}\right) - D\right) \tag{10.12}$$

where $\overline{\phi} = \phi_e^r - \widehat{D_m}\tanh\left(\frac{s}{\varepsilon}\right)$; $\widetilde{\beta} = \widehat{\beta} - \beta$ is the estimated error of β; and $\widetilde{D_m} = \widehat{D_m} - D_m$ is the estimated error of D_m.

Therefore, the adaptation law for backlash model parameters $\widehat{\beta}$ and $\widehat{D_m}$ is

$$\begin{cases} \dot{\widehat{\beta}} = -\delta_1 \overline{\phi} s \\ \dot{\widehat{D_m}} = \delta_2 \beta \tanh\left(\frac{s}{\varepsilon}\right) s \end{cases} \tag{10.13}$$

where δ_1 and δ_2 are positive adaptation gains. The initial values for $\widehat{\beta}$ and $\widehat{D_m}$ are set to be zero.

10.2.2.3 Low-level controller: friction compensation and position control

The low-level control layer is intended to drive the actuation stage by sending the input to the DC motor and to compensate for the nonlinear friction occurring because of the cable sliding along the Bowden sheath. The friction continuously and unpredictably changes according to the curvature of the sheath which moves with the subject arm motion; if not compensated, the torque generated by the actuator is partly lost during operation. An adaptive algorithm compensates for the friction and controls the DC motor in tracking the desired trajectory ϕ_a^d. The actuator with the friction is modeled as follows:

$$J\ddot{\phi}_a + B\dot{\phi}_a + \tau_f = u \tag{10.14}$$

where J and B represent the inertia and damping coefficient of the actuation stage, τ_f the friction torque, and u is the control output to be sent to the DC motor. The dynamic parameters of the actuation stage comprising the DC motor and the

Bowden cable and the friction variability are unknown: the Lugre model for the dynamic friction compensation [44], allows to express τ_f as follows:

$$\tau_f = B_v\dot{\phi}_a + \tau_z(\phi_a, \dot{\phi}_a, z) \tag{10.15}$$

where z is a variable in the Lugre model; $B_v\phi_a$ represents the viscous friction; and $\tau_z(\phi_a, \dot{\phi}_a, z)$ represents the dynamic friction depending on z. Before designing the control signal u, we define the tracking error e_1, reference motion $\dot{x}_a^r$, and sliding surface s_1 for the actuator as

$$\begin{cases} e_1 = \phi_a - \phi_a^d \\ s_1 = \dot{e}_1 + \lambda_1 e_1 \\ \dot{\phi}_a^r = \dot{\phi}_a^d - \lambda_1 e_1 \end{cases} \tag{10.16}$$

where ϕ_a and ϕ_a^d denote the measured and desired rotation of the DC motor, respectively; and λ_1 is an arbitrary positive constant. The control signal u for the DC motor is designed as

$$u = \widehat{J}\ddot{\phi}_a^r + \left(\widehat{B} + \widehat{B}_v\right)\dot{\phi}_a - \widehat{\tau_{zm}}\tanh\left(\frac{s_1}{\varepsilon_1}\right) - \kappa_1 s_1 \tag{10.17}$$

where $\widehat{J}$ is the estimated value of J; $\widehat{B}$ and $\widehat{B}_v$ denote, respectively, the estimated values of B and B_v; $\widehat{\tau_{zm}}$ denotes the estimated value of τ_{zm}; and ε_1 and κ_1 are two positive constants. Substituting (10.15) and (10.17) to (10.14), and replacing $B_t = B + B_v$ result in the dynamics of the sliding surface s_1 as (Table 10.2)

$$J\dot{s}_1 + \kappa_1 s_1 = \widetilde{J}\ddot{\phi}_a^d + \widetilde{B}_t\dot{\phi}_a - \widetilde{\tau_{zm}}\tanh\left(\frac{s_1}{\varepsilon_1}\right) - \tau_{zm}\tanh\left(\frac{s_1}{\varepsilon_1}\right) - \tau_z(\phi_a, \dot{\phi}_a, z) \tag{10.18}$$

where $\widetilde{J} = \widehat{J} - J$ is the estimated error of J; $\widetilde{B}_t = \widehat{B}_t - B_t$ is the estimated error of B_t; and $\widetilde{\tau_{zm}} = \widehat{\tau_{zm}} - \tau_{zm}$ is the estimated error of τ_{zm}. Therefore, the unknown parameters $\widehat{J}$, $\widehat{B}_t$, and $\widehat{\tau_{zm}}$ are updated at each period by

$$\begin{cases} \dot{\widehat{J}} = -\delta_3\ddot{\phi}_a^r s_1 \\ \dot{\widehat{B}}_t = -\delta_4\dot{\phi}_a s_1 \\ \dot{\widehat{\tau_{zm}}} = \delta_5\tanh\left(\frac{s_1}{\varepsilon_1}\right)s_1 \end{cases} \tag{10.19}$$

where δ_3, δ_4, and δ_5 are positive adaptation gains. Table 10.2 summarizes the identified control parameters averaged on three subjects.

Table 10.2 Average identified control model parameters

Parameters	**Values [unit]**
a, b, R	0.05 [m], 0.10 [m], 0.06 [m]
m, l, l_c	1.55 ± 0.12 [kg], 0.24 ± 0.03 [m], 0.2 ± 0.02 [m]
b_e	1.50 [N m/(rad/s)]
$\alpha_0, \alpha_1, \varepsilon_a$	0.5, 0.5, 0.1
$\lambda, \kappa, \varepsilon, \sigma_1, \sigma_2$	2, 10, 0.01, 2, 2
$\lambda_1, \kappa_1, \varepsilon_1, \sigma_3, \sigma_4, \sigma_5$	1.5, 5, 0.01, 1, 1, 1

10.2.3 Evaluation

The elbow exosuit was evaluated on three unimpaired subjects (average age: 26.6 ± 1.5), with the controller described above. The experiment aims at demonstrating that the proposed hierarchical controller is accurate, stable, and provides a smooth intervention when assistance is requested by the users, modulating the amount of torque at the elbow depending on the capacity of motion of the subject and decreasing the muscular effort during load manipulation.

During the experiment the subjects were instructed to lift their forearm, following the motion of a virtual avatar on a screen (as shown in Figure 10.6) while holding a 1-kg load in their left hand, in two distinct task phases: (1) performing 10 repetitive flexion/extension movements with and without assistance delivered by the exosuit to prove that the use of the proposed device/controller effectively decrease the muscular activity, helping the wearer to complete the task; (2) performing 10 repetitive flexion/extension movements at two different speeds ($\dot{\phi}_e^1 = 0.3$ rad/s and $\dot{\phi}_e^2 = 0.6$ rad/s), to show that the intervention of the exosuit is based on the subject's capacity of motion. It is commonly accepted that kinematics in stroke subjects is dramatically jeopardized [45,46] and a lower elbow speed is associated to a reduced voluntarily capacity of motion. For this reason, asking subjects to move at a lower and higher speed implies that the proposed controller should interpret a lower speed as a reduced motion capacity and consequently increase the level of assistive torque, respectively, as described in Section 10.2.2.1.

Muscular effort was estimated from the root mean square (RMS) of the EMG activity [47] of the main muscle involved in performing elbow flexion movements, i.e., the biceps brachii of the left arm. The raw EMG was acquired using Trigno wireless EMG sensors (Delsys Inc.) and was preprocessed in MATLAB™ Simulink® using a full-wave rectification, followed by a low-pass second-order Butterworth filter with a 8-Hz cut-off frequency. The elbow joint angle ϕ_e was acquired during the experiment and used for control purposes using a resistive flex sensor (Spectrasymbol, USA). Data acquisition and motor control were performed using the Quanser Quarc® real-time workstation running at 1-kHz refresh rate.

Results of the first task comparing the EMG activity of one subject with and without the assistance are depicted in Figure 10.7(a), showing a clear decrease in the amplitude of the EMG activity when the exosuit was assisting the motion. Analysis of the RMS value of the EMG signal, averaged over repetitions and subjects, is shown in Figure 10.7(a). The latter shows an average drop in RMS of 48.3% between the nonassisted and assisted case. Figure 10.7(b) shows the estimated assistive torque, as computed from (10.6) and the recorded EMG activity of a single flexion/extension task from one of the participants. This results in a trajectory profile as shown in Figure 10.7(b), with a tracking error smaller than 0.3 rad and a fast response time.

The second experiment examines the efficacy of the controller in adapting its contribution to the user's capacity of motion. The subjects performed the elbow movement at two different angular velocities ($\dot{\phi}_e^1 = 0.3$ rad/s and $\dot{\phi}_e^2 = 0.6$ rad/s) and the exosuit modulates the degree of assistance, i.e., the assistive torque, based on the speed of the subjects. Figure 10.8(a) and (b) shows the EMG activities and

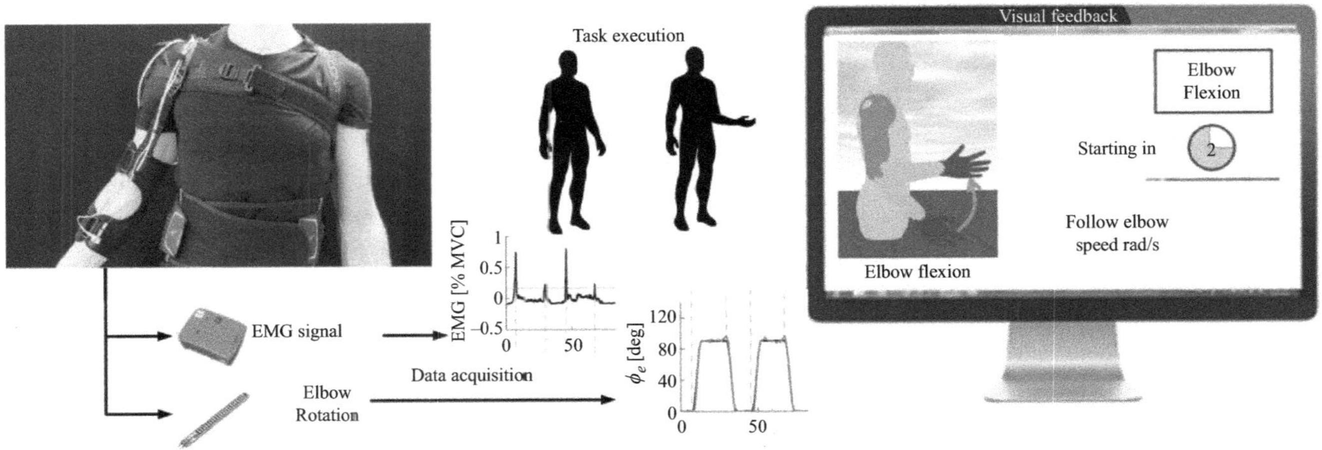

Figure 10.6 *The experiment was run by instructing the subjects to replicate the movements of a virtual avatar on the screen trying to match the elbow speed. EMG and elbow flexion ϕ_e were acquired and post processed to compare the muscular activities*

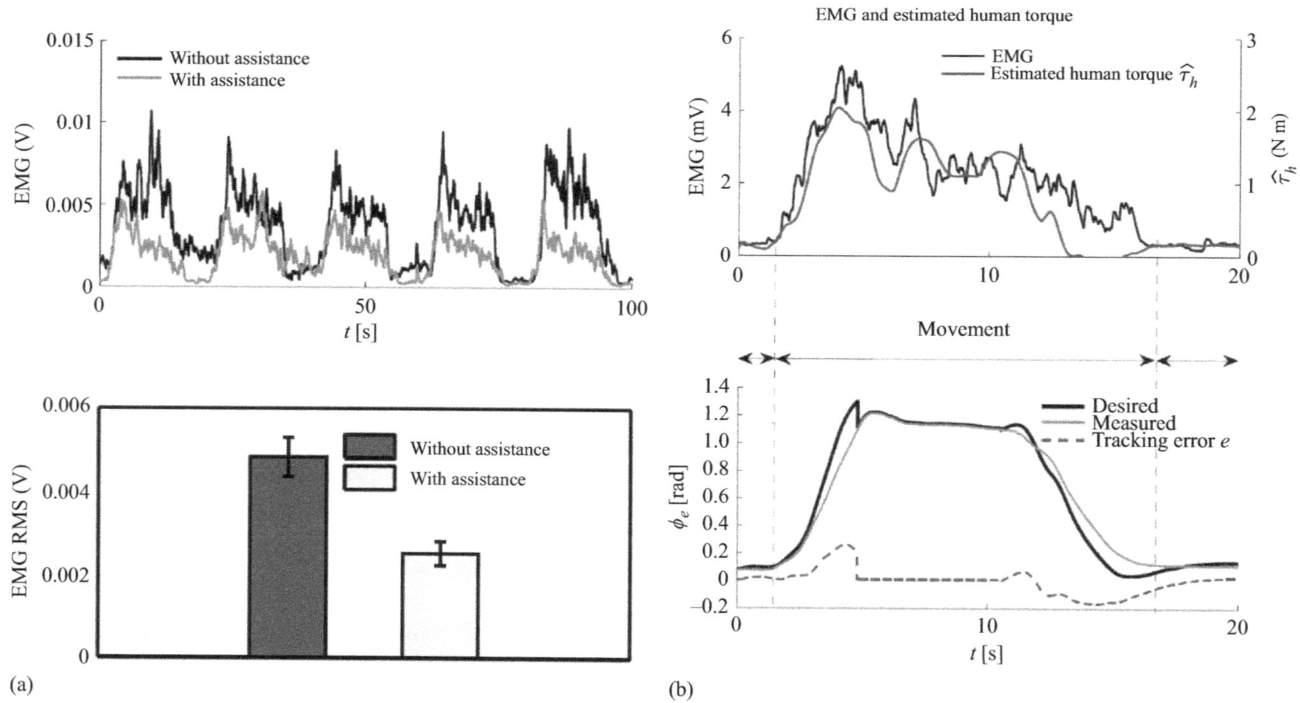

Figure 10.7 *Evaluation of the performance and assistance level of the proposed control paradigm. (a) The top plot shows the EMG amplitude comparison on a single subject with and without the exosuit's assistance, the bottom bar plot shows the mean and standard deviation of the RMS of the EMG activity, averaged over repetitions and subjects. (b) EMG and estimated human torque and trajectory tracking accuracy. The top plot shows the amplitude of the EMG signal (black) during a single repetition of a subject performing an elbow flexion/extension task and estimated human torque (gray) as computed from (10.6). The bottom plot shows the trajectory tracking accuracy during a single flexion/extension task*

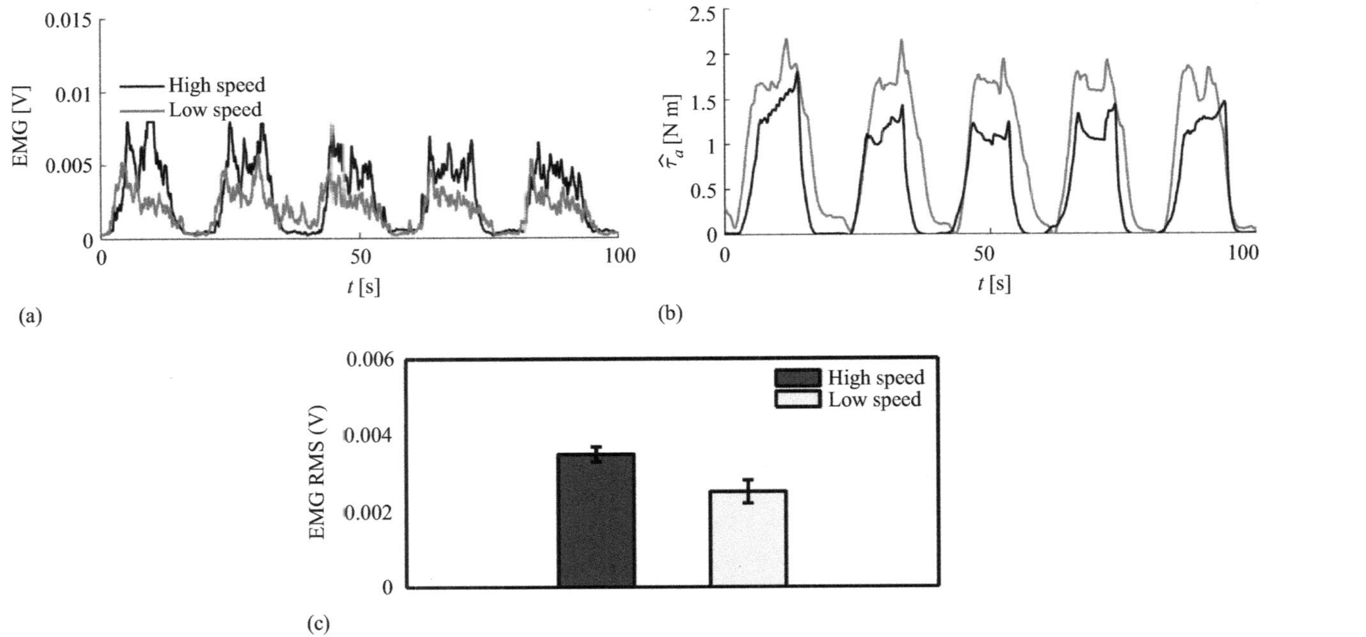

Figure 10.8 *Evaluation of the effectiveness of the assistance modulation at different elbow speeds. (a) The amplitude of the EMG activities is shown when a single subject performed the elbow motions at two different velocities. When the subject moves slower, (b) the assistive torques at a lower (light gray) and higher (dark gray) elbow speed motion. (c) Comparison of the EMG activity at low and high speed of movement with the proposed controller*

the delivered assistive torques $\hat{\tau}_a$, respectively, for one typical subject at the two execution speeds: it is observable that at lower speed (lower capacity of motion) the controller estimates a higher assistive torque than when the subject moves at the higher speed (which corresponds to a higher capacity of motion). The same result was observed for the whole group of subjects. An analysis of the EMG signals supports these results: Figure 10.8(c) shows that the mean RMS value of the EMG signal, averaged over subjects and trials, is lower at a lower speed of motion, confirming that the controller has adapted to the lower motor ability of the subjects by providing higher assistance.

10.2.4 Discussion

Despite the unquestionable advances achieved in the last 50 years in wearable assistive devices, current technologies are still far from being used on a daily basis. This is mostly due to their limitations in terms of portability, safety, ergonomics and, energy-wise, autonomy. Moreover, the cost of most of the developed exoskeleton make them prohibitive but for the most affluent users.

In this section, we presented the design and testing of a soft wearable exosuit for assisting elbow movements and hand grasping. Using fabrics and Bowden cables instead of traditional rigid transmissions would potentially result in cheaper devices, moreover making the device low-profile, lightweight, compliant, and less restrictive to the wearer's motion. We based our design on a set of documented force and motion requirements and kept the weight and size of the actuators as low as possible. Finally, we introduced a novel, hierarchical control paradigm that exploits sensory data to adapt its model of the system, adapting to the wearer's retained motor capacity and compensating for the nonlinear phenomena that make a simple linear control insufficient.

Despite having multiple advantages, exosuits rely on the wearer's skeletal structure to transmit compressive forces and are thus limited in the amount of assistance they can provide, especially if the wearer suffers from bone weakness caused by disuse osteoporosis, a common comorbidity of neuromuscular impairments [48]. This suggests that their effectiveness might be strongly dependent on the degree of retained motor ability of the patient. This point needs to be experimentally assessed: to the authors' knowledge, the only clinical criteria for the use of a soft wearable robot have been defined for the SEM Glove (Bioservo Ltd), which is recommended for patients with and action research arm test (ARAT) score between 10 and 35 and a stroke upper limb capacity scale (SULCS) score of 4–7 [49]. We are confident that our glove could prove to be useful for a slightly larger population since it actuates both flexion and extension of the fingers, while the SEM glove only aids gripping strength. No documented criteria, on the other side, exist for deciding the level of impairment that a soft elbow sleeve would be suitable for, thus tests with patients are of paramount importance for identifying the contribution of our technology.

In conclusion, while there is still a great need for improvement in the design, control and knowledge of their contribution, soft wearable devices for assistance have the potential of becoming a valid and cost-effective solution for increasing independence and quality of life of patients suffering from motor disorders.

10.3 Supernumerary limbs

Long-term disabilities of the upper limb affect millions of stroke survivors [50]. More than 80% of individuals who experience severe hemiparesis after stroke cannot completely recover hand and arm use [51]. The improvement of the paretic hand functionality plays a key role in the functional recovery of stroke patients with a paretic upper limb [52,53]. Different motor impairments can affect the hand both at motor execution and motor planning/learning level, including weakness of wrist/finger extensors, increased wrist/finger flexors tone and spasticity, cocontraction, impaired finger independence, poor coordination between grip and load forces, inefficient scaling of grip force and peak aperture, and delayed preparation, initiation, and termination of object grip [54]. In the last two decades, several rehabilitation teams have started integrating robotic-aided therapies in their rehabilitation projects. Such treatments represent a novel and promising approach in rehabilitation of the poststroke paretic upper limb. The use of robotic devices in rehabilitation can provide high-intensity, repetitive, task-specific, and interactive treatment of the impaired upper limb, and can serve as an objective and reliable means of monitoring patient progress [55–57]. Most of the proposed devices for hand and arm rehabilitation are designed to increase functional recovery in the first period after the stroke when, in some cases, biological restoring and plastic reorganization of the central nervous system take place [58]. However, even after extensive therapeutic interventions in acute rehabilitation, the probability of regaining functional use of the impaired hand is low [59]. For this reason, we recently started studying robotic devices for the compensation of hand function in chronic stroke patients, and in [60–62], we introduced a wearable robotic extra finger that can be used as an active compensatory tool for grasping objects.

The working principle of the proposed extra finger is rather simple and intuitive. The device can be worn on the paretic forearm by means of an elastic band, so that the robotic finger and the paretic hand can act like the two parts of a gripper working together to hold an object, see Figure 10.9. This solution represents the

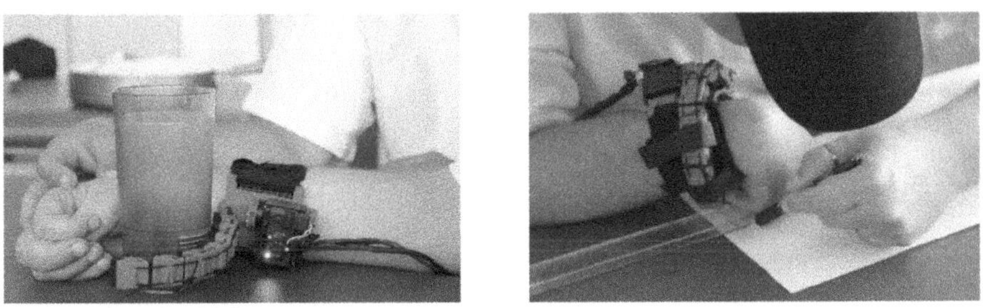

Figure 10.9 Working principle of the robotic extra finger. It cooperates with the paretic limb to compensate for hand grasp functionality (left), and it can be shaped into bracelet when not used (right)

minimum robotic complexity necessary to perform grasping tasks. To further improve its wearability, the finger can be shaped into bracelet when being not used. The user can control its flexion/extension through an EMG interface placed on the patient forehead [63] or through wearable switches [64], as described in Section 10.3.2. In [65], we showed how the supernumerary finger can be used in ADL involving common bimanual tasks such as opening cans and jars with different closing systems and shapes. Sections 10.3.1 and 10.3.2 present the principles that can be used to design and control supernumerary limbs and, in particular, focus on the choices that were made for the Soft-SixthFinger. The reader can refer to Section 10.3.5 for details on the application and on the performance evaluation of the device.

10.3.1 Design and actuation

Wearable assistive robotic devices used for clinical applications must meet specific human factors and performance criteria. The general guidelines which can be found in the literature include: durability, energy efficiency, low encumbrance, ease of use, error tolerance, and configurability [66].

Two key characteristics for supernumerary limbs are the wearability and the acceptability for the users, and, in order to improve them, the design must satisfy a number of conditions related to ergonomics and functionality [67]. The specific indexes to be used and the human ergonomics strongly depend on the actual patient conditions and needs [68,69]. Several experiments with patients, conducted in cooperation with a rehabilitation team, and reported in [62,63,65], led us to the development of the Soft-SixthFinger.

This supernumerary finger has been designed to be wearable, robust, and capable of adapting to different objects. In general, the robustness plays a 2-fold role: it makes the device capable of withstanding large impacts and other forces due to the unintended contact with the environment; it enables the robotic device to reliably grasp objects in presence of sensory uncertainty and unpredictable environment features. In previous works, robustness and soft interaction are achieved either by regulating the compliance of the robotic joints [70] or by tuning the intrinsic softness, acting on the passive characteristics of the robot bodyware [71–73]. The former approach is based on complex and bulky variable impedance actuators that are more suitable for other applications. Differently, our device is inspired by the latter approach in order to be simple, lightweight and compact. In particular, passive compliant joints and cable driven actuation guarantee robustness and safety during the interaction with the environment. These two characteristics not only assure that the device can endure collisions with hard objects and even strikes from a hammer without breaking into pieces, but also allow it to be highly adaptable to different object shapes, as it typically happens in underactuated and compliant hands [74]. In such types of robotic hands, in fact, transmission solutions that allow the motion of the other joints to continue after a contact occurs on a coupled link enable the device to passively adapt to objects using a reduced set of control parameters and in presence of uncertainties in sensing and actuation [75,76].

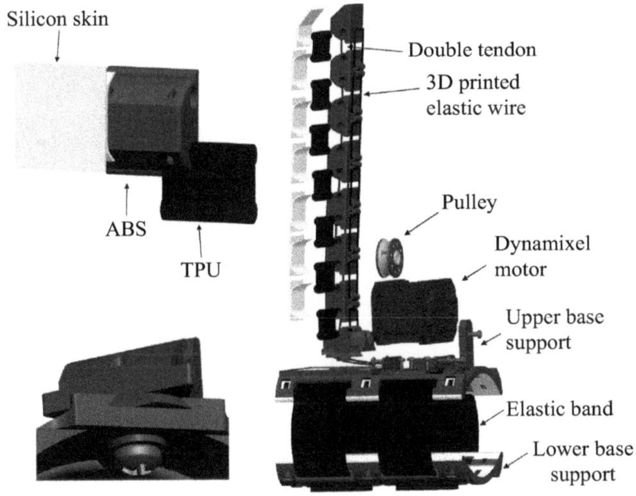

Figure 10.10 *The CAD exploded view of a module (top left), the passive locking mechanism (bottom left), and the complete Soft-SixthFinger (right) with its tendon holes, support base, and actuator*

As shown in Figure 10.10, the Soft-SixthFinger is composed of two main parts: a flexible finger and a support base, coupled through a passive locking mechanism needed to switch between working and rest position (Figure 10.9). The flexible finger is built with a modular structure. Each module consists of a rigid 3D printed link made of ABS (Acrylonitrile Butadiene Styrene, ABSPlus, Stratasys, USA) and covered with a silicon skin, and a flexible 3D printed joint made of thermoplastic polyurethane (NinjaFlex®). We selected polyurethane for the flexible parts because the high elongation of this material allows for repeated movement and impact without wear or cracking, proving also an excellent vibration reduction. Reasons for adding passive elements are manifold, including storing elastic energy, avoiding tendon slackness and ensuring the uniqueness of the position of the extra finger when not in contact with the object [77]. The modules are connected one to each other by sliding the thermoplastic polyurethane part inside the ABS one. This method allows to assemble the device in an easy way, without using any screw to combine the modules.

The device is developed by combining two different manufacturing processes: rapid prototyping 3D printing for the structure and molding for the silicon skin. The molding process shapes the raw material using a solid frame of a particular shape, called a pattern. We used 3D printed skeletons to hold the liquid silicon in the desired shape until it turned solid. We realized closed-top molds which are used for more complex part geometries. We pored the silicon mixture over the skeletons of the modules and used other mold's parts to constrain the liquid silicon to achieve the desired geometry and shape of the skin. Metal tubes were inserted into the module holes so to avoid silicon to fill the tendon holes. The silicon used is Fast

Rubber FR-18 which is bicomponent curing at room temperature. The mixing ratio of components is 100 g of base per 5 g of catalyst. It has viscosity of 30 Pa s and the final hardness is 17 ± 2 *shore A*. The silicon skin on the rigid links aiming to increase the friction at the possible contact areas.

The support base of the finger has been designed to guarantee a firm grip on the arm of the user and the wearability of the entire device. It consists of two parts coupled with Velcro strips to facilitate the wearing process and assure the adaptation to different arm sizes. The upper part contains the actuation system of the Soft-SixthFinger that consists of a single actuator and two tendons running in parallel through the modules of the finger. The cables (polyethylene Dyneema fiber, Japan) run through the finger and are attached on one side to the fingertip and on the other one to a pulley rigidly connected to the actuator shaft. When the motor is actuated, the tendon wires are wound on the pulley reducing the length of the wire and thus flexing the finger. As the motor is rotated in the opposite direction, the extension of the finger is achieved thanks to the elastic force stored in the flexible joints. Currently, we are using as actuator a Dynamixel MX-28T (Robotis, South Korea) driven by an ArbotiX-M Robocontroller [78].

10.3.2 Control

One of the major challenges in the development of extra limbs lies in finding suitable control interfaces for the integration of the robotic devices with the human. The choice of the control strategy to adopt, strictly depends on the design of the supernumerary limb and on its purpose. In [79], the authors present a control algorithm enabling a human hand, augmented with two robotic fingers, to share the task load together and adapt to diverse task conditions. Postural synergies were found for the seven-fingered hand comprised of two robotic fingers and five human fingers through the analysis of measured data from grasping experiments. Prattichizzo *et al.* [60] describe a mapping algorithm able to transfer to an arbitrary number of robotic extra fingers the motion of the human hand. The algorithm is based on the method proposed in [80], in which the mapping procedure was developed to reproduce human hand postural synergies on robotic hands with a dissimilar kinematics. In [60], this procedure is extended to the case of a human hand augmented with robotic extra fingers, introducing an interesting and, to the best of our knowledge, still not very exploited framework, in which the human part and the robotic one share a common workspace. The human hand and the supernumerary extra fingers define a higher level kinematic structure that we referred to as *augmented hand*. The mapping algorithm is based on the definition of a virtual object obtained as a function of a set of reference points placed on the augmented hand, and allows to move the extra fingers according to the human hand motions, without requiring any explicit command from the user. This type of control strategy was successfully applied to a fully actuated robotic finger used by an healthy human, and was found to be useful in tasks that are typically rather difficult, such as grasping a large object or holding a bottle and unscrewing its cap with the same hand, see Figure 10.11.

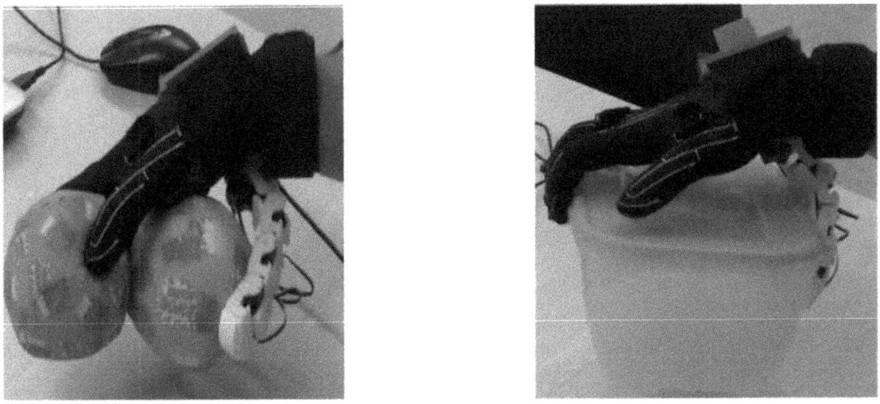

Figure 10.11 Fully actuated extra finger to enhance the capabilities of an healthy human hand. Left: grasping multiple objects in one augmented hand (ulnar grasp). Right: grasp of a big object (anatomically impossible grasp)

Both control approaches presented in [60,79] used an instrumented glove to track the human hand. However, in applications where the coordination between the human hand and the robotic extra limb is not a feasible solution, e.g., when extra limbs are used for patients that are not able to control their hand motion due to an upper limb paralysis, alternative solutions must be explored. A possibility is to use wearable control interfaces designed to be as intuitive and easy-to-use as possible. These features become paramount, for example, in chronic stroke patients that may also be affected by some cognitive deficit, possibly limiting their compliance during a demanding learning phase.

Next subsections describe two of the most suitable interfaces that were used to control the Soft-SixthFinger: the wearable cutaneous device called hRing and the EMG interface called frontalis muscle cap.

10.3.3 The hRing

All the above-mentioned works for the design of novel supernumerary limbs have shown incredibly promising results, reenabling impaired users to grasp objects they would not be able to grasp otherwise. Despite this, most of them do not provide any information about the forces exerted by the supernumerary limb on the environment, which is known to be useful [81–83]. Often this is not necessary, as operators use their own bodies to counterbalance the forces exerted by the supernumerary limb. For example, in Figure 10.9, the patient through his own hand can feel the force applied by the robotic finger through the object [65]. However, this is not always the case. In fact, many poststroke patients suffer from tactile anesthesia, or anaphia [84,85], in the hand contralateral to the stroke, which is the one that should provide force information about the supernumerary limb. For this reason, there have been attempts to restore the sense of touch in patients affected by this deficit.

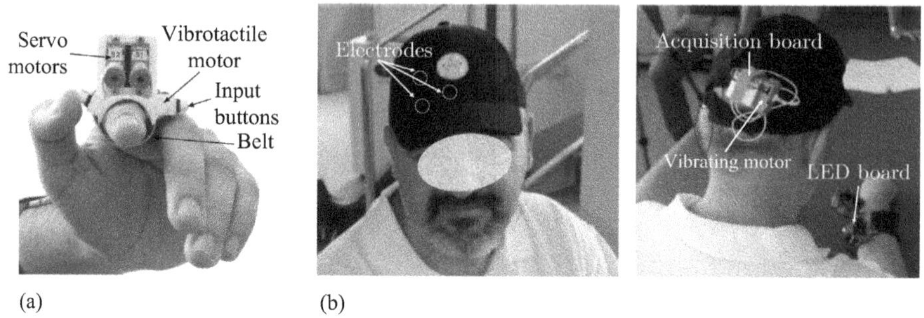

Figure 10.12 Possible interfaces for supernumerary limbs. (a) The hRing consists of two servo motors, a vibrotactile motor, two pairs of push buttons, and a belt. The frontalis muscle cap is an EMG interface where electrodes are placed inside the cap at front side to be positioned on the patient's forehead. The acquisition board and a vibrating motor are placed in a box on the back of the cap

In this respect, the group at the University of Siena proposed a wearable cutaneous interface with the purpose of relying information about the force exerted by the robotic finger. It can be worn on the proximal finger phalanx of the healthy hand and it is called hRing.

A prototype of the hRing is shown in Figure 10.12(a), and Hussain *et al.* [65] tested it together with the Soft-SixthFinger described in Section 10.3.1. It consists of two servo motors, a vibrotactile motor, two pairs of push buttons, and a belt. The servo motors move the belt in contact with the user's finger skin. When the motors spin in opposite directions, the belt presses into the user's finger, while when the motors spin in the same direction, the belt applies a shear force to the skin. The device structure is symmetrical and each side features two push buttons, enabling the hRing to be used both on the left and on the right hand. Through the moving belt, the hRing is able to provide normal and shear stimuli at the fingertip. Through the vibrotactile motor, it can provide vibrotactile transients information, such as the making and breaking contact with the grasped object. Moreover, featuring two push buttons, users can close the robotic finger and increase the grasping force (blue button in Figure 10.12(a)), and reduce the grasping force and open the robotic finger (yellow button in Figure 10.12(a)).

Two chronic stroke patients took part to the experimental evaluation of this interface for supernumerary limbs [64]. Each patient used the Soft-SixthFinger and the hRing for bimanual tasks, such as unscrewing a cap of a tomato jar, opening a popcorn bag, or opening a can of beans. Both patients were able to carry out several ADL bimanual tasks that would have not been possible without the supernumerary robotic system. Moreover, they both appreciated the haptic feedback. The demonstration of this integrated system (Soft-SixthFinger and hRing) was awarded with the Best Demonstration Award at the 2016 IEEE Haptics Symposium in Philadelphia, USA [86]. The same haptic device was also used for other applications, not paired with a supernumerary limb. Pacchierotti *et al.* [87], for example, used it to provide haptic feedback in a virtual environment. The authors showed that providing tactile feedback through the device improved the performance and perceived effectiveness of the virtual interaction task of 20% and 47% with respect to not providing any force feedback, respectively.

10.3.4 The frontalis muscle cap

The use of the hRing to control the motion of a robotic finger requires the involvement of the healthy hand. To cope with this issue, we proposed the frontalis muscle cap [65], an EMG-based wireless interface which maintains the principle of simplicity of a switch without interrupting the patient activities and without the involvement of the healthy hand during task execution.

Several EMG interfaces have been already successfully adopted for the control of prosthesis [88] and exoskeletons [89]. The EMG electrodes are usually placed either in the muscles coupled with the robot (exoskeleton) or in muscles where amputees still have the phantom of functions and hence they are able to generate a repeatable EMG pattern corresponding to each of the functions (prosthesis). For chronic stroke patients, it is generally difficult to generate repeatable EMG patterns in their paretic upper limb due to the weakness in muscle contraction control. For this reason, we coupled the flexion/extension motion of the robotic device with the contraction of the frontalis muscle. The user can contract this muscle by moving the eyebrows upwards. This muscle is always spared in case of a motor stroke either of the left or of the right hemisphere, due to its bilateral cortical representation [90].

The frontalis muscle cap is a wearable wireless EMG interface where electrodes, acquisition, and signal conditioning boards, as well as a vibration motor to provide haptic feedback are embedded in a cap, see Figure 10.12. This solution allows the patients to autonomously wear the interface using only their healthy hand. The vibration motor provides the feedback about the robotic finger status in terms of contact/no contact with the grasped object and in terms of force exerted by the device. The acquired EMG signal is sampled at 1 kHz (double EMG band) to avoid aliasing and a wireless communication is realized by a pair of Xbee modules (Series 1). The transmitter is embedded in the frontalis muscle cap while the receiver is placed on the actuator controller unit (specifications of the EMG acquisition board are summarized in Table 10.4). The reference values of received EMG signals were normalized using maximum voluntary contraction (MVC) technique [91]. This solution avoids the problems related to the high influence of detection condition on EMG signal amplitude. In fact, amplitude can greatly vary between electrode sites, subjects, and even day-to-day measures of the same muscle site. We implemented an auto-tuning procedure based on the MVC in order to better match the user-dependent nature of the EMG signal. The motion of the compensatory robotic device is controlled by using a finite state machine based on a trigger signal generated by using the hRing or the frontalis muscle cap. When using the EMG control interface, the trigger signal is obtained by using a single-threshold value defined as the 50% of the MVC, a level that was repeatable and

Table 10.3 Motion control of soft sixth finger using trigger signal

Trigger signal	Associated action
Single trigger	Move/stop
Double trigger	Change direction

sustainable for the subject without producing undue fatigue during the use of the device [65]. The state machine works under the simple control strategy presented in Table 10.3.

10.3.5 Evaluation

This section explains two aspects of the evaluation of a supernumerary finger. On the one side (Section 10.3.6), there is the need of assessing the characteristics of the robotic device from a quantitative point of view using standard indexes devised for robotic grippers (Table 10.4), and from a qualitative point of view, evaluating its shape adaptability to a standard database of objects [92]. On the other (Section 10.3.7), it is important to check whether the device works as expected when used by patients.

10.3.6 Performance evaluation

The performances of the Soft-SixthFinger were evaluated through a subset of the tests proposed in [93]. In particular, we measured the maximum fingertip force, the maximum payload and maximum horizontal grasp resistive force. The results of the experiments are summarized in Table 10.4. The maximum fingertip force of the device was recorded while fixing its support base on a table with the finger perpendicular to the table surface. The initial configuration of the finger was fully extended and it was commanded to close at the maximum torque. The hook of a dynamometer (Vernier, USA) was rigidly coupled with the fingertip of the device so to measure the force in the vertical direction. The maximum horizontal grasp resistive force was measured by grasping an object (diameter=65 mm, weight=400 g) with the robotic device and the arm while resting the arm on the table. The object was slowly pulled horizontally w.r.t. the table surface. To check the maximum payload, the operator's arm was stabilized on a table while grasping a cylindrical object (diameter=65 mm, weight=400 g) with the aid of the Soft-SixthFinger driven at the maximum actuator's torque. The grasped object was slowly pushed down using the dynamometer's bumper. The maximum pushing force was recorded when the object started to slip. The maximum payload of the device is the sum of the weight of the grasped object and the load due to the pushing force.

In order to prove the grasping ability of the device and its shape adaptability to different objects, we used a subset of the objects of the YCB grasping toolkit [92]. This toolkit is intended to be used to facilitate benchmarking in prosthetic design, rehabilitation research, and robotic manipulation. The objects in the set are designed to cover a wide range of aspects of the manipulation problem. It includes

Table 10.4 Technical details of the complete system

Dimensions	
Module	$20 \times 31 \times 12$ mm^3
Total length of finger (on arm)	180 mm
Support base	$110 \times 63 \times 3.5$ mm^3
Actuator control unit box	$71 \times 71 \times 45$ mm^3
Weights	
Module	4 g
Actuator control unit box	146 g
Actuator	
Maximum torque	3.1 N m @ 12 V
Pulley radius	8 mm
Maximum current	1.4 A @ 12 V
Continuous operating time	3.5 h @ stall torque
Maximum operating angles	300 deg, endless turn
Maximum nonloaded velocity	684 deg/s
The SSF performances	
Maximum Force at fingertip	40 N
Maximum payload	2.4 kg
Maximum horizontal resistive force	13 N @ dia = 65 mm
Total: finger + support base	180 g
Diameter smallest graspable obj.	14 mm
Motion control	
Single trigger	Move/stop
Double trigger	Change direction
Specifications of EMG acquisition board	
EMG acquisition box dimensions	$35 \times 31 \times 45$ mm^3
EMG acquisition box weight	46 g
Principle	Differential voltage
Number of electrodes	3
Bandwidth	10–400 Hz
Gain	1,000
Input impedance	100 GΩ
CMRR	110 dB
Operating voltage	V_{cc}=3.3 V

objects of daily life with different shapes, sizes, textures, weight, and rigidity. In general, grasp success is greatly affected by the ability of the robotic device to adapt to the shape of the object and the environment in response to contact forces. This adaptation increases the contact area and thereby the robustness of the grasp. In literature, the improvements on grasping of unknown object obtained through shape adaptation are well known [72,76,94,95]. We tested the device with different objects to evaluate how the robotic finger can adapt to the shape of the objects to realize a stable enveloping grasp. The tests were performed by a healthy subject wearing the device. This assured to evaluate only shape adaptation of the Soft-SixthFinger, avoiding possible grasp failures which could occur due to the low residual mobility of patients' arms. The finger was able to adapt itself to the shape of the grasped objects due to its intrinsic compliance [65].

10.3.7 Tests with chronic stroke patients

We tested the proposed device with six chronic stroke patients (five males, one female, age 40–62) to see how it can be used for hand grasping compensation in subjects showing a residual mobility of the arm. For being included in the experimental phase, patients had to score ≤ 2 when their motor function was tested with the National Institute of Health Stroke Scale (NIHSS) [96], item 5 "paretic arm." Moreover, the patients had to show the following characteristics: normal consciousness (NIHSS, item 1a, 1b, 1c = 0), absence of conjugate eyes deviation (NIHSS, item 2 = 0), absence of complete hemianopia (NIHSS, item 3 $\leq$ 1), absence of ataxia (NIHSS, item 7 = 0), absence of completely sensory loss (NIHSS, item 8 $\leq$ 1), absence of aphasia (NIHSS, item 9 = 0), absence of profound extinction and inattention (NIHSS, item 11 $\leq$ 1). The goal of the tests was to verify how quickly the patients can learn to use the device and its control interface. We performed qualitative tests composed of the Frenchay Arm Test [97], and some ADL bimanual tasks, where the paretic limb and the robotic finger worked together to constrain the motion of the object while the healthy hand manipulated it (e.g., constrain the motion of a jar while unscrewing its cap). Figure 10.13 shows two examples of tasks performed by the patients, whereas detailed results are presented in [64,65].

10.3.8 Discussion

In this part of the chapter, we presented the newly introduced *framework* of SRLs and shared our experience regarding the development of supernumerary robotic fingers that can be used for augmenting and compensating the human manipulation abilities. In particular, the new generation of robotic fingers can be used by the stroke patients to recover their missing grasping abilities and by healthy

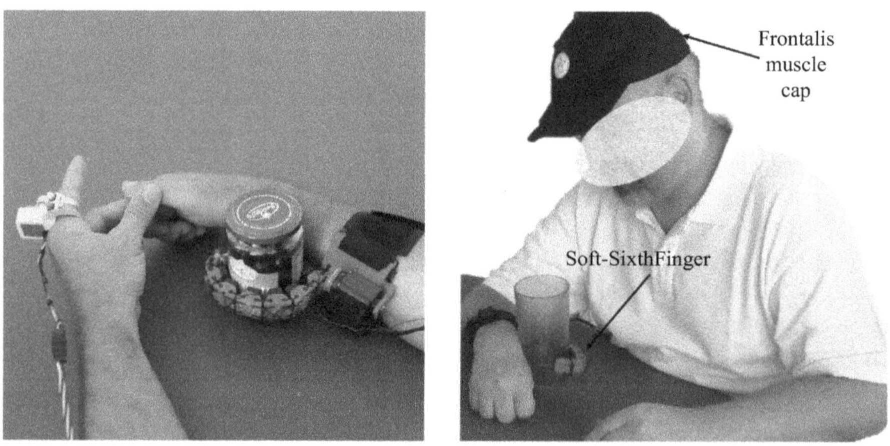

*Figure 10.13 Example of a bimanual tasks (*left*) and a task of Frenchay Arm Test (*right*), performed by patients using two different control interfaces: the hRing (*left*) and the frontalis muscle cap (*right*), are shown*

subjects to enhance their manipulation capabilities. For people with paretic upper limb, most of the attention of the community has been focused on exoskeletons [98] which are very difficult to use specially in chronic stage because of their limitation in accommodating the subjects anatomical variations due to impairment. Moreover, the poor wearability, in terms of weight and size, make them difficult to use in ADL. One of the biggest challenge of rehabilitation and assistive engineering is to develop technology to practice intense movement training at home [99]. The creation of a functional grasp by means of the supernumerary fingers enables patients to execute task-oriented grasp and release exercises and practice intensively using repetitive movements. Supernumerary robotic fingers can increase patients' performances, with a focus on objects manipulation, thereby improving their independence in ADL, and simultaneously decreasing erroneous compensatory motor strategies for solving everyday tasks. The idea of wearable supernumerary limbs as assistive devices is different in nature than other approaches used in rehabilitation and assistive robotics. Supernumerary limbs will provide novel opportunities to recover missing abilities, resulting in improvements of patients' quality of life. One of the more important aspects that has to be taken into account when designing supernumerary limbs is their close interaction with the human body. For this reason, the design guiding principles are safety, wearability, ergonomics, and user comfort. Although, until now the supernumerary robotic fingers are mainly being used for grasping compensation but there is a good expectation in using these devices to rehabilitate at least the arm.

Moreover, we also believe that extra fingers can play a role even in hand rehabilitation. Patients with hemiparesis often have limited functionality in the left or right hand. The standard therapeutic approach requires the patient to attempt to make use of the weak hand even though it is not functionally capable, which can result in feelings of frustration. The aim is to provide patients with a sense of purpose and accomplishment during ADL training, even during the early phase of treatment when the task can only be partially completed. We hope that the proposed devices can facilitate therapist-guided ADL training and encourage patients to continue exercising the affected limb. From the neuroscientific point of view, the development of this type of devices opens a series of interesting questions on how the supernumerary limbs are perceived by human cognitive system that needs to be investigated. For example, Hoyet *et al.* [100] measured humans' sense of ownership of a six-fingers hand avatar controlled in a virtual reality scenario. Participants responded positively to the possibility of controlling the six-fingers hand, despite the structural difference with respect to their own hand. In this respect, we are interested in evaluating the sense of ownership of our extra finger.

Acknowledgments

The research has received founding from European Union's Horizon 2020 Research and Innovation Programme, Grant Agreement No. 688857 (SoftPro) and from the European Union Seventh Framework Programme FP7/2007-2013, Grant Agreement No. 601165 (WEARHAP).

References

[1] Pratt GA, Williamson MM. Series elastic actuators. In: IEEE/RSJ Int. Conf. Intell. Robot. Syst. Human Robot Interact. Coop. Robot. vol. 1. IEEE Comput Soc Press; 1995. p. 399–406.

[2] Rouse EJ, Mooney LM, Herr HM. Clutchable series-elastic actuator: implications for prosthetic knee design. Int J Rob Res. 2014;33:1611–1625.

[3] Noritsugu T, Yamamoto H, Sasakil D, Takaiwa M. Wearable power assist device for hand grasping using pneumatic artificial rubber muscle. SICE 2004 Annu. Conf. 2004;1:420–425.

[4] Aguirre-Ollinger G, Colgate JE, Peshkin MA, Goswami A. Active-impedance control of a lower-limb assistive exoskeleton. In: 2007 IEEE 10th International Conference on Rehabilitation Robotics; 2007. p. 188–195.

[5] Herr H. Exoskeletons and orthoses: classification, design challenges and future directions. J Neuroeng Rehabil. 2009;6:21.

[6] Karavas N, Ajoudani A, Tsagarakis N, Saglia J, Bicchi A, Caldwell D. Tele-impedance based assistive control for a compliant knee exoskeleton. Rob Auton Syst. 2015;73:78–90.

[7] Asbeck AT, De Rossi SMM, Holt KG, Walsh CJ. A biologically inspired soft exosuit for walking assistance. Int J Rob Res. 2015;34(6):744–762.

[8] Panizzolo F, Galiana I, Asbeck AT, *et al.* A biologically-inspired multi-joint soft exosuit that can reduce the energy cost of loaded walking. J Neuroeng Rehabil. 2016 Jan; submitted(1):43.

[9] Quinlivan BT, Sangjun L, Malcolm P, *et al.* Assistance magnitude vs. metabolic cost reductions for a tethered multiarticular soft exosuit. Sci Robot. 2016;2(2):1–17.

[10] In H, Kang BB, Sin M, Cho KJ. Exo-Glove: a wearable robot for the hand with a soft tendon routing system. IEEE Robot Autom Mag. 2015 Mar;22(1):97–105.

[11] Robinson DW. Design and analysis of series elasticity in closed-loop actuator force control. PhD thesis, Massachusetts Institute of Technology, Boston, Massachusetts; 2000.

[12] Asbeck AT, De Rossi SMM, Galiana I, Ding Y, Walsh CJ. Stronger, smarter, softer: next-generation wearable robots. IEEE Robot Autom Mag. 2014 Dec;21(4):22–33.

[13] Rossi SMMD, Bae J, Donnell KEO, *et al.* Gait improvements in stroke patients with a soft exosuit. In: Proc Gait Clin Mov Anal Soc Meet. 2015. p. 2–3.

[14] Ding M, Ueda J, Ogasawara T. Pinpointed muscle force control using a power-assisting device: system configuration and experiment. In: Proc. 2nd Bienn. IEEE/RAS-EMBS Int. Conf. Biomed. Robot. Biomechatronics, BioRob 2008. IEEE; 2008. p. 181–186.

[15] Ueda J, Hyderabadwala M, Krishnamoorthy V, Shinohara M. Motor task planning for neuromuscular function tests using an individual muscle control technique. In: IEEE Int. Conf. Rehabil. Robot. IEEE; 2009. p. 133–138.

[16] Lee SW, Landers KA, Park HS. Development of a biomimetic hand exotendon device (BiomHED) for restoration of functional hand movement post-stroke. IEEE Trans Neural Syst Rehabil Eng. 2014;22(4):886–898.

[17] In H, Cho Kj. Exo-Glove: soft wearable robot for the hand using soft tendon routing system. IEEE Robot Autom. 2015;22(Mar 2015):97–105.

[18] Nef T, Guidali M, Riener R. ARMin III – arm therapy exoskeleton with an ergonomic shoulder actuation. Appl Bionics Biomech. 2009 Jul;6(2): 127–142.

[19] Carignan C, Tang J, Roderick S. Development of an exoskeleton haptic interface for virtual task training. In: IEEE/RSJ Int. Conf. Intell. Robot. Syst. IROS. IEEE; 2009. p. 3697–3702.

[20] Ren Y, Park HS, Zhang LQ. Developing a whole-arm exoskeleton robot with hand opening and closing mechanism for upper limb stroke rehabilitation. In: IEEE Int. Conf. Rehabil. Robot. ICORR. IEEE; 2009. p. 761–765.

[21] Ball SJ, Brown IE, Scott SH. MEDARM: a rehabilitation robot with 5DOF at the shoulder complex. In: IEEE/ASME Int. Conf. Adv. Intell. Mechatronics, AIM. IEEE; 2007. p. 1–6.

[22] Lo HS, Xie SQ. Exoskeleton robots for upper-limb rehabilitation: state of the art and future prospects. Med Eng Phys. 2012 Apr;34(3):261–268.

[23] Kiguchi K, Rahman MH, Sasaki M, Teramoto K. Development of a 3DOF mobile exoskeleton robot for human upper-limb motion assist. Rob Auton Syst. 2008 Aug;56(8):678–691.

[24] Perry JC, Rosen J, Burns S. Upper-limb powered exoskeleton design. IEEE/ ASME Trans Mechatron. 2007;12(4):408–417.

[25] Ueki S, Nishimoto Y, Abe M, *et al.* Development of virtual reality exercise of hand motion assist robot for rehabilitation therapy by patient self-motion control. Conf Proc IEEE Eng Med Biol Soc. 2008 Jan;2008(1):4282–4285.

[26] Carmeli E, Peleg S, Bartur G, Elbo E, Vatine JJ. HandTutor™ enhanced hand rehabilitation after stroke – a pilot study. Physiother Res Int. 2011 Dec;16(4):191–200.

[27] Chiri A, Vitiello N, Giovacchini F, Roccella S, Vecchi F, Carrozza MC. Mechatronic design and characterization of the index finger module of a hand exoskeleton for post-stroke rehabilitation. IEEE/ASME Trans Mechatron. 2012 Oct;17(5):884–894.

[28] Jarrassé N, Morel G. Connecting a human limb to an exoskeleton. IEEE Trans Robot. 2012 Jun;28(3):697–709.

[29] Jarrassé N, Morel G. A formal method for avoiding hyperstaticity when connecting an exoskeleton to a human member. In: Proc. – IEEE Int. Conf. Robot. Autom.; 2010. p. 1188–1195.

[30] Stienen A, Hekman E, van der Helm FCT, *et al.* Self-aligning exoskeleton axes through decoupling of joint rotations and translations. IEEE Trans Robot. 2009;25(3):628–633.

[31] Fontana M, Dettori A, Salsedo F, Bergamasco M. Mechanical design of a novel hand exoskeleton for accurate force displaying. In: Proc. – IEEE Int. Conf. Robot. Autom.; 2009. p. 1704–1709.

[32] Toya K, Miyagawa T, Kubota Y. Power-assist glove operated by predicting the grasping mode. J Syst Des Dyn. 2011;5(1):94–108.
[33] Kadowaki Y, Noritsugu T, Takaiwa M, Sasaki D, Kato M. Development of soft power-assist glove and control based on human intent. J Robot Mechatron. 2011;23(2):281–291.
[34] Polygerinos P, Wang Z, Galloway KC, Wood RJ, Walsh CJ. Soft robotic glove for combined assistance and at-home rehabilitation. Rob Auton Syst. 2015;73:135–143.
[35] Kobayashi H, Hiramatsu K. Development of muscle suit for upper limb. In: IEEE Int. Conf. Robot. Autom. 2004. Proceedings. ICRA '04. 2004. vol. 3. IEEE; 2004. p. 3–8.
[36] Asbeck AT, Dyer RJ, Larusson AF, Walsh CJ. Biologically-inspired soft exosuit. In Rehabilitation Robotics (ICORR), 2013 IEEE International Conference on (pp. 1–8). IEEE.
[37] Asbeck AT, Schmidt K, Walsh CJ. Soft exosuit for hip assistance. Rob Auton Syst. 2015;73:102–110.
[38] In H, Lee H, Jeong U, Kang BB, Cho Kj. Feasibility study of a slack enabling actuator for actuating tendon-driven soft wearable robot without pretension. In: ICRA. Seattle, WA; 2015. p. 1229–1234.
[39] Tsai LW. Robot Analysis and Design; New York, NY, USA: John Wiley & Sons, Inc., 1999.
[40] Murray RM, Li Z, Sastry SS. A mathematical introduction to robotic manipulation. vol. 29. Boca Raton, Florida: CRC Press; 1994.
[41] Kong K, Tomizuka M. Control of exoskeletons inspired by fictitious gain in human model. IEEE/ASME Trans Mechatron. 2009 Dec;14(6):689–698.
[42] Weinger MB, Wiklund ME, Gardner-Bonneau DJ. Handbook of human factors in medical device design; 2010. Available from: http://www.crcnetbase.com/doi/abs/10.1201/b10439-16.
[43] Ikhouane F, Rodellar J. Systems with hysteresis: analysis, identification and control using the Bouc–Wen model. West Sussex, England: John Wiley & Sons; 2007.
[44] Olsson H, strm KJ, de Wit CC, Gfvert M, Lischinsky P. Friction models and friction compensation. Eur J Control. 1998;4(3):176–195. Available from: http://www.sciencedirect.com/science/article/pii/S094735809870113X.
[45] Cirstea MC. Compensatory strategies for reaching in stroke. Brain. 2000; 123(5):940–953. Available from: http://brain.oxfordjournals.org/content/123/5/940.abstract.
[46] Berardelli A, Hallett M, Rothwell JC, *et al.* Single-joint rapid arm movements in normal subjects and in patients with motor disorders. Brain. 1996;119 (Pt 2):661–674.
[47] Challis RE, Kitney RI. Biomedical signal processing (in four parts). Med Biol Eng Comput. 1990;28(6):509–524. Available from: http://dx.doi.org/10.1007/BF02442601.
[48] Takata S, Yasui N. Disuse osteoporosis. J Med Invest. 2001;48(3–4): 147–156.

[49] Nilsson M. A Helping Hand On Innovations for Rehabilitation and Assistive Technology; PhD thesis, KTH Royal Institute of Technology, Stockholm, Sweden, 2013.

[50] Go AS, Mozaffarian D, Roger VL, *et al.* Heart disease and stroke statistics – 2014 update: a report from the American Heart Association. Circulation. 2014;129(3):e28.

[51] Nakayama H, Jorgensen HS, Raaschou HO, Olsen TS. Compensation in recovery of upper extremity function after stroke: the Copenhagen Stroke Study. Arch Phys Med Rehabil. 1994;75(8):852–857.

[52] Kwakkel G, Kollen B. Predicting improvement in the upper paretic limb after stroke: a longitudinal prospective study. Restor Neurol Neurosci. 2007;25(5):453–460.

[53] Faria-Fortini I, Michaelsen SM, Cassiano JG, Teixeira-Salmela LF. Upper extremity function in stroke subjects: relationships between the international classification of functioning, disability, and health domains. J Hand Ther. 2011;24(3):257–265.

[54] Balasubramanian S, Klein J, Burdet E. Robot-assisted rehabilitation of hand function. Curr Opin Neurol. 2010;23(6):661–670.

[55] Volpe BT, Krebs HI, Hogan N. Is robot-aided sensorimotor training in stroke rehabilitation a realistic option? Curr Opin Neurol. 2001;14(6):745–752.

[56] Masiero S, Celia A, Rosati G, Armani M. Robotic-assisted rehabilitation of the upper limb after acute stroke. Arch Phys Med Rehabil. 2007;88(2): 142–149.

[57] Chiri A, Vitiello N, Giovacchini F, Roccella S, Vecchi F, Carrozza MC. Mechatronic design and characterization of the index finger module of a hand exoskeleton for post-stroke rehabilitation. IEEE/ASME Trans Mechatron. 2012;17(5):884–894.

[58] Lum PS, Godfrey SB, Brokaw EB, Holley RJ, Nichols D. Robotic approaches for rehabilitation of hand function after stroke. Am J Phys Med Rehabil. 2012;91(11):S242–S254.

[59] Kwakkel G, Kollen BJ, van der Grond J, Prevo AJ. Probability of regaining dexterity in the flaccid upper limb impact of severity of paresis and time since onset in acute stroke. Stroke. 2003;34(9):2181–2186.

[60] Prattichizzo D, Salvietti G, Chinello F, Malvezzi M. An object-based mapping algorithm to control wearable robotic extra-fingers. In: Proc. IEEE/ASME Int. Conf. on Advanced Intelligent Mechatronics. Besançon, France; 2014.

[61] Prattichizzo D, Malvezzi M, Hussain I, Salvietti G. The Sixth-Finger: a modular extra-finger to enhance human hand capabilities. In: Proc. IEEE Int. Symp. in Robot and Human Interactive Communication. Edinburgh, United Kingdom; 2014.

[62] Hussain I, Salvietti G, Meli L, Pacchierotti C, Prattichizzo D. Using the robotic sixth finger and vibrotactile feedback for grasp compensation in chronic stroke patients. In: Proc. IEEE/RAS-EMBS International Conference on Rehabilitation Robotics (ICORR); 2015.

[63] Salvietti G, Hussain I, Cioncoloni D, Taddei S, Rossi S, Prattichizzo D. Compensating hand function in chronic stroke patients through the robotic Sixth Finger. IEEE Transactions on Neural Systems and Rehabilitation Engineering 25.2 (2017): 142–150.
[64] Hussain I, Spagnoletti G, Pacchierotti C, Prattichizzo D. A wearable haptic ring for the control of extra robotic fingers. In: Proc. Asia Haptics. Chiba, Japan; 2016.
[65] Hussain I, Salvietti G, Spagnoletti G, Prattichizzo D. The Soft-SixthFinger: a wearable EMG controlled robotic extra-finger for grasp compensation in chronic stroke patients. IEEE Robot Autom Lett. 2016 Jul;1(2):1000–1006.
[66] Meng Q, Lee MH. Design issues for assistive robotics for the elderly. Adv Eng Inform. 2006;20(2):171–186.
[67] Pons JL. Rehabilitation exoskeletal robotics. IEEE Eng Med Biol Mag. 2010;29(3):57–63.
[68] Stanger CA, Anglin C, Harwin WS, Romilly DP. Devices for assisting manipulation: a summary of user task priorities. IEEE Trans Rehabil Eng. 1994;2(4):256–265.
[69] Miguelez J, Miguelez M, Alley R. Amputations about the shoulder: prosthetic management. Altas of Amputations and Limb Deficiencies – Surgical, Prosthetic, and Rehabilitation Principles Rosemont, IL: American Academy of Orthopaedic Surgeons. 2004; p. 263–273.
[70] Vanderborght B, Albu-Schäffer A, Bicchi A, *et al.* Variable impedance actuators: a review. Robot Autonomous Syst. 2013;61(12):1601–1614.
[71] Manti M, Hassan T, Passetti G, d'Elia N, Cianchetti M, Laschi C. An Under-Actuated and Adaptable Soft Robotic Gripper. In: Wilson PS, Verschure FMJP, Mura A, Prescott JT, editors. Biomimetic and Biohybrid Systems: 4th International Conference, Living Machines 2015, Barcelona, Spain, Jul 28–31, 2015, Proceedings. Cham: Springer International Publishing; 2015. p. 64–74. Available from: http://dx.doi.org/10.1007/978-3-319-22979-9_6.
[72] Dollar AM, Howe RD. Joint coupling design of underactuated hands for unstructured environments. Int J Robot Res. 2011;30(9):1157–1169.
[73] Laschi C, Cianchetti M. Soft robotics: new perspectives for robot bodyware and control. Front Bioeng Biotechnol. 2014;2–3.
[74] Birglen L, Lalibertè T, Gosselin C. Underactuated Robotic Hands. vol. 40 of Springer Tracts in Advanced Robotics. New York City, New York: Springer, 2008.
[75] Dollar AM, Howe RD. The highly adaptive SDM hand: design and performance evaluation. Int J Robot Res. 2010;29(5):585–597.
[76] Eppner C, Brock O. Grasping unknown objects by exploiting shape adaptability and environmental constraints. In: Intelligent Robots and Systems (IROS), 2013 IEEE/RSJ International Conference on; 2013. p. 4000–4006.
[77] Catalano MG, Grioli G, Farnioli E, Serio A, Piazza C, Bicchi A. Adaptive synergies for the design and control of the Pisa/IIT SoftHand. Int J Robot Res. 2014;33(5):768–782.

[78] ArbotiX. ArbotiX-M Robocontroller, open source; 2012. On-line: http://www.trossenrobotics.com/p/arbotix-robot-controller.aspx.

[79] Wu F, Asada H. Bio-artificial synergies for grasp posture control of supernumerary robotic fingers. In: Proceedings of Robotics: Science and Systems. Berkeley, USA; 2014.

[80] Gioioso G, Salvietti G, Malvezzi M, Prattichizzo D. Mapping synergies from human to robotic hands with dissimilar kinematics: an approach in the object domain. IEEE Trans Robot. 29.4 (2013): 825–837.

[81] Meli L, Pacchierotti C, Prattichizzo D. Sensory subtraction in robot-assisted surgery: fingertip skin deformation feedback to ensure safety and improve transparency in bimanual haptic interaction. IEEE Trans Biomed Eng. 2014;61(4):1318–1327.

[82] Pacchierotti C, Prattichizzo D, Kuchenbecker KJ. Cutaneous feedback of fingertip deformation and vibration for palpation in robotic surgery. IEEE Trans Biomed Eng. 2016;63(2):278–287.

[83] Pacchierotti C. Cutaneous haptic feedback in robotic teleoperation. Springer Series on Touch and Haptic Systems. New York City, New York: Springer International Publishing; 2015.

[84] Son SM, Kwon YH, Lee NK, Nam SH, Kim K. Deficits of movement accuracy and proprioceptive sense in the ipsi-lesional upper limb of patients with hemiparetic stroke. J Phys Ther Sci. 2013;25(5):567.

[85] Lima NMFV, Menegatti KC, Yu É, *et al.* Sensory deficits in ipsilesional upper-extremity in chronic stroke patients. In: Arquivos de Neuro-psiquiatria. 2015; p. 1–6.

[86] Best demonstration award winners, IEEE Haptics Symposium, 2016, Philadelphia, USA.

[87] Pacchierotti C, Salvietti G, Hussain I, Meli L, Prattichizzo D. The hRing: a wearable haptic device to avoid occlusions in hand tracking. In Haptics Symposium (HAPTICS), 2016 IEEE (pp. 134–139).

[88] Zecca M, Micera S, Carrozza M, Dario P. Control of multifunctional prosthetic hands by processing the electromyographic signal. Crit Rev Biomed Eng. 2002;30(4–6).

[89] Kiguchi K, Tanaka T, Fukuda T. Neuro-fuzzy control of a robotic exoskeleton with EMG signals. IEEE Trans Fuzzy Syst. 2004;12(4):481–490.

[90] Brodal A. Neurological anatomy in relation to clinical medicine. Oxford University Press, USA; 1981.

[91] Farina D, Merletti R. Comparison of algorithms for estimation of EMG variables during voluntary isometric contractions. J Electromyogr Kinesiol. 2000;10(5):337–349.

[92] Çalli B, Walsman A, Singh A, Srinivasa S, Abbeel P, Dollar AM. Benchmarking in Manipulation Research: The YCB Object and Model Set and Benchmarking Protocols. CoRR. 2015; abs/1502.03143. Available from: http://arxiv.org/abs/1502.03143.

[93] Falco J, Van Wyk K, Liu S, Carpin S. Grasping the performance: facilitating replicable performance measures via benchmarking and standardized methodologies. IEEE Robot Autom Mag. 2015;22(4):125–136.

[94] Ciocarlie M, Allen P. A design and analysis tool for underactuated compliant hands. In: Intelligent Robots and Systems, 2009. IROS 2009. IEEE/RSJ International Conference on; 2009. p. 5234–5239.

[95] Krut S, Bégoc V, Dombre E, Pierrot F. Extension of the form closure property to underactuated hands. IEEE Trans Robot. 2010;26:853–866.

[96] Brott T, Adams H, Olinger CP, *et al.* Measurements of acute cerebral infarction: a clinical examination scale. Stroke. 1989;20(7):864–870.

[97] Heller A, Wade D, Wood VA, Sunderland A, Hewer RL, Ward E. Arm function after stroke: measurement and recovery over the first three months. J Neurol Neurosurg Psychiatry. 1987;50(6):714–719.

[98] Bogue R. Exoskeletons and robotic prosthetics: a review of recent developments. Ind Robot. 2009;36(5):421–427.

[99] Stefanov DH, Bien Z, Bang WC. The smart house for older persons and persons with physical disabilities: structure, technology arrangements, and perspectives. IEEE Trans Neural Syst Rehabil Eng. 2004;12(2):228–250.

[100] Hoyet L, Argelaguet F, Nicole C, Lécuyer A. "Wow! I have six Fingers!": Would You accept structural changes of Your hand in Vr? Frontiers. 2016;3(27):1.

Chapter 11

Walking assistive apparatus for gait training patients and promotion exercise of the elderly

Eiichiro Tanaka[1], Keiichi Muramatsu[2], Keiichi Watanuki[2], Shozo Saegusa[3], and Louis Yuge[4]

Abstract

This chapter presents authors development work on walking assistant exoskeletons for gait training of patients and promotion exercise of the elderly. The first exoskeleton is a whole leg assisting device using a special parallel link mechanism. For gait training of motor palsy patients, a weight bearing lifter was attached, and an impedance control was tuned by using frequency entertainment. For the elderly, a torque controller taken into account the dynamics of both a user and the apparatus with real time acceleration data is developed. The second exoskeleton is a whole body assisting suit was developed by adding arm assistance to the whole leg apparatus. By assisting not only legs but also swinging arms, an increased cerebral activity of all areas can be expected via rehabilitation with the whole body exoskeleton. Finally, we developed a close-fitting type exoskeleton assisting only ankle joint, which leads to a product RE-Gait® in 2016. By utilizing the structure of bi-articular muscle and physiological phenomenon of stretch reflex, the user's leg can be raised assisting only ankle joint. Experiments with by hemiplegic patients are enclosed, which show that abduct variation of the hip joint is decreased while the stride length is increased by using the device.

Keywords: Walking assistance; gait training; promote exercise

11.1 Introduction

For many apoplexy patients (over 1 million people per year in Japan) recovering and regaining an independent life through neuro-rehabilitation is highly demanded.

[1]Faculty of Science and Engineering, Waseda University, Japan
[2]Graduate School of Science and Engineering, Saitama University, Japan
[3]Faculty of Business Administration, Shujitsu University, Japan
[4]Graduate School of Biomedical & Health Sciences, Hiroshima University, Japan

If the patients start training immediately after the incident, they have a high possible to recover compared to no training. In the case of walking, when the ideal leg gait motion for a motor palsy patient is input by using a gait assistive device, the subject regains his/her motor function, and the motion activates the subject's brain in the undamaged area. Then the neural network of the brain is reconstructed.

Neuro-rehabilitation, which repairs the neural circuits by directly inputting movement into the body, requires programable training device to perform these movement and treatment. Novel training machines and methods have been developed in recent year.

NIRS (Near-Infrared Spectroscopy) is a useful approach to evaluate cerebral activity during walking and hypothesized that the alternating leg movements seen during gait are controlled by the leg regions of the primary motor cortex (PMC) and other related motor areas [1]. Suzuki *et al.* showed that the prefrontal and premotor cortices are involved in adapting walking and running speed on the treadmill using NIRS [2]. They also indicated that preparation for walking cued by a verbal instruction enhanced frontal activations both during the preparation and execution of walking as well as walking performance [3]. Furthermore, Yano *et al.* developed a footpad-type locomotion interface and by using NIRS, they confirmed it activated user's brain more effectively than walking on a treadmill [4]. Therefore, it is more effective to use the motion assistive device than normal training. However, most of these machines require the motor palsy patient to remain stationary during use (e.g., [5]). Many conventional devices (e.g., [6–8]) are fixed to arms and legs tightly, and most of these devices do not equip the actuator for assisting the ankle joint.

Figure 11.1 shows the comparison between ideal gait and hemiplegic gait. Ideal gait has four points of the feature; straight posture, long stride, heel contact by dorsiflexion, and kicking ground by plantarflexion. On the other hand, hemiplegic gait also has four features, namely, the droopy posture, toe contact by equinus foot, inadequacy of plantarflexion, and circumductive foot. Of them, equinus foot is dangerous for walking, because it is easy to stumble and falling. It is most

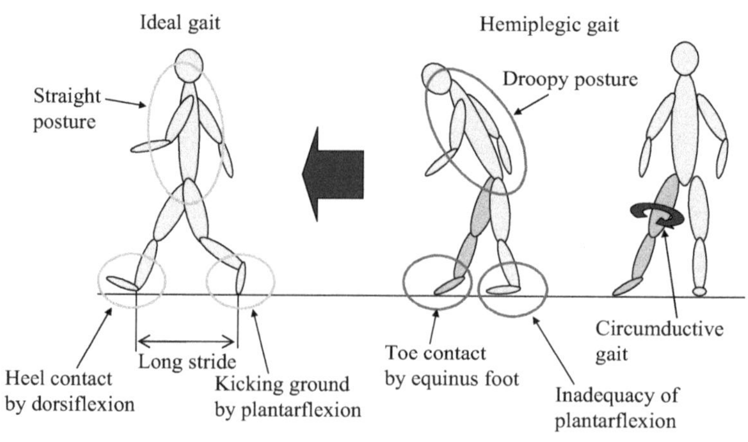

Figure 11.1 Comparison between ideal gait and hemiplegic gait. ©, with permission, from [12]

important for hemiplegic patient to recover heel contact, therefore, it is necessary to develop the ankle motion support type of walking assistive apparatus.

Our research is focused also on the problem about rapid aging population. Most of the elderly's muscles become weak with age. Especially, TA muscle (Tibialis Anterior muscle), used in the dorsiflexion of ankle joint, is the most fatigable muscle. It is more likely to stumble and fall down even though there may only be a slight step. Therefore, they lose confidence in walking alone, if they don't walk, they will become to be easy to fracture their bone, and finally, they will be bedridden.

To address the needs and problems noted above, we have developed several exoskeletons to meet the different training requirements. We made a whole leg assisting type of walking assistive apparatus using a special parallel link mechanism [9]. For gait training of motor palsy patients, the weight bearing lifter was attached, and an impedance control was tuned by using frequency entertainment. For the elderly, a torque controlled taken into account the dynamics both a user and the apparatus and using real time acceleration data [10]. Furthermore, we developed a whole body motion support type mobile suit. By utilizing NIRS, we evaluated the cerebral activity while walking, and compared the difference assisting only legs or not only legs but also swinging arms, and walking on a treadmill or in a corridor [11]. However, patients hesitated to use this apparatus, because it was so bulky and afraid of going out of control. For gait training of hemiplegic patients, it is necessary to reduce the weight of the apparatus which is attached to the patient's leg. To address this problem, we developed a close-fitting type of walking assistive apparatus which was attached a motor drive mechanism to an Ankle Foot Orthosis for assisting the ankle motion [12]. It was improved and sold as a product, RE-Gait® in 2016 [13]. This is very light weight, low cost, and very small as it can be hidden in the hem of pants. By using this apparatus, both dorsiflexion and plantar flexion increased and the gait was improved. RE-Gait® is able to be used for not only gait training of patients but also promotion exercise of walking. We developed the prototype for promotion exercise of the able-bodied elderly [14]. However, even though the user is assisted with the apparatus physically, he/she should keep motivation for the exercise mentally. Therefore, we suggested to evaluate the user's emotion two-dimensionally [15], and developed the apparatus controlled by using the relation between two-dimensional emotion map and walking condition map [16,17]. In this chapter, these developed apparatuses are introduced.

11.2 Whole leg assisting type of walking assistive apparatus

We proposed a walking-assistance method in which flat steps structurally follow the sole of the foot. That is, the user places the legs on flat steps and is assisted by the walking movement of the entire leg from the sole of the foot, as shown in Figure 11.2(b). Key joints and corresponding procedures used in walking are as follows:

1. Hip joint: Swings back and forth parallel to the direction of walking movement
2. Knee joint: Swings in a back-and-forth direction
3. Ankle joint: Swings in a back-and-forth direction

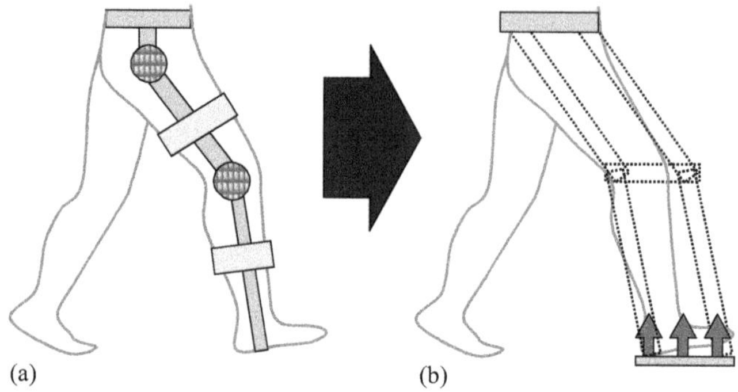

Figure 11.2 The difference between conventional and new proposal structures of the power assistance apparatus: (a) Conventional structure, (b) New proposal structure. ©, with permission, from [9]

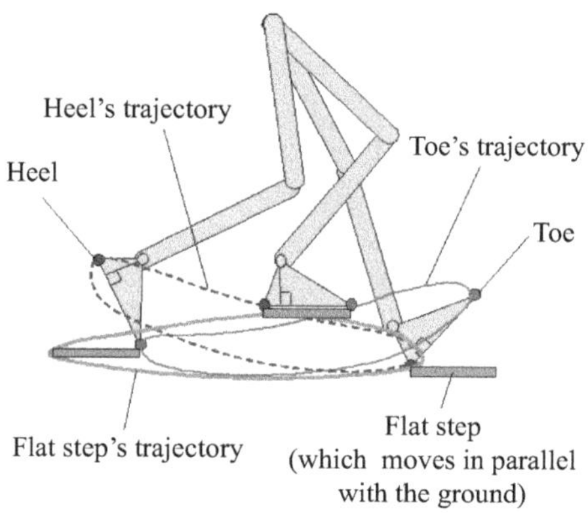

Figure 11.3 Trajectory of flat step of apparatus. ©, with permission, from [9]

Specifically, the apparatus we developed supports the back-and-forth swinging above, which influences the direction of movement. As shown in Figure 11.3, flat steps are designed to remain parallel to the ground so that the three joints can be supported simultaneously. In the stance phase of walking, the flat step rests on the ground, while in the swing phase, the flat step is driven according to the lowest position of the sole of each foot. Because posture is controlled when the user stands, this helps prevent stumbling by equipping the apparatus.

We proposed a spatial parallel link mechanism to enable the walking-assistance method to be applied. This mechanism connects two parallelograms lengthwise on the front and the side of the user. As shown in Figure 11.4, the flat step maintains translational motion in the three degrees of freedom (DOF) of the user. As shown in Figure 11.5, this mechanism is applied to the walking-assistive

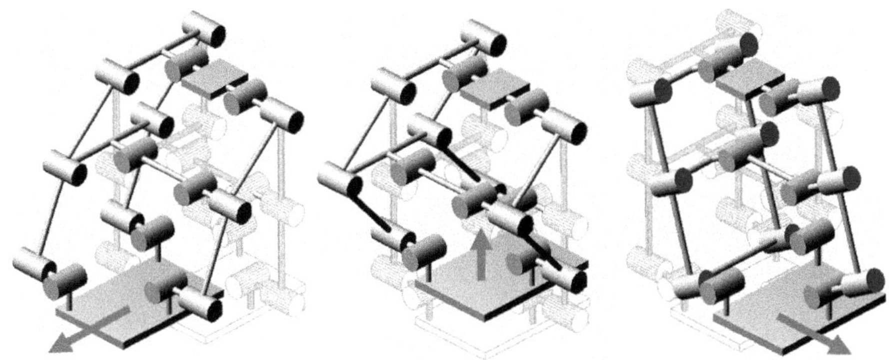

Figure 11.4 Three DOFs supported by spatial parallel link mechanism shown for right leg. ©, with permission, from [9]

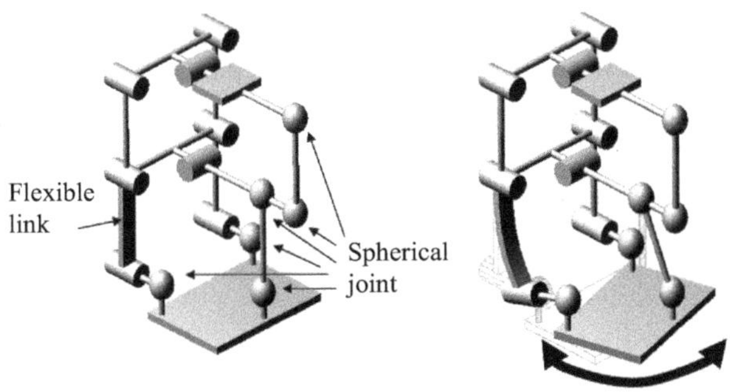

Figure 11.5 Flexible link and spherical joints in mechanism enabling twisting, shown for right leg. ©, with permission, from [9]

apparatus; the structure links the parallelograms corresponding to the femoral and crural parts of the leg. To appropriately consider twisting of the ankle, we use a flexible link and spherical joints, as indicated in Figure 11.5. This link is constructed from a stainless steel plate and rubber to bend and twist with moderate elasticity without buckling because the bending strength of the flexible link has strength equivalent to the aluminum square pipe utilized for other crural links. This makes it durable enough to be pliable in the direction of the link length and strong enough to maintain proper posture.

Figure 11.6 shows a prototype of the structure. Our control system has the SH-4 board as shown in Figures 11.7 and 11.8. The operating system of the computer is ART-LINUX with a control interval of 1 ms. We used three lithium-ion batteries as the power supply for the apparatus as shown in Figures 11.7 and 11.8.

To estimate the walking phase of the user's leg, pressure sensors were attached under the thenar eminence and the heel. The pressure variation at each sensing point was measured. The sensor was constructed by sandwiching pressure-sensitive conductive rubber 0.5 mm thick within thin copper sheet. Its output voltage was

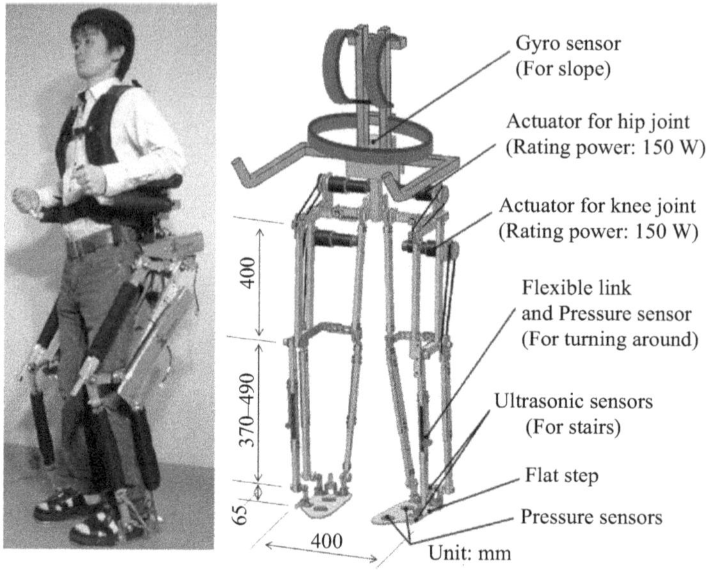

Figure 11.6 Structure of mechanism. ©, with permission, from [9]

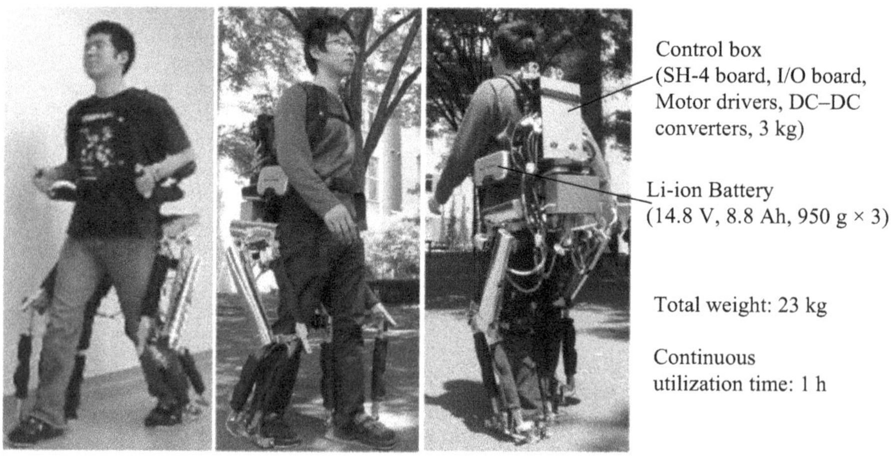

Figure 11.7 Example of apparatus utilized as vehicle and self-contained system. ©, with permission, from [9]

0–5 V, which is obtained using a partial pressure circuit. Because the sensor is attached to the insole of the shoe, pressure can only be measured with footgears such as a sandal. Figure 11.9 illustrates pressure conditions under the thenar eminence and the heel in the walking phase. The stance phase is estimated by dividing the walking phase into the four distinct steps shown in the figure. By specifying the

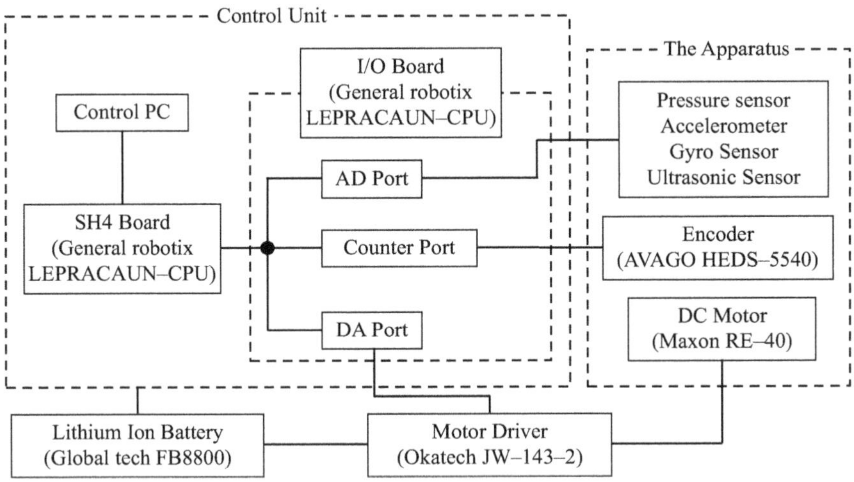

Figure 11.8 System constitution of the control device for the apparatus. ©, with permission, from [10]

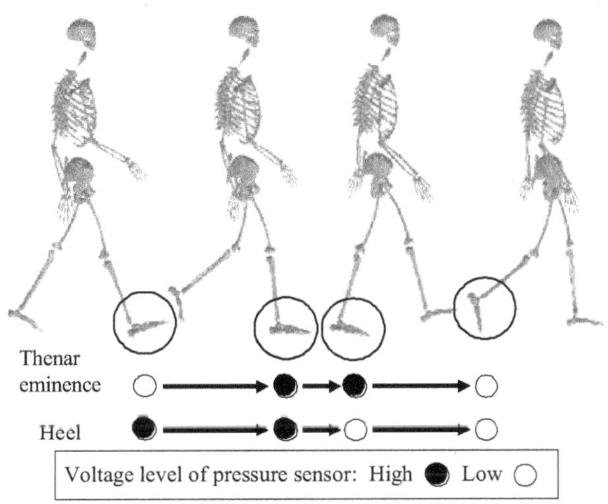

Figure 11.9 Measurement of walking phase. ©, with permission, from [9]

target value of the step according to each phase, control corresponds to the behavior of the user, which in turn improves safety. In the case of a system malfunction or an emergency, a direct stop switch was also incorporated into the apparatus.

To determine the direction in which the user was walking, pressure sensors were also attached to the flexible crural link. Given the parallel link mechanism, a gyro sensor attached to the back of the apparatus measured the slope angle of the surface on which the user was standing. To measure the distance between flat steps and stairs, ultrasonic sensors were attached to the toe and heel of the flat step. If the foot in the swing phase is about to hit the edge of a stair, the sensor detects it. Hitting the stair is prevented by compensating the trajectory of the flat step according to the direction of the stair is detected. The user can thus walk on stairs without fear of stumbling or falling.

Users, including the elderly who can walk independently can use the apparatus freely. In contrast, it is difficult for those with ambulation difficulty or unable to walk independently to use the apparatus without assistance by others, because such users are more likely to stumble, making it necessary to support the user's posture. A weight-bearing lift is effective for using apparatus power effectively. As shown in Figure 11.10, we developed a weight-bearing lift with two arms, which can bear the combined weight of the user and the apparatus. This lift is controlled with a joystick that specifies where the helper or the user wants to go.

To control the apparatus, we made a calculation model as shown in Figure 11.11, and developed two control methods; one uses the impedance control (11.1) and targeted trajectory is inputted.

$$\begin{aligned}\tau &= M(\boldsymbol{\theta})\ddot{\boldsymbol{\theta}} + h(\dot{\boldsymbol{\theta}},\boldsymbol{\theta}) + g_{ap}(\boldsymbol{\theta}) \\ &\quad + \alpha g_{\text{human}}(\boldsymbol{\theta}) - J^{\mathrm{T}}\left[\left\{M_d\left(J\ddot{\boldsymbol{\theta}} + \dot{J}\dot{\boldsymbol{\theta}}\right) + D_d\dot{\boldsymbol{p}}_e + K_d\boldsymbol{p}_e\right\} + \boldsymbol{f}_r\right] \in \mathrm{R}^2 \end{aligned} \tag{11.1}$$

where M and h are total (apparatus and user's leg) values. g_{ap} is the gravity compensation for one leg of the apparatus, g_{human} is the gravity compensation for one leg, α is the gravity compensation ratio for one leg, J is the Jacobian matrix from the hip and knee joints of the apparatus to the flat step, M_d, D_d, K_d are the desired mass, damping, stiffness matrixes when the user moves, $\boldsymbol{p}_e$ is the deviation from the target, $\boldsymbol{f}_r$ is the floor reaction force in the stance phase.

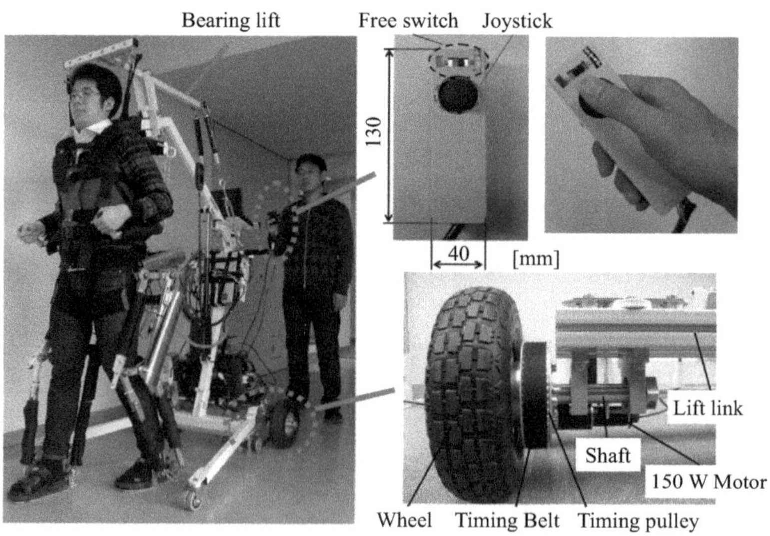

Figure 11.10 Photos of the apparatus and weight bearing lift. ©, with permission, from [9]

Angle variations at each joint (while an able-bodied person is walking) are described in various studies. By using angle variation data for each joint, i.e., hip, knee, and ankle, and a user's height, we calculated the user's trajectory with the leg model by direct kinematics.

Another is the torque control (11.4) and by using (11.2) and (11.3), real time measured acceleration data is inputted in $\ddot{\boldsymbol{p}}$; therefore, even though the user behaves freely, the apparatus can assist adequately. Both methods were taken into account the dynamics of the apparatus and the legs of the user, as well as the assist ratio α for the user.

$$\begin{aligned}\boldsymbol{\tau} &= M_{ap}(\boldsymbol{\theta})\ddot{\boldsymbol{\theta}} + h_{ap}(\dot{\boldsymbol{\theta}},\boldsymbol{\theta}) + g_{ap}(\boldsymbol{\theta}) - J^T f_r \\ &\quad + \alpha(M_{\text{human}}(\boldsymbol{\theta})\ddot{\boldsymbol{\theta}} + h_{\text{human}}(\dot{\boldsymbol{\theta}},\boldsymbol{\theta}) + g_{\text{human}}(\boldsymbol{\theta})) \in \mathrm{R}^2 \end{aligned} \tag{11.2}$$

$$\ddot{\boldsymbol{p}} = J\ddot{\boldsymbol{\theta}} + \dot{J}\dot{\boldsymbol{\theta}} \in \mathrm{R}^3 \tag{11.3}$$

$$\begin{aligned}\boldsymbol{\tau} &= M_{ap}(\boldsymbol{\theta})J^{-1}(\ddot{\boldsymbol{p}} - \dot{J}\dot{\boldsymbol{\theta}}) + h_{ap}(\dot{\boldsymbol{\theta}},\boldsymbol{\theta}) + g_{ap}(\boldsymbol{\theta}) - J^T f_r \\ &\quad + \alpha(M_{\text{human}}(\boldsymbol{\theta})J^{-1}(\ddot{\boldsymbol{p}} - \dot{J}\dot{\boldsymbol{\theta}}) + h_{\text{human}}(\dot{\boldsymbol{\theta}},\boldsymbol{\theta}) + g_{\text{human}}(\boldsymbol{\theta})) \in \mathrm{R}^2 \end{aligned} \tag{11.4}$$

Equations (11.1) and (11.2) are for one leg based on the center of the hip joint, and the component of the angle vector $\boldsymbol{\theta}$ is the hip joint and the knee joint. When the user wants to carry out gait training, to get the correct gait motion, the impedance control method is selected. On the other hand, if the user wants to have walking assistance, the torque control method is selected.

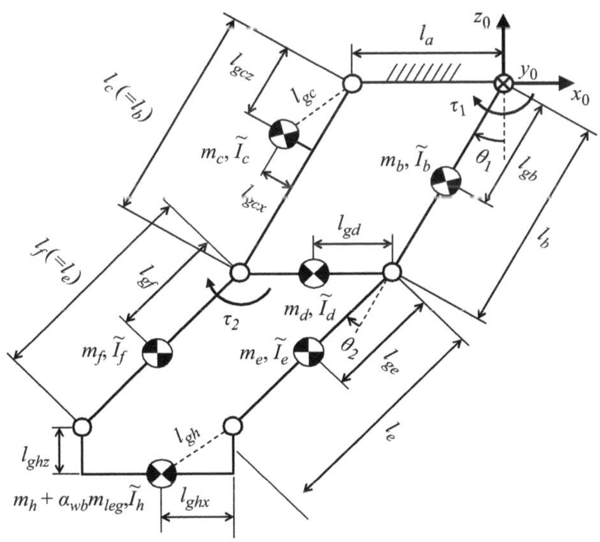

Figure 11.11 Apparatus model to make equation (side view). ©, with permission, from [9]

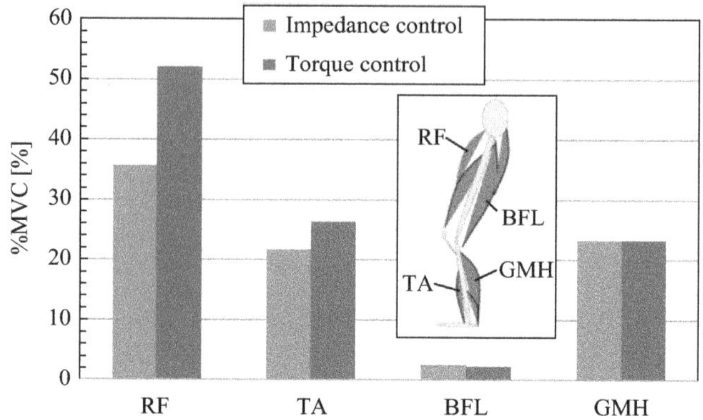

Figure 11.12 Example data of %MVC of one unaccustomed subject while walking. ©, with permission, from [10]

To verify the validity of the control methods of the apparatus, we performed interviews and measuring experiments of electromyography (EMG) during assisted walking, which was controlled using (11.1) and (11.2). Subjects were five able-bodied men (age: 21–24, height: 165–172 cm), and three subjects were unaccustomed to the apparatus, while two subjects were accustomed to the apparatus. The two accustomed subjects had previously carried out the training smoothly for about 1 h.

At first, all subjects walked wearing the apparatus with both control methods on a flat floor, and compared each feeling, and measured EMG. EMG was measured with FREEEMG 300 (BTS) while walking every 10 steps. The measured data was collected from four points on the right leg (Tibialis Anterior muscle: TA, Medial Head of Gastronemius muscle: GMH, Rectus Femoris muscle: RF, Long Head of Biceps Femoris muscle: BFL). The distance between the electrodes of the EMG sensor was 34 mm; the signal was converted from analog to digital signal whose sampling time was 1 kHz, sent the signal wirelessly, and recorded it with a computer. IEMG (integrated EMG) at intervals of 0.1 s was calculated from the EMG data, and the maximum values of IEMG in each step was averaged, and the averaged value was compared with MVC (maximum voluntary contraction), therefore the %MVC (percentage of MVC) of each control method was derived.

In the case of the unaccustomed subjects, they felt difficulty to walk smoothly in both control methods. An example data of %MVC of one unaccustomed subject while walking is shown in Figure 11.12. From this result, especially the front muscle; RF and TA were hyper activated, because it was hard to adapt their own motion to the apparatus immediately.

On the contrary, Figure 11.13 shows an example data of %MVC of one accustomed subject while walking. From this figure, the activity of RF and TA of the torque control was lower than the data of Figure 11.13. As subjects become more accustomed to the apparatus, torque control method assists the leg raising-up motion adequately. However, in the case of BFL and GMH in Figure 11.13, the

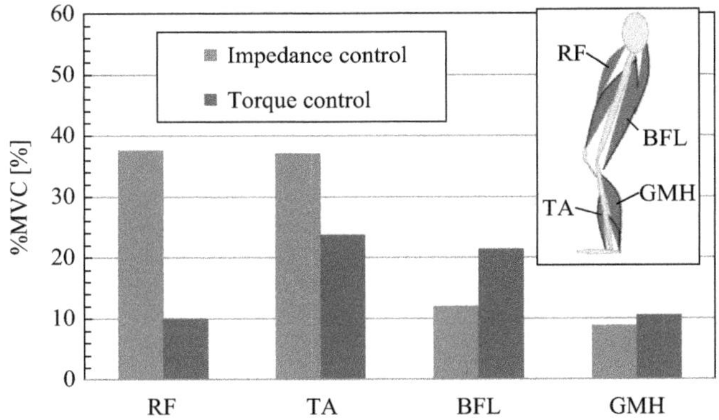

Figure 11.13 Example data of %MVC of one accustomed subject while walking. ©, with permission, from [10]

results of the impedance control were lower than the data of the torque control. Because the impedance control method prepared ideal trajectory of walking, so the kicking motion was carried out correctly. Therefore, we suggested to select the control methods for different applications:

For walk promotion: Torque control method is better.
For gait training: Impedance control method is better.

Furthermore, we carried out the walking experiment on various floors; slope, step, stairs, while wearing the apparatus of the torque control method. Subject was an accustomed, able-bodied man (age: 24, height 172 cm). For example, Figure 11.14 shows the photos of walking and climbing up a step continuously, and Figure 11.15 shows from the downstairs to the flat floor. As shown in these figures, as a result, he could walk smoothly on various floors. For future work, we will research decreasing muscle activity while walking on various floors.

11.3 Whole body motion support type mobile suit

We developed a whole body motion support type mobile suit as shown in Figure 11.16. This suit can be used separately for supporting the upper and/or lower limbs as shown in Figures 11.6 and 11.17. The assistive apparatus for upper limbs (Figure 11.17) is designed for the patients who can control their finger but they cannot lift up their arms themselves, especially myopathy patients. The mechanism of the assistance is utilized the differential gears to lose the weight and volume of the mechanical arm, and that enabled to configure three motors to drive two DOFs for the shoulder and one DOF for the elbow around the root of the mechanical arm.

This suit and the lifter (Figure 11.10) can be used by motor palsy patients, people who have suffered from strokes, patients with spinal-cord-injury, and people

Figure 11.14 Experiment of wearing the apparatus on a step (0.14 s/photo). ©, with permission, from [10]

with central nerve disorders. Using this device, these patients can recover normal gait with no risk of falling. This suit is equipped with 10 motors, and the total power is 950 W. The control system of the suit is shown in Figure 11.18. In this system, a self-contained computer (LEPRACAUN-CPU, General Robotics Inc.) controls 10 motors in a lump and control interval is 1 ms.

It is important to activate the brain extensively, not only PFC (prefrontal cortex), but also other areas such as PSC (primary somatosensory cortex), PMA (premotor area), PMC, and SMA (supplementary motor area), which are associated

Figure 11.15 Experiment of wearing the apparatus from the downstairs to the floor (0.27 s/photo). ©, with permission, from [10]

to motor learning. To verify the effectiveness of our developed suit for Neuro-Rehabilitation, the cerebral activity during walking using the suit and normal gait without the suit are measured with NIRS and compared. We used ETG-4000 (Hitachi Medical Corporation) and measured the variation of Oxy-Hb with 44 probes at tetragonal intervals of 3 mm. The measured areas are PFC (associated with consideration), PMA (associated with generate motion with sense), SMA (associated with generate motion with memory), PMC (associated with the output signal of motion), and PSC (associated with the received signal of somatesthesia) as shown in Figure 11.19. The experimental block design is as follows:

{Rest (Pre) 20 [s], Task 40 [s], Rest (Post) 40 [s]} *5 [sets].

Each group of eight tasks is shown in Table 11.1. By comparing the results of Tasks 1–4, the influence of the difference of walking velocity and swing arms or not are estimated. The walking velocity of each task was determined in response to the value which each subject felt comfortably during walking previously. By comparing the results of Tasks 3–6, the influence of assistance or not and swing arms or not are estimated. By comparing the results of Tasks 7 and 8, the influence of assistance or not for swing arms are estimated. In Task 6–1, subjects walked as well as grasping the grip which was attached the treadmill. In Task 6–2, subjects

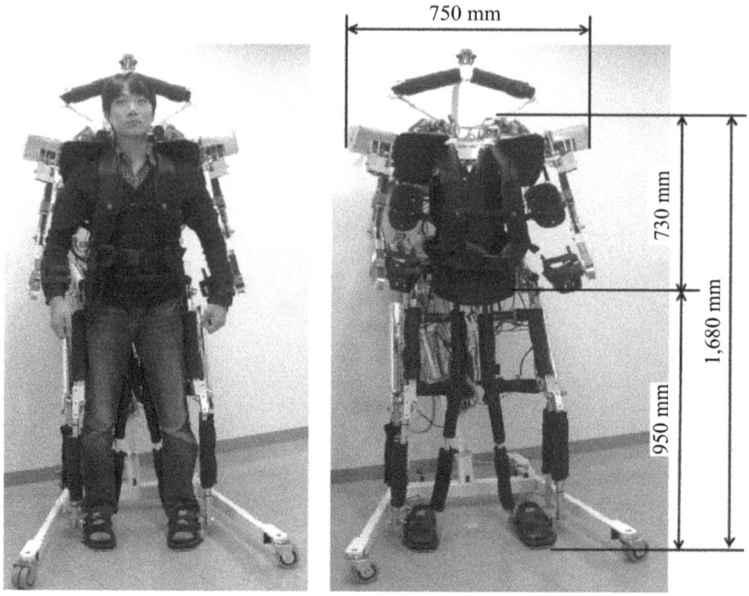

Figure 11.16 Whole body motion support type mobile suit. ©, with permission, from [11]

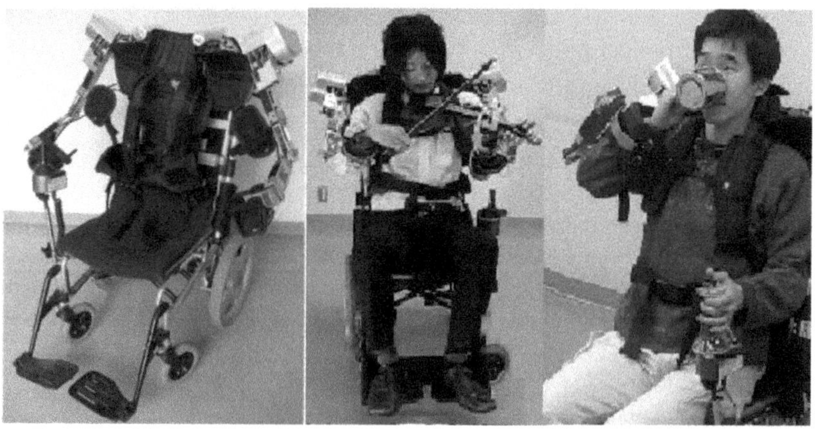

Figure 11.17 ADL assistance apparatus for upper limbs. ©, with permission, from [11]

walk with arms crossed arms. Each task was carried out five times respectively and averaged. Each measured data was processed as the baseline, which takes mean data for durations 10–20 and 90–100 s. Furthermore, to compare the activities in each area and for each subject, effect size [18,19] was calculated by using (11.5).

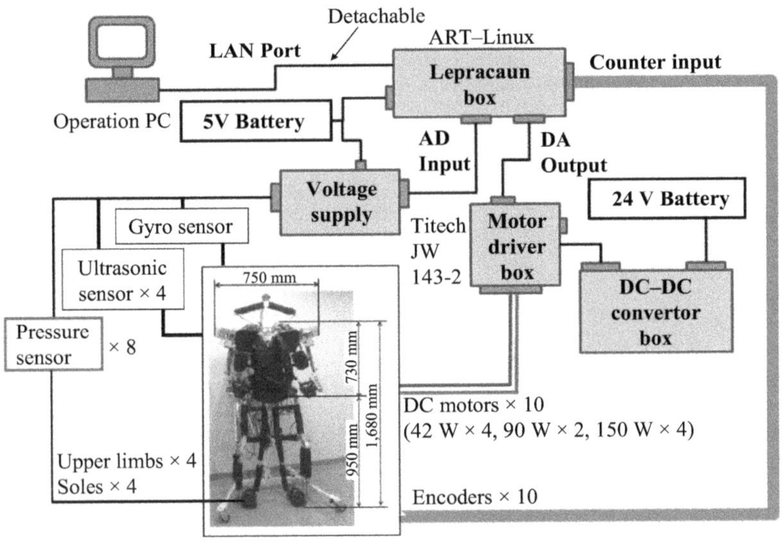

Figure 11.18 Control system of the suit. ©, with permission, from [11]

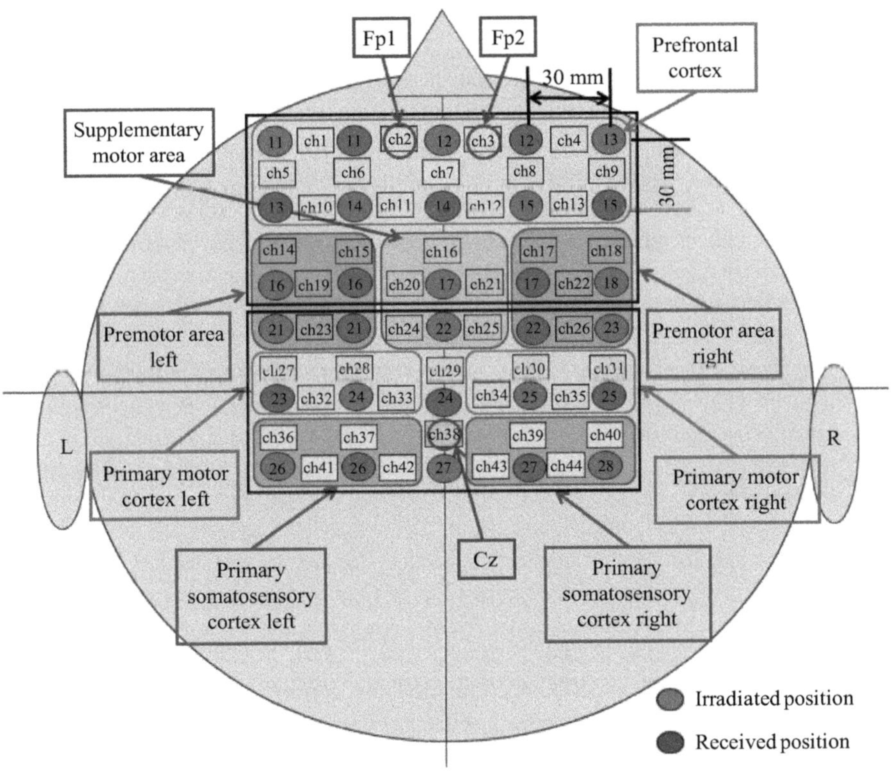

Figure 11.19 Configuration of each probe. ©, with permission, from [11]

Table 11.1 Example of tables. ©, with permission, from [11]

Task no.	Walking velocity (km/h)	Assist	Arm swing
1.	0.8–2.8	None	Yes
2.	0.8–2.8	None	No
3.	0.5–0.6	None	Yes
4.	0.5–0.6	None	No
5.	0.5–0.6	Upper and lower limbs	Yes
6–1.	0.5–0.6	Lower limbs	No (grasp grip)
6–2.	0.5–0.6	Lower limbs	No (cross arms)
7.	0 (Standing)	None	Yes
8.	0 (Standing)	Upper limbs	Yes

$$\text{Effect Size} = (\text{Rmean} - \text{Nmean})/\text{NSD}, \tag{11.5}$$

where Rmean (mM mm) is the mean data of each task, Nmean (mM mm) is the mean data of Task 1, NSD (mM mm) is the standard deviation of the Rest of Task 1. For fear of the fatigue by the influence of each task's turn, the tasks were carried out in order of low fatigue for un-accustomed subjects we assumed, i.e., (1) Only swing arms (Tasks 7 and 8), (2) Walking without the device (Tasks 1–4), (3) Walking with the device (Tasks 5 and 6). Furthermore, we asked all subjects to take an enough rest break between tasks.

The first experiment was carried out on a treadmill as shown in Figure 11.20. To maximize output and to reduce expended power for the compensation of the suit's weight, the weight bearing lifter was used to lift up the suit. However, to treat the condition of the floor reaction force the same for all Tasks, the lift didn't lift up subject's weight. Subjects were six; three un-accustomed subjects (able-bodied men, age: 21–24, height: 170–176 cm, weight: 60–66 kg) and three accustomed subjects (able-bodied men, age: 22–25, height: 170–176 cm, weight: 60–65 kg). Un-accustomed subjects did not have an experience to use the suit, and practice to walk with the suit for only 1 h before measured. Accustomed subjects have an experience to use the suit over one year as a staff of developing the suit. The mean results of the brain activity with each task for three un-accustomed subjects and three accustomed subjects are shown in Figures 11.21 and 11.22.

From Figure 11.21, most results of the brain activity of unaccustomed subjects are activated than the result of Task 1. The results of Tasks 2 and 4 hardly changed, and the result of Task 3 slightly increased. Therefore, it is possible to influence the variation of walk velocity to cerebral activity. The results of Tasks 5 and 6 were also higher than the results of Tasks 3 and 4. PFC is especially increased, therefore the subjects might be more conscious for walking motion to keep own posture, by wearing the apparatus. However, from the result of Task 6–1 (grasp the grip) in Figure 11.22, PFC is decreased. By grasping the grip and receiving lower leg's assistance, subjects did not have to control to keep posture and consider how to behave to walk. From the result of Task 6–2 (cross arms) in Figure 11.22, the activities of PFC, PMA, PMC, and PSC; the areas are concerned about his/her

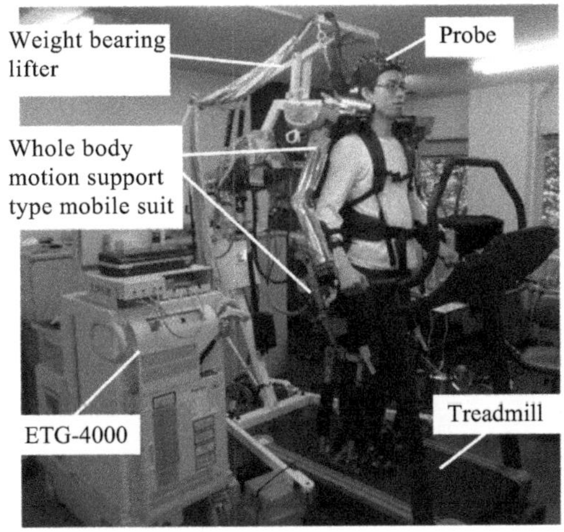

Figure 11.20 Component devices of the experiment for gait training on a treadmill. ©, with permission, from [11]

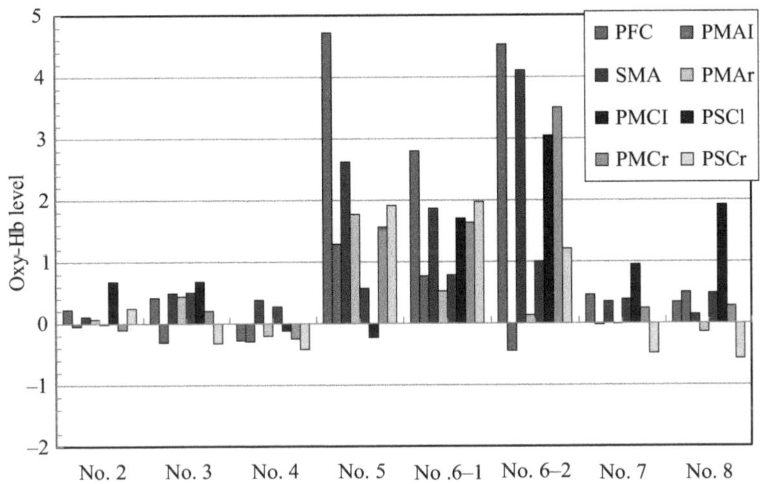

Figure 11.21 Results of mean data of each task on a treadmill (three unaccustomed subjects). ©, with permission, from [11]

motion, are decreased. Therefore, it is hard to gain effective training only while walking relying on the apparatus without swinging arms. The results indicate the subjects accustomed the apparatus and they could walk smoothly, because PMA is concerned about the information about the sensation. However, from the result of Task 5, even though the subjects are accustomed, PFC and SMA are slightly activated. By swinging arms, inactivation of the cerebral activity is prevented, and it is effective for Neuro-Rehabilitation. From the results of Tasks 7 and 8, only

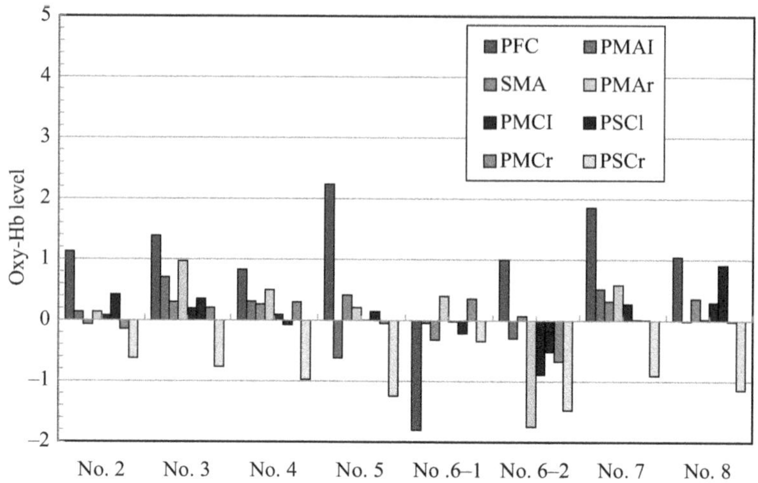

Figure 11.22 Results of mean data of each task on a treadmill (three accustomed subjects). ©, with permission, from [11]

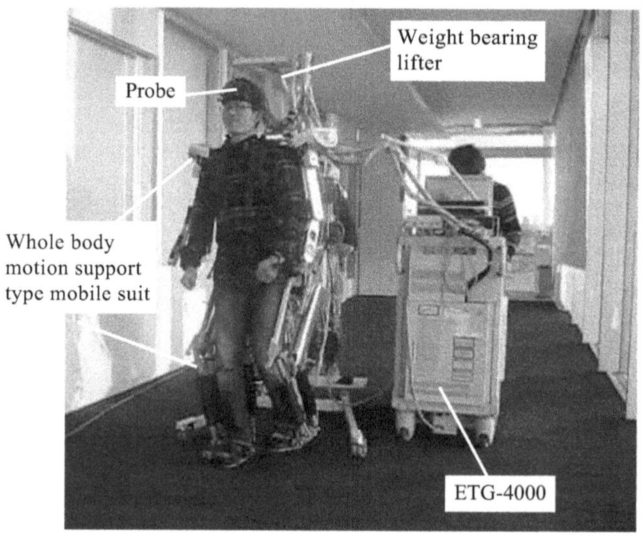

Figure 11.23 Component devices of the experiment for gait training in a corridor. ©, with permission, from [11]

swinging arms activated their brains, even though with the upper limbs assistance and by the accustomed subjects. Especially, SMA is activated. It could also be possible to activate CPG (central pattern generator) which generates rhythmic motion.

Next experiment was carried out in a corridor with an outside view as shown in Figure 11.23. Figure 11.24 shows the mean result of two accustomed subjects.

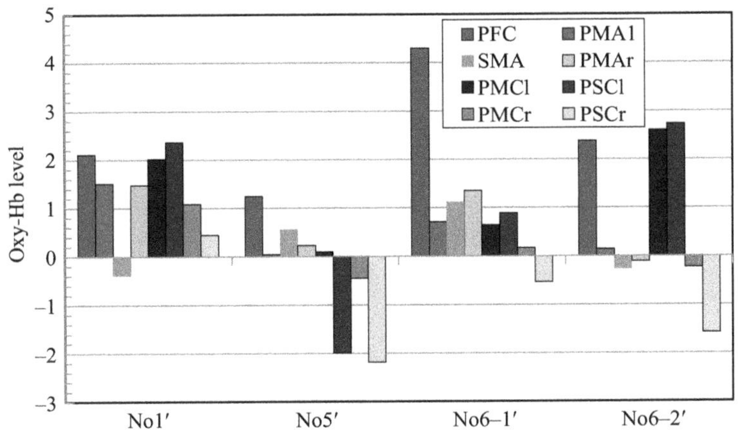

Figure 11.24 Results of mean data of each task in a corridor (two accustomed subjects). ©, with permission, from [11]

To distinguish the experiments between on a treadmill and in the corridor, the task numbers of the experiment in a corridor are labeled with a prime "′" symbol (e.g., Task 1′). From the result of Task 1′, all area is activated except for SMA. Gait training in a corridor is apparently more effective for Neuro-Rehabilitation than the training on a treadmill. In Task 5′, all subjects said that they could walk the most comfortably in all experiments; however, the result is almost the same outcome as the result of Task 5 in Figure 11.22. Therefore, if the patient cannot do gait training only on a treadmill according to his/her condition or environment, it is possible to achieve the effective rehabilitation by swinging arms (even though they are assisted with a suit). In the result of Task 6–1′ (grasp the grip), most of the areas are activated, however, the result of Task 6–2′ (cross arms) was inactivate except for PFC, PMCl, and PSCl. From the subject's comments, even though he didn't swing his arms, by grasping the grip while walking, he felt like as if he drove a kind of vehicle. Consequently, it is most effective for gate training to walk as well as moving from the aspect of motivation.

11.4 Close-fitting type of walking assistive apparatus

For gait training of hemiplegic patients, it is necessary to reduce the weight of the apparatus which is attached to the patient's leg. To address this problem, we developed a walking assistive apparatus using a flexible shaft which can raise the user's leg utilizing stretch reflex to bi-articular muscle of the user, by only assisting the ankle joint as shown in Figure 11.25. Some photos of wearing the apparatus are shown in Figure 11.26.

Figure 11.27 shows major muscles of a leg. The mechanism of raising a leg only by assisting the ankle joint can be explained as follows (Figure 11.28): (1) By the dorsiflexion of the ankle joint by the apparatus, GMH (Medial Head of

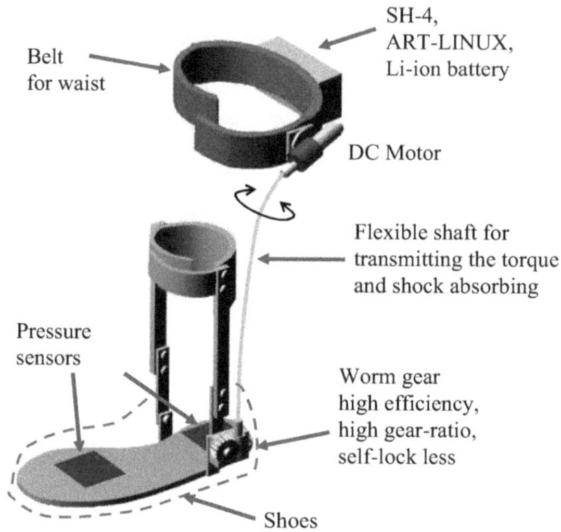

Figure 11.25 Walking assistance apparatus of flexible shaft and worm type. ©, with permission, from [14]

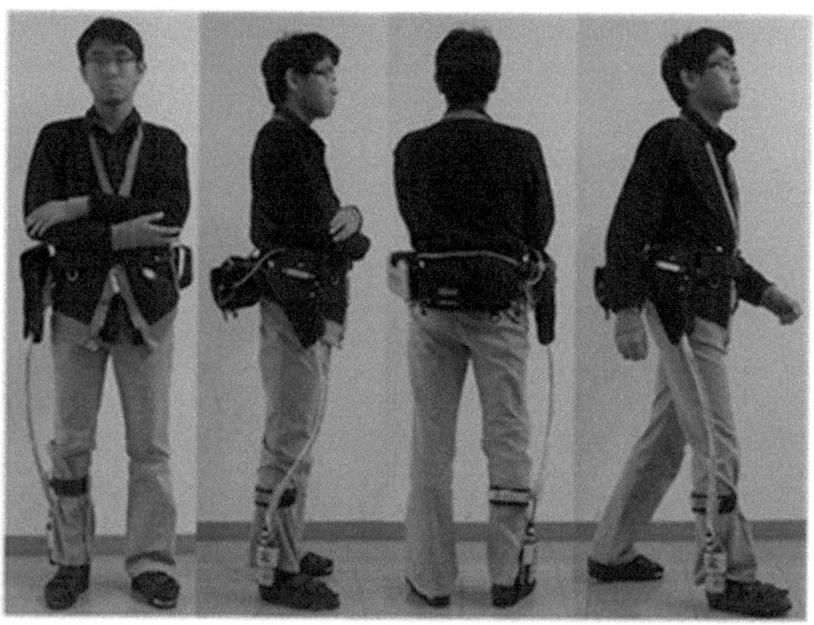

Figure 11.26 Photos of wearing the walking assistance apparatus of flexible shaft and worm type. ©, with permission, from [12]

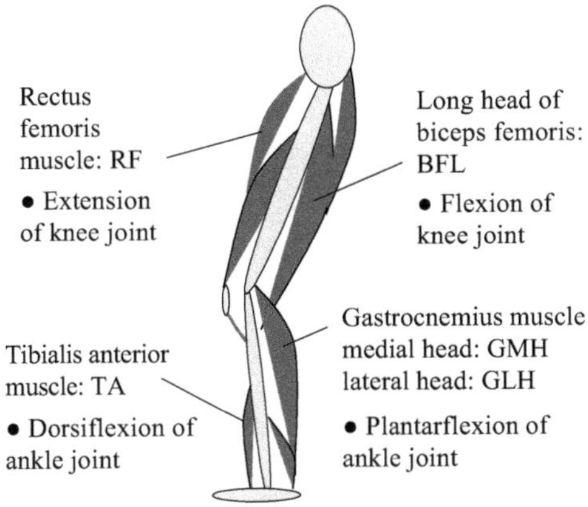

Figure 11.27 Mechanism to raise a foot by dorsiflexion. ©, with permission, from [14]

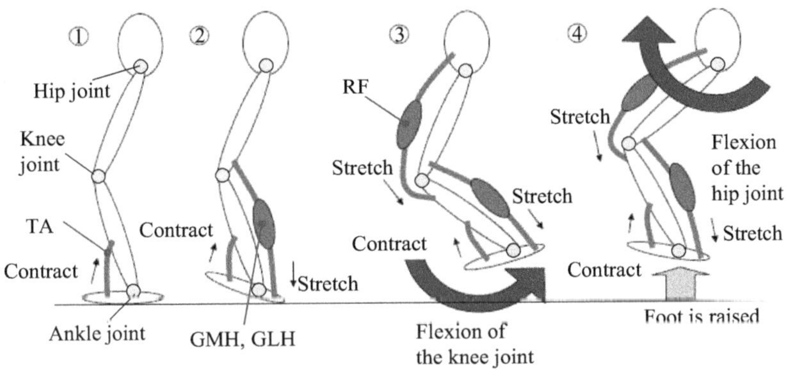

Figure 11.28 Mechanism to raise a foot by dorsiflexion assistance. ©, with permission, from [12]

Gastronemius muscle) of bi-articular muscle is diminished by stretch reflex, and the knee joint is inflected. (2) By inflecting the knee joint, RF of bi-articular muscle is also diminished by stretch reflex, and the foot is raised. By utilizing this method, the actuator for the knee joint does not have to be equipped and the load on the user's leg can be decreased.

As shown in Figures 11.27 and 11.28, the motor of this apparatus is attached on the waist belt of the user, and the output torque of the motor is transmitted by the flexible shaft to the worm gear box on the ankle joint [8].

The flexible shaft can bend but it can also transmit the torque with 95% efficiency. To convert the direction of the shaft rotation to a right angle, we used a worm gear, because the width of the gearbox is smaller than a bevel gear. We used the multiple type worm gear as show in Figure 11.29, to prevent the self-lock phenomenon for the safety of the user. Furthermore, to obtain a high efficiency,

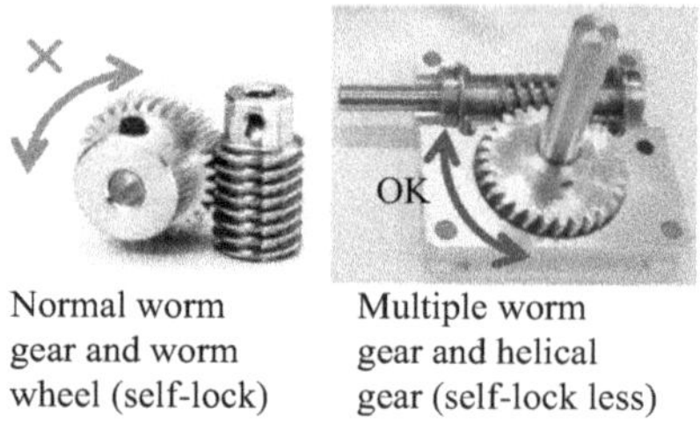

Figure 11.29 Comparison between the normal worm gear (left) and our used worm gear (right). ©, with permission, from [12]

Table 11.2 Subjects for walking experiment

Patient	Sex	Age	Brunnstrom stage (Brs)
A	Male	68	4
B	Male	68	5
C	Male	60	6
D	Female	65	4

a helical gear is used as a worm wheel. By using the helical gear for meshing with the worm gear, the area of contact decreases. The gear we used is at about 70% efficiency. Generally, the efficiency of the normal gear pair of a worm gear and a worm wheel is under 50%, as the area of contact of this gear pair is larger than that of ours. However, our gear pair's stress is larger than normal, therefore our gear pair's lifetime limit is shorter than normal.

This apparatus is controlled with a SH4 computer, whose OS is ART-LINUX. This system can use a hybrid control method, in which torque and angle are controlled with measurement of the output torque from the angle variation data from both ends of the flexible shaft as shown in Figures 11.26 and 11.27. To estimate the walking phase of each leg of the user, pressure sensors were attached under the thenar eminence and the heel of the sock liner as shown in Figure 11.9 and the pressure variation at each sensing point was measured.

We carried out the walking experiment with four hemiplegic patients as shown in Table 11.2. First, they walked for 10 m without wearing the device, and 10 m wearing the device repeatedly with resting in between each walking experiment. The target angle and torque data were only tuned on the first equipped experiment. We recorded video from the side and front angles of walking. We analyzed two angle variations; ankle joint from the side angle and hip joint from the front angle. We measured angle variations of each joints from the video, and divided each cycle from the data of pressure sensors.

Figure 11.30 (left) shows the result data of the measured mean ankle joint data of Patient A. From this figure, he was talipes equius except in the range from 30% to 50% of the walking phase (the data of without device, black line). However, after

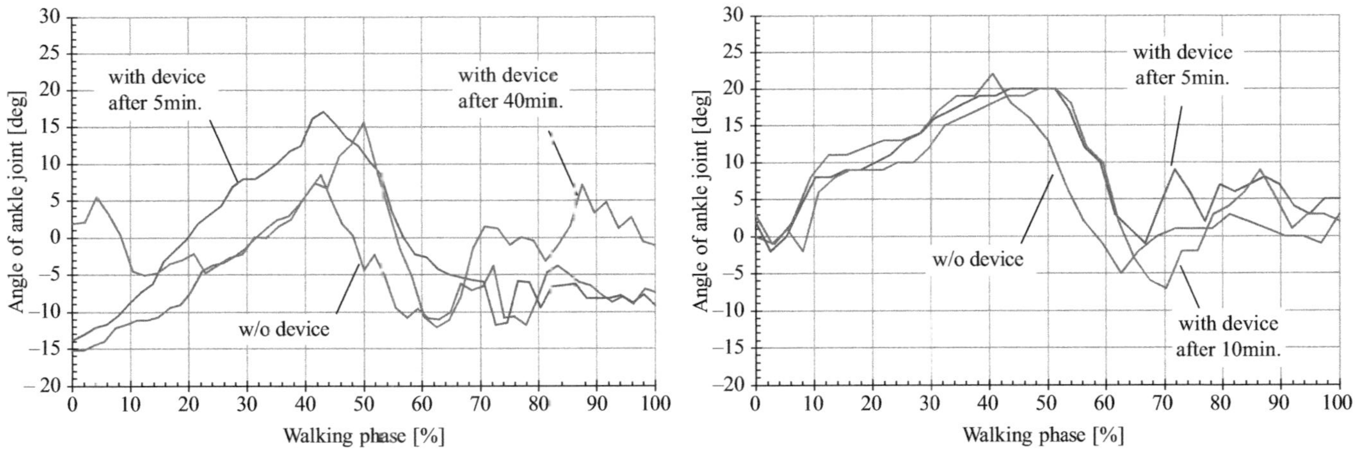

Figure 11.30 *Variation of the angle of the ankle joint (left: hemiplegic patient A, right: hemiplegic patient B). ©, with permission, from [12]*

walking with the apparatus for 5 min, he dorsiflexion range was extended from 20% to 60% (the data of with device, gray line). This range was in stance phase, therefore, the patient could make contact from heel to toe on the ground the whole time in stance phase. Furthermore, after 40 min from the start of walking while wearing the device, at the range from 70% to 100% during swing phase, this patient could increase the dorsiflexion, therefore, he improved to safety gait. The result data of Patient B is shown in Figure 11.30 (right). From the data without the device (black line), the subject could not achieve dorsiflexion in swing phase. However, by attaching the device for only 5 min, in the range from 70% to 100% in swing phase, dorsiflexion increased. Furthermore, after 10 min, at the range from 65% to 85% between stance and swing phase, plantarflexion increased as shown in Figure 11.30 (right). From this improvement, the subject was able to kick the ground strongly, his stride expanded, and his walking cycle diminished, therefore, his gait became closer to an able-bodied person's gait. From these results, we confirmed the effectiveness of this walking apparatus.

Figure 11.31 shows the relation between the training time and the angle of the hip joint for abduction. Especially, Patient A and D were heavy circumductive gait. The variations of the angle of the hip joint before using the device and after using device were compared. In this graph, when it's time to start training with the device, it is defined as 0 min, therefore, the plots before 0 min are the experiments without the device. As shown in Figure 11.31, apparently the angle decreased by using the device. However, after 20 min of starting training with the device, both angle data of patients A and D were slightly increased than former plotted data. From their comments, at the beginning of gait training with device, they were conscious of the gait which was improved by the device, however, according to the continuing the training over 20 min, they accustomed the device and they were likely to walk by usual gait (circumductive gait) again unconsciously. To achieve more effectiveness of this gait training, it is necessary not only to keep motivation

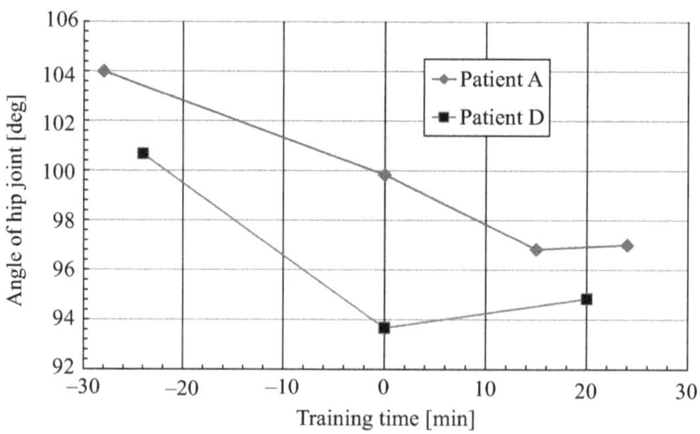

Figure 11.31 Relation between training time and angle of hip joint (hemiplegic patients A and D). ©, with permission, from [12]

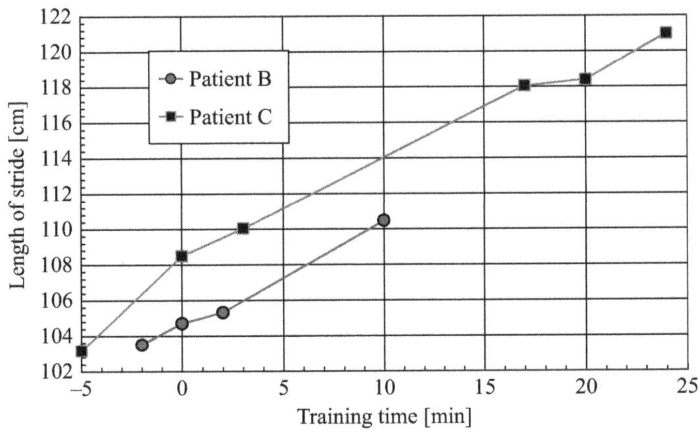

Figure 11.32 Relation between training time and the length of stride (hemiplegic patients B and C). ©, with permission, from [12]

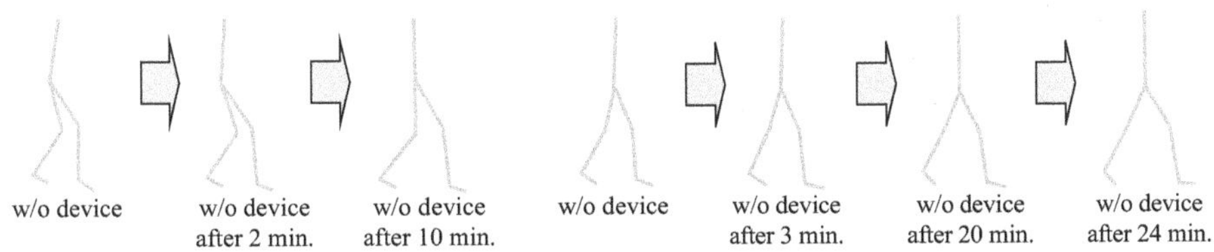

Figure 11.33 Posture while walking (left: hemiplegic patient B, right: hemiplegic patient C). ©, with permission, from [12]

for this training but also to tune of decreasing the targeted cycle and increasing the targeted angle according to the condition of the subject.

Figure 11.32 shows the relation between the training time and the length of stride. Figure 11.33 shows the posture while walking. From the recorded video from the side angle of walking, the pictures of each subject at almost same phase (start of double stance phase) were extracted. These lines were drawn by connecting the markers on subjects. By using the device, the gait of Patients B and C improved in a moment and his droopy posture changed to straight posture. Especially, in the case of Patient C, the length of stride extended obviously according to the training time and the posture made a strong recovery.

From these results, by using our developed device, the gait of the patients improved and we confirmed the effectiveness of the device.

11.5 Walking support robot "RE-Gait®" and "RE-Gait®Light"

The previous type of our developed apparatus (Figure 11.26) had a flexible shaft and worm gear was relatively heavy weight (5.6 kg) for patients. Therefore, we

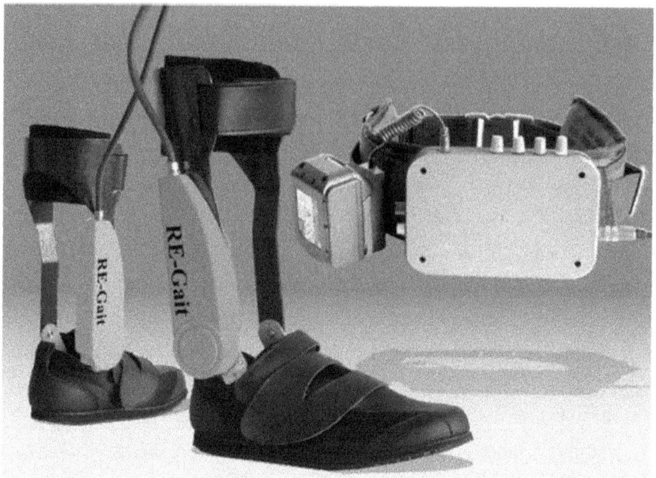

Figure 11.34 Walking support robot RE-Gait®

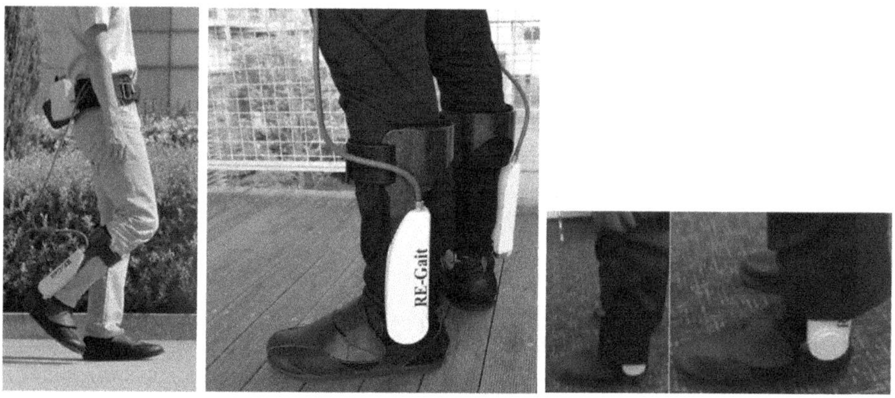

Figure 11.35 Photos of wearing RE-Gait®

developed RE-Gait®; a new type of the apparatus as shown in Figure 11.34, which can be used not only for neuro-rehabilitation but also for the promotion of exercise, because it is very small (able to wear under the pants as shown in Figure 11.35 (right)), light weight (1 ft: 1 kg). The photos of wearing RE-Gait® are shown in Figure 11.35. This apparatus assists the ankle joints of the equipped person with the motor controlled with a microcomputer. The insole is attached to pressure sensors, and the apparatus can grasp the phase of the contact condition between the sole and the ground as shown in Figure 11.9. In the reference [20], motion intention estimation is important and it requires the human's control method for locomotion and the sensing to understand the state and intent. It is necessary to prepare the hierarchical controller. However, if the robot has a complex controller, the equipped

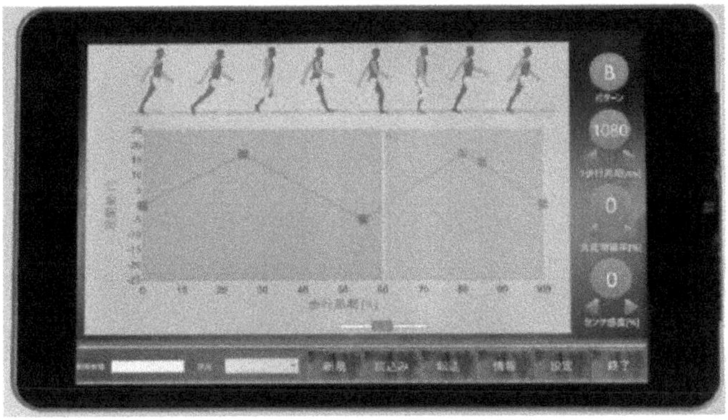

Figure 11.36 Setting program of the tablet PC

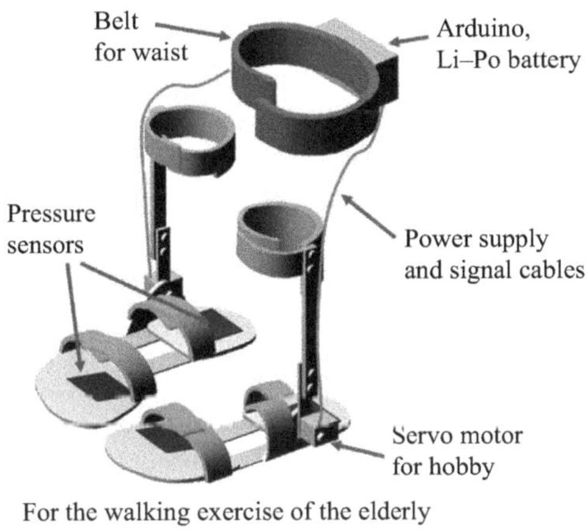

Figure 11.37 Servo motor type of the assistance apparatus and its parts. ©, with permission, from [14]

person may be hard to estimate the motion of the robot. Especially, the main targeted user is the elderly and operator is Physical Therapist. Therefore, in our device, the motion is controlled previously input for each user adequately with the tablet PC as shown in Figure 11.36, according to the contact condition of the pressure sensors. This setting program is very easy to use, because only shifting five points of the angle variation line, the data is automatically set to the apparatus wirelessly.

Furthermore, we are also developing a smaller type of the apparatus (RE-Gait®Light) as shown in Figures 11.37 and 11.38, which can be used not only for neuro-rehabilitation but also for the promotion of exercise, because it is able to put on the user's shoes, very small, lightweight (both feet assistance type: 1.6 kg), and low cost (initial cost is under $1,000). This apparatus assists the ankle joints of the equipped person with the servo motors directly according to the posture of

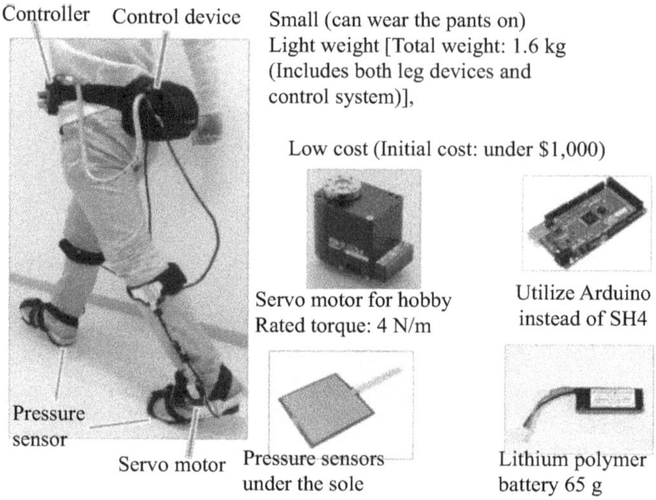

Figure 11.38 Photos of wearing the servo motor type of the assistance apparatus and its parts

him/her and selected from four conditions as shown in Figure 11.40 while not only walking, but also getting up, sitting, and standing (all round mode). These servo motors are controlled with angle control method or angle variation control method, and these target angles of each leg are fine-tuned according to the condition of the equipped person. The control system is so simple and cost effective, because we utilized the Arduino microcomputer, Arduino I/O ports and the servo motors which are hobby parts. It is easy to create, repair and distribute easily. Our goal is for the elderly or apoplexy patients to promote extended exercise after finishing the neuro-rehabilitation.

The apparatus can be equipped very easily and quickly, on his/her both feet while wearing shoes. The structure of the apparatus has only one frame on the outside of the leg, to enable the internal and external rotation of the ankle joint. Furthermore, the plate where the footrest is attached to pressure sensors which are under the sole, and the apparatus can grasp the phase of the contact condition between the sole and the ground as shown in Figure 11.9. For example, when the equipped person wants to stand up from the sitting posture, the center of pressure under the sole shifts from the heel toward the toe. Then the apparatus assists the plantarflex motion of the ankle joint as shown in Figure 11.39, and the lower legs are supported while standing up motion (all round mode). Another program is our conventional type which was programed with ideal angle variation data of the ankle joint as a target value previously, for gait training of the elderly or hemiplegic patients (walk mode). The difference of the targeted angle was shown in Figure 11.40.

Finally, we carried out the measurement experiment of the variation of the ankle joint while walking on a treadmill. Subjects were same as previous section.

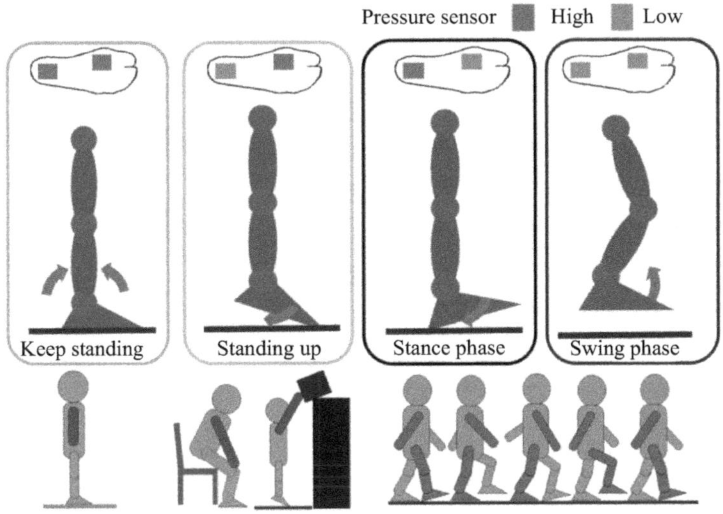

Figure 11.39 All-round mode of the control method of the apparatus. ©, with permission, from [14]

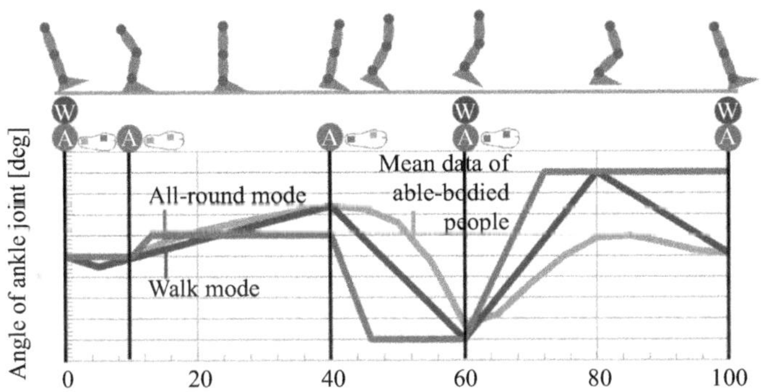

Figure 11.40 Comparison between the example of the targeted angle variation data (all round mode and walk mode) and mean data of able-bodied people. ©, with permission, from [14]

At first, they walked on a treadmill without using the apparatus. Next, they walked using the apparatus by all-round mode after enough rest. Third, they walked using the apparatus by Walk mode after enough rest. The velocity of treadmill was same, 1.7 km/h. We recorded video from the side angle of walking. We measured angle variation of ankle joint from the video, divided by each cycle from the data of pressure sensors, and calculated the average values of each phase.

Measured results of each subject are shown in Figure 11.41. The result of Subject A is Figure 11.41(a). All data can be found the motions of dorsiflexion and

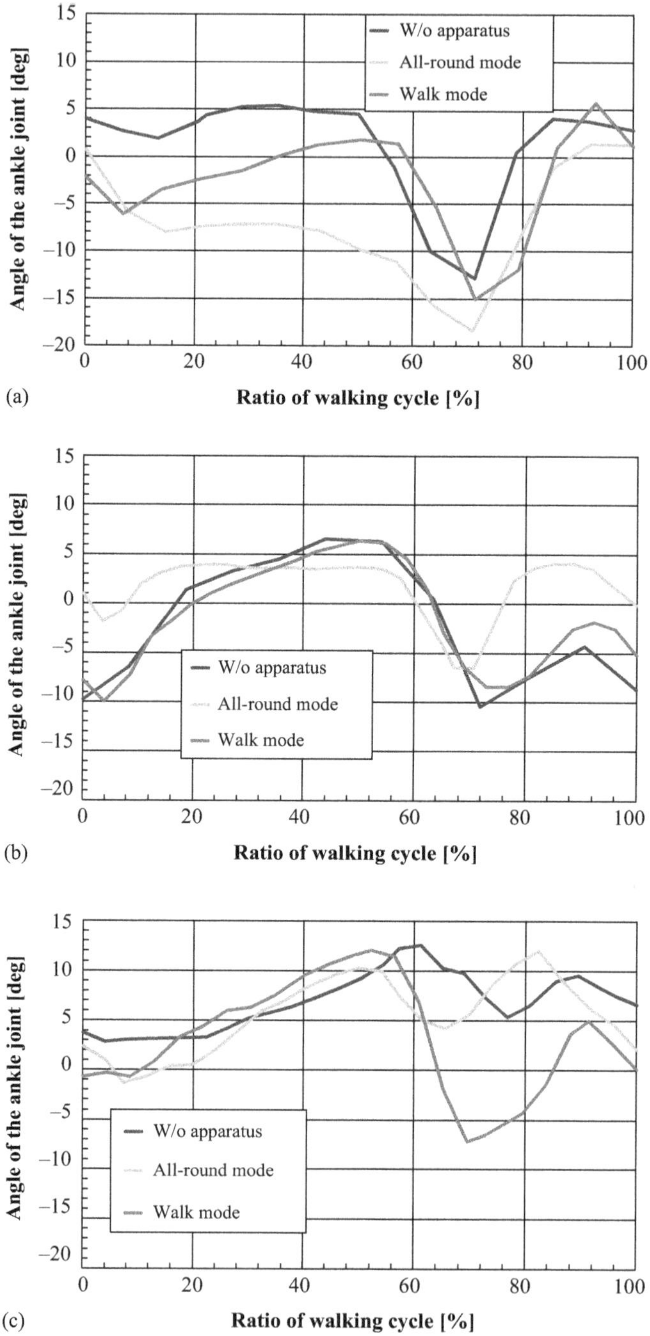

Figure 11.41 Measurement results of angle variation of the ankle joint: (a) Subject A (age: 22, able-bodied man), (b) Subject B (age: 42, able-bodied man), (c) Subject C (age: 66, able-bodied woman). ©, with permission, from [14]

plantarflexion. Especially, in the data of all-round mode, plantarflexion was larger than the other results. From the comment of the Subject A, all-round mode was most comfortable for him. Therefore, all-round mode is effective for relatively young people.

Subject B felt walk mode was most comfortable for him. The result of Subject B is shown in Figure 11.41(b), and the data without using the apparatus and walk mode were almost the same. Therefore he felt Walk mode did not interfere in his walk. However, it can be seen that dorsiflexion is increased in the data of all-round mode while swing phase. Therefore, all-round mode has ability of improving the gait for middle age people.

The data of Subject C is shown in Figure 11.41(c). In the data of without using the apparatus and all-round mode, dorsiflexion and plantarflexion were relatively small. However, in the data of Walk mode, both motions are improved and she felt Walk mode was most comfortable for her. Therefore, Walk mode is effective for the Elderly.

11.6 Control method of two-dimensional emotion map and future work

We proposed a method that promotes walking by using a two-dimensional emotion map. The two-dimensional map is made with two axes, the horizontal axis is either pleasant or un-pleasant, with the vertical axis being arousal or not, the same as the circumplex model of affect [21]. Figure 11.42 shows our assumed emotional shift while walking on the two-dimensional emotion map. When people walk for gait training as a rehabilitation, their emotion might be un-pleasant and their emotion vectors exist in the second and third quadrants. When the emotion vector exists in the first quadrant, a walking person will be promoted by his/her emotion and walking velocity will be accelerated. However, if the heart beat is too high, the

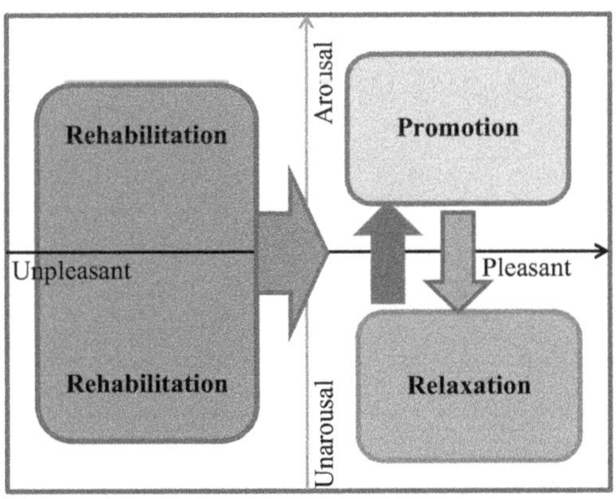

Figure 11.42 Assumption of the emotional shift while walking on the two-dimensional emotion map. ©, with permission, from [15]

emotion vector will be shifted toward the second or third quadrant. On the other hand, when the emotion vector exists in the fourth quadrant, a walking person will be relaxed by his/her emotion and walking velocity will be decelerated. However, if the heart beat is too low, he/she might stop walking. Therefore, to keep motivation and walking for a long time, it is important to control the walking person's emotion.

We also proposed a method that promotes walking by using a two-dimensional walking condition map. The two-dimensional map is made with two axes, the horizontal axis is the cadence (walking frequency) (step/min) which is the inverse number of the walking cycle, with the vertical axis being the stride length (m/step), same as [22]. Figure 11.43 shows our assumed walking condition shift on the two-dimensional map. From this reference, when the walking ratio is 0.006, which is the gradient value in this map, it is better for an able-bodied person's walk. In this work, this line is defined as an ideal walk line. When a patient tries to improve his/her gait, the point of the current location will be shifted toward the ideal walk line on the two-dimensional walking condition map. When the walking person walks more slowly with a smaller stride length than the current walk, the point will shift lower left along the ideal walk line in the two-dimensional walking condition map, and he/she will feel relax. On the other hand, when the walk person walks more fast with a longer stride length than the present walk, the point will shift upper right along the ideal walk line in the two dimensional walking condition map, and he/she will feel active. If some parameters of walking are able to adjust, and if the tendency of the point shift has some relations between the two-dimensional walking condition map and emotion map, the emotion while walking will be change to keep motivation and walking for a long time. Therefore, our purpose of this study is to confirm the relation of the tendency of the shift between the emotion vector on the emotion map and the point on the walking condition map.

We found the relation of the tendency of the shift in two maps. We developed the automatic control system according to the shift in the emotion map, which can

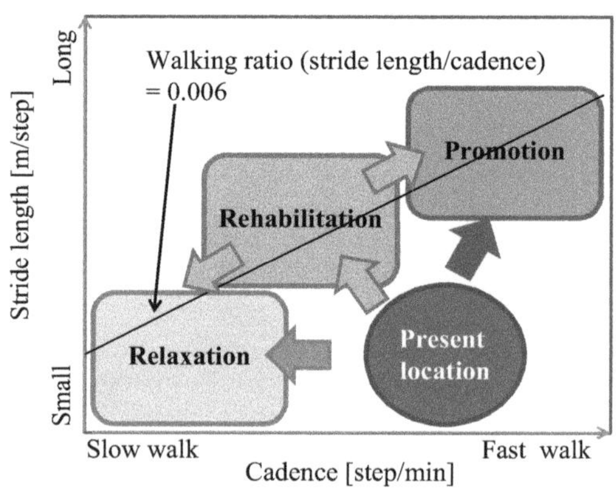

Figure 11.43 Assumption of the walking condition shift on the two-dimensional walking condition map. ©, with permission, from [15]

realize that the user can walk long time keeping the good heart beat value. At first, we evaluated the emotion from the measured body data, and two dimensional emotion map was made. A pleasant feeling can be evaluated from the face EMG or face temperature data. An arousal state can be evaluated from the variation of the ratio low frequency/high frequency of the heart beat or the variation of alpha wave of EEG. Second, when the user wants to promote exercise, the targeted cadence and stride length are increased using the previously measured relation ratio. Third, if the user's heart beat raises extremely, to shift to the relax emotion area, the targeted cadence and stride length are decreased. Now we are carrying out so many experiments [15–17]. As a future work, we will install this automatic emotion control system to RE-Gait® and RE-Gait®Light, and we hope so many patients recover their gait early and so many elderly can promote exercise continuously, and keep their health.

11.7 Conclusions

We have been developing walking assistive apparatus for gait training patients and promotion exercise of the elderly. One of them is already utilized by many patients to improve their gait. Furthermore, now we are developing automatic emotion evaluation system and by connecting to the walking assistive device, many elderly will be able to promote exercise easily. They will regain their independent life.

References

[1] Miyai, I., *et al.*, Cortical mapping of gait in humans: a near-infrared spectroscopic topography study, NeuroImage, 14, (2001), pp. 1186–1192.

[2] Suzuki, M., Miyai, I., Ono, T., *et al.*, Prefrontal and premotor cortices are involved in adapting walking and running speed on the treadmill: an optical imaging study, NeuroImage 23, (2004), 1020–1026.

[3] Suzuki, M., Miyai, I., Ono, T., Kubota, K., Activities in the frontal cortex and gait performance are modulated by preparation. An fNIRS study, NeuroImage 39, (2008), 600–607.

[4] Yano, H., Nakajima, Y., Iwata, H., Evaluation of effect of walking using near-infrared spectroscopy, Journal of the Virtual Reality Society of Japan, Vol. 12, No. 1, (2007), pp. 67–74. (In Japanese).

[5] Colombo, G., Joerg, M., Schreier, R., Dietz, V., Treadmill training of paraplegic patients using a robotic orthosis, Journal of Rehabilitation Research and Development, Vol. 37, No. 6, (2000), pp. 693–700.

[6] Zoss, A. B., Kazerooni, H., Chu, A., Biomechanical design of the Berkeley lower extremity exoskeleton (BLEEX), IEEE/ASME Transactions on Mechatronics, Vol. 11, No. 2, (2006), pp. 128–138.

[7] Yonetake, J., Toyama, S., Development of the Ultrasonic Motor-Powered Assisted Suit System, Proc. of the 1st KSME-JSME Joint Int. Conf. on

Manufacturing, Machine Design and Tribology (ICMDT), Seoul, Korea, CD-ROM, (2005).
[8] Sankai, Y., HAL: Hybrid Assistive Limb Based on Cybernics, Robotics Research, The 13th International Symposium ISRR, (2010), pp. 25–34.
[9] Tanaka, E., Ikehara, T., Yusa, H., *et al.*, Walking-assistance apparatus as a next-generation vehicle and movable Neuro-Rehabilitation training appliance, Journal of Robotics and Mechatronics, Vol. 24, No. 5, (2012), pp. 851–865.
[10] Tanaka, E., Suzuki, T., Saegusa, S., Yuge, L., Walking Assistance Apparatus Able to Select the Control Method According to the Purpose of the User, Proceedings of 14th International Symposium on Robotics and Applications (ISORA 2014), the World Automation Congress 2014 (WAC2014), IEEE SMC Society, August 3–7 2014 Waikoloa Hilton Village, Kona, Big Island of Hawaii, (2014). DOI: 10.1109/WAC.2014.6936036.
[11] Tanaka, E., Saegusa, S., Yuge, L., Development of a whole body motion support type mobile suit and evaluation of cerebral activity corresponding to the cortical motor areas, Journal of Advanced Mechanical Design, Systems, and Manufacturing, Vol. 7, No. 1, (2013), pp. 82–94.
[12] Tanaka, E., Muramatsu, K., Watanuki, K., Saegusa, S., Yuge, L., Walking assistance apparatus enabled for Neuro-Rehabilitation of patients and its effectiveness, Mechanical Engineering Letter, Vol. 1, No. 15-00530, (2015–12), pp. 1–15. DOI: http://doi.org/10.1299/mel.15-00530.
[13] Space Bio-Laboratories Co., Ltd., Development of walking support robot "RE-Gait®", http://www.spacebio-lab.com/03development/index-en.html.
[14] Tanaka, E., Muramatsu, K., Watanuki, K., Saegusa, S., Yuge, L., Development of a walking assistance apparatus for gait training and promotion of exercise, IEEE International Conference on Robotics and Automation Stockholm (ICRA 2016), Sweden, May 16–21, (2016), pp. 3711–3716. DOI: 10.1109/ICRA.2016.7487557.
[15] Tanaka, E., Muramatsu, K., Osawa, Y., Saegusa, S., Yuge, L., Watanuki, K., A Walking Promotion Method using the Tuning of a Beat Sound Based on a Two-Dimensional Emotion Map, *Proceedings of the AHFE 2016 International Conference on Affective and Pleasurable Design*, July 27–31, (2016), Walt Disney World®, Florida, USA, pp. 519–525. DOI: 10.1007/978-3-319-41661-8_50.
[16] Tanaka, E., Osawa, Y., Muramatsu, K., Watanuki, K., Saegusa, S., Yuge, L., Study of RE-Gait® as the Device that Promotes Walking using a Two-Dimensional Emotion Map, ROMANSY 21 -Robot Design, Dynamics and Control (Romansy 2016), *Proceedings of the 21st CISM-IFToMM Symposium*, June 20–23, Udine, Italy, (2016), pp. 369–376. DOI: 10.1007/978-3-319-33714-2_41
[17] Tanaka, E., Osawa, Y., Muramatsu, K., Watanuki, K., Saegusa, S., Yuge, L., Evaluation of a device that promotes walking using a two-dimensional emotion map, *Proceedings of the World Automation Congress 2016*

(WAC2016), July 31–August 4, Wyndham Grand Rio Mar Beach Resort & Spa, Puerto Rico (2016). DOI: 10.1109/WAC.2016.7582962.

[18] Schroeter, M. L., Zysset, S., Kruggel, F., von Cramon, D. Y., Age dependency of the hemodynamic response as measured by functional near-infrared spectroscopy, NeuroImage, Vol. 19, (2003), pp. 555–564.

[19] Suzuki, M., Miyai, I., Ono, T., Kubota, K., Activities in the frontal cortex and gait performance are modulated by preparation. An FNIRS study, NeuroImage, Vol. 39, (2008), pp. 600–607.

[20] Tucker, M. R., Olivier, J., Pagel, A., *et al.*, Control strategies for active lower extremity prosthetics and orthotics: a review. Journal of NeuroEngineering and Rehabilitation, Vol. 12, No. 1, (2015), pp. 1–29.

[21] Russell, J. A., A circumplex model of affect, Journal of Personality and Social Psychology, Vol. 39, No. 6, (1980), pp. 1161–1178.

[22] Diedrich, F.J., Warren, W.H., Why change gaits? Dynamics of the walk-run transition, Journal of Experimental Psychology: Human Perception and Performance, Vol. 21, No. 1, (1995), pp. 183–202.

Section 4

Commercialization issues

Chapter 12: Regulatory issues for exoskeletons
Chapter 13: Test methods for exoskeletons—lessons learned from industrial and response robots
Chapter 14: Ekso bionics

This section focuses on commercialization issues for wearable exoskeletons. For this, there are many business, management, finance and commercial details which need to be considered to ensure the research results achieved in developing technically innovative prototypes so that they can be taken forward to realize commercially viable products. Many of these relate to business decisions such as how the R&D team can be morphed into a commercialization team where instead of focusing on technical innovation, the attention turn to design for manufacturing, marketing and after sales activities. It is not the intention to focus on these issues here but instead to ensure the technical innovations will be able to satisfy regulatory requirements and hence have the potential for commercialization.

In this respect, mandatory safety requirements must be complied with and these are presented in Chapter 12 from machinery and medical device perspectives. Wearable exoskeletons are classified as different types of robots and as such will be regulated as robots. Traditional robots have been regulated as industrial robots where safety was ensured by essential keeping humans and robots separated; the safety requirements were presented in ISO 10218-1 and -2 but recently, new types of service robots have been emerging which are designed to allow close human–robot interaction, and also allow robot–human contact. In fact, medical applications of robots have also been initiated and these allow robots to invasively enter human bodies to perform a variety of clinical functions. It turns out that wearable exoskeletons will fall into new markets which do not yet exist but when they do, they will be governed by machinery and medical device regulations. The new needed regulations have started to be developed within ISO and IEC and the key Technical committees carrying out the standardization work are ISO 299 (Robotics) and IEC TC 62 (Electrical equipment in medical practice). New standards are beginning to emerge and the chapter gives an up-to-date summary by experts leading the standardization work within ISO and IEC from medical and non-medical robot application viewpoints.

When new regulations are formulated for new emerging markets, it takes time for the new requirements to be understood. Manufacturers need to ensure they understand the rationale used by the standardization experts in specifying the requirements and regulators need to understand how to test them. Without close coordination and open discussion between the stakeholders, it is difficult to get consensus on how normative new test methods can be developed so that consistent results are obtained. In this respect, Chapter 13 presents a viable approach that has been developed at NIST, USA for developing the urgently needed new methods for testing the exoskeletons that are beginning to emerge. It is proposed that lessons learned from established sectors of industrial and response robots be used to develop suitable new test methods. Key issues related to identifying performance metrics, sensory systems to be used for the human-related and wearable robot-related measurements. The ISO/IEC working groups are also developing test methods for the new standards but these are not yet developed sufficiently and hence possible test approaches which could be used are urgently needed from the research community.

Chapter 14 presents a case study from Ekso Bionics which has emerged as a market leader commercializing lower body exoskeletons for a variety of healthcare, industrial, military and consumer markets. Able-bodied people and persons with physical disabilities are targeted by the company for the exoskeletons being developed. The rehabilitation market is focused upon in particular, and the Ekso GT is being assessed by The Food and Drug Administration (FDA). The case study gives an excellent overview of the sector and how business activities can evolve rapidly in evolving markets. It is clear from the presentation that getting good discussions with regulatory bodies is essential so that the certification process can be carried out quickly and efficiently to get the new products commercialized.

Chapter 12
Regulatory issues for exoskeletons

Koen Chielens, Burkhard Zimmerman**, and Gurvinder S. Virk***,†*

Abstract

The chapter presents the safety standards which apply to wearable exoskeleton robots. Wearable robots are either regulated as machines or as medical electrical equipment. The machine regulations are formulated to comply with the Machinery Directive and the harmonised ISO 13482 safety standard has been recently published to present the safety requirements for personal care robots and which includes physical assistant robots. Medical exoskeletons are regulated under the Medical Device Directive and apply to patients needing medical help in carrying out human motion tasks. Both machine and medical regulations of wearable robots are described.

Keywords: Safety standards; medical and non-medical exoskeletons; machinery directive; medical device directive; compliance

12.1 Introduction

The body-worn exoskeletons discussed in this book can have a wide variety of application domains depending on the type of user and how the wearable robot will be used to help the wearer perform some task. The main exemplar classes are as follows:

- Class 1: Exoskeleton could be worn by a soldier to amplify human strength for carrying heavy equipment over rough terrain so that the destination is reached in good physical condition for engaging in combat as required in military scenarios. Gains of such military exoskeletons could be in excess of 5–10 or more, implying a soldier can have 5–10 times normal human strength.

*New Approach Consultant for Machinery Directive, Vinçotte, Belgium
**Quality Management, Hocoma AG, Switzerland
***Technical Director, InnotecUK, United Kingdom
†Trustee, CLAWAR Association Ltd, United Kingdom

- Class 2: Exoskeleton could be worn by workers to reduce physical burden in various manual tasks. Gains of such exoskeletons for helping manual workers could be in the range 2–5 so reasonably heavy tasks could be carried out by average ability persons.
- Class 3: Exoskeletons could be worn by healthy elderly persons for performing daily living activities. Gains of such assistive exoskeletons could be in the range 0–1 so supplementary support can be provided as needed to persons whose normal physical capabilities are diminishing due to being tired or because of ageing.
- Class 4: Exoskeletons could be used by disabled persons who have lost their functionality due to disease or other trauma and includes amputees. Gains of such exoskeletons should be around unity or just over so that 'normal' functionality can be provided by the external actuators to move human joints as required to perform desired motions.
- Class 5: Exoskeletons could be used by patients to perform a variety of operations such as rehabilitation, or assessment of capabilities by identifying impairments, or compensate for identified impairments, or alleviation of symptoms following impairment.

It is important to identify these classes of exoskeletons so that appropriate regulations can be applied when they are commercialized. Class 1 is military; whereas Classes 2 and 3 are machines and Classes 4 and 5 are medical electrical equipment type systems (or medical devices as stated in the EC's Medical Device Directive) which are covered by the, for example, International Electrotechnical Commission (IEC) 60601 family of standards. Medical and non-medical products have their own regulations which need to be satisfied and both will be considered here as exoskeletons can arise in both areas. For convenience, military regulations will not be covered but instead we can group them as special high-powered exoskeleton machines which will be discussed in this chapter under Class 2 exoskeletons as well discussing Classes 4 and 5 exoskeletons for medical applications.

We will focus here on free-walking body-worn exoskeletons; stationary exoskeletons for upper and lower extremities are also available especially for assistive and medical applications.

The main international standardization organizations responsible for developing standards are ISO (International Organization for Standardization [1]) and IEC [2]. ISO and IEC have jointly produced a number of guides providing advice on how to deal with wide range of issues related to standardization and the list of guides can be found at the ISO website [3], e.g. Guide 51 for how to include safety in standards [4]. There are also a number of guides on more specialized topics that are designed for use by committees working in particular sectors – for example, rules for drafting and presentation of safety standards for machinery (ISO Guide 78 [5]) or guidance on the inclusion of safety aspects in standards for medical devices (ISO/IEC Guide 63 [6]). The collection of guides is continually evolving.

We start the discussions by looking at key regulatory issues for medical and non-medical exoskeletons.

12.1.1 Exoskeletons: medical–non-medical applications

The key starting issues here are some fundamental definitions as to what is a machine, and what is a medical device (or medical electrical equipment). Although there are various definitions, it is appropriate to adopt those which are presented in appropriate international standards:

- Machine/Machinery: Assembly, fitted with or intended to be fitted with a drive system consisting of linked parts or components, at least one of which moves, and which are joined together for a specific application (from ISO 12100:2010 [7], Definition 3.1; see also the EC Machinery Directive [8]).
- Medical device regulation is complex and part of the complexity is that there are no harmonized definitions of key terms. Two important definitions need to be considered here namely medical devices and medical electrical equipment which are defined as follows:
  - Medical devices are defined in the EC Medical Directive 93/42/EEC [9] to mean any instrument, apparatus, appliance, software, material or other article, whether used alone or in combination, including the software intended by its manufacturer to be used specifically for diagnostic and/or therapeutic purposes and necessary for its proper application, intended by the manufacturer to be used for human beings for the purpose of
    - diagnosis, prevention, monitoring, treatment or alleviation of disease,
    - diagnosis, monitoring, treatment, alleviation of or compensation for an injury or handicap,
    - investigation, replacement or modification of the anatomy or of a physiological process,
    - control of conception,

  and which does not achieve its principal intended action in or on the human body by pharmacological, immunological or metabolic means, but which may be assisted in its function by such means.
  - Medical electrical equipment is defined in IEC 60601-1 ed. 3.1 [10] as electrical equipment having an applied part or transferring energy to or from the patient or detecting such energy transfer to or from the patient and which is
  - provided with not more than one connection to a particular supply mains supply; and
  - intended by its manufacturer to be used
    - in the diagnosis, treatment, or monitoring of a patient or
    - for compensation or alleviation of disease, injury or disability.

Using these important definitions, it is possible to see if the specific exoskeleton being considered falls under the machinery or medical regulations by identifying if the exoskeleton will be used by a healthy person or by a patient for some clinical function, respectively. It is most important to get this correct so that the correct regulations can be applied. In most cases, it is clear if the exoskeleton is medical or non-medical because it is based on the manufacturer's intended use for his product so that all necessary design and testing protocols from regulatory perspectives are completed but there is potential for misuse and current standards require manufacturers to mitigate foreseeable misuse situations. These are issues which are likely to evolve as wearable exoskeletons are relatively new and the markets for medical and non-medical applications have not yet been developed in any significant form. In addition, the boundary issues between them are also not clarified well enough especially when it comes to support activities of elderly persons for normal daily living.

As the final result of using any exoskeleton is to apply external force to a human joint, it is conceivable that medical and non-medical exoskeletons can be used interchangeably even though their regulations are quite different. As medical regulations demand more severe requirement, medical exoskeletons are likely to be more expensive than non-medical exoskeletons. This mismatch in pricing is likely to lead to non-medical assistive products to be used in medical applications and issues of misuse are likely to arise. If these are deemed to be 'foreseeable', manufacturers have a responsibility to address any potential issues arising from the misuse. In some cases, medical professionals have used products which do not formally comply with medical regulations and this can only be acceptable if the use has involved carrying out an acceptable risk–benefit analysis for the patient to warrant the use for approved clinical trials according to national laws. If individuals chose to buy the cheaper non-medical products from retail outlets rather than go to get them as prescribed by clinicians, the situation could lead to harm when the persons have significant medical problems needing addressing. Most of the national requirements include very strict roles also for professionals in the health care sector to use only approved or registered medical devices for their patients. How society and regulators will deal with such issues will be important to ensure the supply and demand for the needed systems is appropriate and can be nurtured and developed in a sensible manner. The main application which is effected is assistive exoskeletons for helping elderly persons perform normal daily living activities. It is important to note that ageing is normal and if assistance is needed by an elderly person to supplement any deficiency due to ageing, this should not be treated as a medical condition. However, normal healthy ageing is often accompanied with the onset of many medical conditions and so it is important to know what the wearable system is supplementing and if this has been affected in any individual case to decide if the exoskeleton needs to comply with medical or non-medical regulations.

12.1.2 Machinery exoskeletons

Machinery exoskeletons have to be designed and their risk assessment and risk reduction carried out according to ISO 12100:2010 [7] which presents the general principles for design machines to achieve mandatory safety requirements to ensure that the residual risk is acceptable. The main steps to be followed are as follows:

1. Try to achieve the safety requirements by means of inherently safe design
2. If inherently safe designs are not possible, then try to achieve the requirements by means of safeguarding or protective measures
3. If neither of these solutions is possible, then provide information for use to the operator (warnings, instructions) to assist the operator in achieving acceptable safety.

12.1.3 Medical exoskeletons

In contrast to machine safety for medical appliance, the process requires manufacturers to use ISO 14971:2007 [11] which deals with the processes for managing risks, primarily to the patient, but also to the operator, other persons, other equipment and the environment. The risk management comprises systematic application of management procedures and practices to the tasks of

1. Risk analysis + risk evaluation + risk control
2. Evaluation of overall residual risk acceptability (risk–benefit balance)
3. Risk management report
4. Production and post-production information (monitoring).

The key differences between the medical and non-medical regulation is that it is necessary to carry out a risk–benefit balance for the patient to define basic safety requirements and essential performances for medical exoskeletons.

In addition, for medical exoskeletons, the technical requirements are as varied as the individual patterns of disease of patients affected in the medical disciplines which are covered, e.g. by different standards in the IEC 60601 family of standards and others. At present, it does not matter whether these medical devices are simple or complex in structure or whether they offer active or passive support. Although the sector is relatively new, medical exoskeletons do not contain any life-supporting functions, which is an important aspect for defining the regulations. However, the medical exoskeletons are always in very close contact with patients, who differ from one another in many ways. A few factors are listed here, but they are not exhaustive

- Age of the patients (from child to advanced old age)
- Low/medium serious impairment
- From no cognitive disorders to severe impairment
- From absence of spasticity to severe spasticity
- From no autonomic disorders to severe impairment
- From self-administration by the patient up to specially trained (para-)medical users specifically trained in the use of the devices
- From the application in special clinical situations such as monitoring rooms up to application in the patient's home healthcare environment
- From early rehabilitation immediately following the occurrence to late or long-term rehabilitation up to daily live alleviation systems
- Free-walking exoskeletons or stationary exoskeletons together with treadmills and/or with body weight support system
- Medical exoskeletons for upper or lower extremities.

All these aspects should be noted at the conception and design phase of the wearable medical exoskeletons. Since the therapy of the patient is at the forefront when utilizing technical appliances, we are dealing with wearable medical devices which should be safe and effective. Only after that, the company can specify the following in a certified quality management system

- The specific design of the medical exoskeleton products is safe and effective, taking into consideration the purpose and intended use: indications,

contra-indications and essential performances based on the clinical functions of the wearable medical device
- The patient population considered
- The operational surroundings of the use case scenario considered
- The usability for the user and/or the patient
- The product risk management and usability engineering
- Clinical evaluation or trial results
- Software life-cycle management detail
- Biocompatibility issues where relevant.

The state of the art for the development of wearable medical devices is based on a design process. Good development practices differentiate the whole process in different phases with milestones. It could be helpful to define for each milestone a checklist summarizing all aspects which should be fulfilled to pass from one phase to the next. From experience, it is helpful to integrate also the necessary steps for fulfilling the applicable tests and standards. In an early phase of development, the applicable standards should be defined. The risk management and usability processes should be started in parallel with the first steps of soft- and hardware development. As soon as applied parts and the material are defined, biocompatibility could start with first steps like material data analyses. First, usability engineering topics could also start in parallel with the first drafts of the (graphic) user interface and/or available prototypes. This approach helps to start and finalize the evidence of applicable standard in accordance with the whole project management for devolving a wearable medical device. It is obvious that final decision for the release of the new or changed wearable medical device is based on the result of completed evidence of the applicable standards (and other aspects).

More detailed information about the controlling of the design of wearable medical devices could be found in the Food and Drug Administration (FDA) guidance document 'Design control guidance for medical device manufacturers' [12].

These considerations must meet the cost demands of today's nationally very varying, health systems. Thus, not only investment and maintenance costs, space and infrastructure costs but also personnel costs will have a decisive influence on the reimbursement of expenses by the service providers in the health system. Requirements on medical devices are regulated in national laws and regulations and are to be fulfilled prior to putting them on the market. They must often be verified, and placement on the market must be approved. But the scope and form are subject to a certain range of variability. Basically, these rules follow the purpose of patient safety by making the risk–benefit analysis efficient as well as sufficiently detailed to protect both users and third parties. Medical device regulatory systems often classify 'medical devices' based on their risk potential in several classes.

At the same time the concept of 'medical devices' is defined in different ways on a national regulatory level. One international organization, the International Medical Device Regulators Forum (IMDRF), formerly known as Global Harmonization Task Force (GHTF), which is aimed at national legislative bodies, becomes increasingly important, when it is a question of 'medical devices' and their regulations.

The IMDRF defines 'medical device' as follows (with several notes): any instrument, apparatus, implement, machine, appliance, implant, in vitro reagent or calibrator, software, material or other similar or related article

1. Intended by the manufacturer to be used, alone or in combination, for human beings, for one or more of the specific purpose(s)
   - (i) Diagnosis, prevention, monitoring, treatment or alleviation of disease
   - (ii) Diagnosis, monitoring, treatment, alleviation of or compensation for an injury
   - (iii) Investigation, replacement, modification or support of the anatomy or of a physiological process
   - (iv) Supporting or sustaining life
   - (v) Control of conception
   - (vi) Disinfection of medical devices
   - (vii) Providing information for medical or diagnostic purposes by means of in vitro examination of specimens derived from the human body.

2. Which does not achieve its primary intended action in or on the human body by pharmacological, immunological or metabolic means but which may be assisted in its intended function by such means
   - (i) Aids for disabled/handicapped people
   - (ii) Devices for the treatment/diagnosis of diseases and injuries in animals
   - (iii) Accessories for medical devices (see Note 3)
   - (iv) Disinfection substances
   - (v) Devices incorporating animal and human tissues, which may meet the requirements of the above definition but are subject to different controls.

Note 1: The definition of a device for in vitro examination includes, for example, reagents, calibrators, sample collection and storage devices, control materials, and related instruments or apparatus. The information provided by such an in vitro diagnostic device may be for diagnostic, monitoring or compatibility purposes. In some jurisdictions, some in vitro diagnostic devices, including reagents and the like, may be covered by separate regulations.

Note 2: Products may be considered to be medical devices in some jurisdictions but for which there is not yet a harmonized approach.

Note 3: Accessories, intended specifically by manufacturers, to be used together with a 'parent' medical device to enable that medical device in achieving its intended purpose, should be subject to the same IMDRF procedures as applied to the medical device itself. For example, an accessory will be classified as though it is a medical device in its own right. This may result in having a different classification for the accessory than for the 'parent' device.

Note 4: Components to wearable medical devices are generally controlled by the manufacturer's quality management system and the conformity assessment procedures for the device. In some jurisdictions, components are included in the definition of a 'medical device'.

The manufacturer of the wearable medical exoskeleton plays an important role by specifying, among other things, the intended use of the medical device for which he has to verify safe and effective utilization. Also, the term 'manufacturer' is not defined in a coherent sense worldwide. Important to understand is the fact that the manufacturer is responsible not only for the design and manufacture process. Medical device registration and all the aspects of product life cycle and quality management processes are also his responsibilities. Actual technologies today support the patient to such an extent that the wearable medical devices can be individually adapted or are adaptable to the performance and the necessary degree of support the individual patient situation requires. For this purpose, sensory and associated control systems, which can control the essential ability to detect and adapt in a given situation, are necessary.

Such systems are also used outside wearable medical devices area since several decades. Under the catchword 'robot', there are very diverse products on the market, which have been in industrial use for a long time, but recently are also becoming established in the 'service and personal care sectors'. All application sectors have their own standards and safety mechanisms, which will be briefly addressed in the following sections. The main difference between industrial, service and person care robots and medical devices which are using robotic technologies is the involvement of patients. Only medical devices are treating patients with all their limitations.

12.2 Legislation applicable for wearable exoskeletons (medical/non-medical)

12.2.1 In Europe

As we have seen in the introduction, exoskeletons fall under the scope of several European regulations; the European system of legislation is based on directives and regulations.

12.2.1.1 What are directives?

Directives are addressed at the member states of the European Union. The member states have to transpose the directives into their national regulation in order to implement the directive on their territory. For this transposition, a transition period is foreseen in the directive. This is 2 years in most cases. The national authorities must make sure that the requirements of the directive are achieved in the member state. In order to avoid the delay that is created by the transition periods, more and more regulations are used.

What are regulations?

In the case of a regulation, the step of transposition into national regulation is skipped. Regulations are in other words immediately applicable in all member states.

How do directives and regulations work?

European Directives are considered as the law. This means that the products in the scope must comply with it before being put on the market, or put into service. Each directive contains a list of essential requirements which are considered essential for putting the product on the market. Each product in the scope of a directive must comply with this list of essential requirements. This is a mandatory rule.

In the frame of the compliance of a product with the essential requirements, the (European harmonized) standards play a significant role. European harmonized standards are specifically written in relation with a directive, based on a mandate by the European Commission to CEN-CENELEC. European standards are more and more based on ISO and IEC standards (through the Vienna Agreement between ISO and CEN and the Dresden Agreement between IEC and CENELEC), which are then adopted as European Standards (or EN standards). In some cases, the EN standards will contain 'common modifications' meant to adapt the international standard to meet the European directive. This is avoided as much as possible (no common modifications) so that only one standard is produced (ISO/IEC = EN).

The advantage for manufacturers is obvious: only one standard can cover (theoretically) most parts of the world. The standards are written by a Technical Committee within the standardization body (CEN/CENELEC/ISO/IEC). The development of the standards is based on a consensus model between all stakeholders (users, manufacturers, testing bodies, authorities, etc.).

A European standard becomes 'harmonized' by publishing its references in the official journal of the European Union. At that moment, the standard will provide a presumption of conformity to the relevant directive. In order to comply with the essential requirements, European harmonized standards are preferably used. If these standards are followed, they provide a legal presumption of conformity with the aspects (essential requirements) that are covered by the standard. Although application of the European harmonized standards is not mandatory, it is important for manufacturers to take them into account. If a manufacturer chooses not to follow a standard, no presumption of conformity will be present and compliance of the solution with the essential requirement will have to be documented. This can happen if a manufacturer has adopted a solution which is not described in the standards, but which can also be valid and must be in compliance with the essential requirements of the directive.

It is important to mention that the link between the standard and the directives that are covered is always mentioned in Annex Z of the standard.

In order for the manufacturer to obtain the presumption of conformity, the following basic principles apply:

- There must be compliance with all the normative elements of the standard, i.e. the body of the text and the normative annexes. This includes also the references that are made to other standards. The informative annexes and notes are only there for clarification or as examples.
- The machine must be completely covered by the standard. If this is not the case, the presumption of conformity will not be complete.
- The presumption of conformity is limited to what is written in the scope and Annex Z. This informative annex describes in detail which essential requirements are covered. It is important that a manufacturer carefully examines this annex.
- Some aspects may be excluded from the standard, because no agreement was reached, or no standardized solution exists, etc.

12.2.2 In other parts of the world

In other parts of the world, no system similar to the European system exists. In most cases, standards will be the reference to follow. As we will see further, there is an international standard available which includes exoskeletons in its scope, i.e. ISO 13482:2014 [13], this international standard is 'harmonized' in Europe, which means that it provides a legal presumption of conformity with European directives. This is not the case in the rest of the world, where no specific legal basis exists.

Each part of the world has its own regulations which can be quite different. Also market-driven rules can make it difficult for manufacturers to put their products in the market. Normally, the ISO and IEC standards should also be applicable here, but with possible different interpretations depending on how specific international standards have been implemented nationally.

Some examples are the following:

- In the US, there is Occupational Safety and Health Administration and the Underwriters Laboratories which will provide a set of rules for machinery, whereas medical devices are covered by FDA regulations.
- In China, there is the China Compulsory Certificate mark, commonly known as a CCC mark. It is a compulsory safety mark for many products that are imported, sold or used on the Chinese market. China has a complex system of standards. For medical devices, other national regulations also apply.
- Korea has the Korea Certification mark. It signifies compliance with Korea's product safety requirements for electrical and electronic equipment and components that use AC power between 50 and 1,000 V. Again, other national regulations apply for medical devices.

12.3 The European directives: application on exoskeletons (non-medical)

12.3.1 The Machinery Directive (2006/42/E)

DIRECTIVE 2006/42/EC OF THE EUROPEAN PARLIAMENT AND OF THE COUNCIL of 17 May 2006 on machinery, and amending Directive 95/16/EC (recast) [7].

The Machinery Directive has the following structure:

Article 1	Scope
Article 2	Definitions
Article 3	Specific directives
Article 4	Market surveillance
Article 5	Placing on the market and putting into service
Article 6	Freedom of movement
Article 7	Presumption of conformity and harmonized standards
Article 8	Specific measures

(*Continues*)

(*Continued*)

Article 9	Specific measures to deal with potentially hazardous machinery
Article 10	Procedure for disputing a harmonized standard
Article 11	Safeguard clause
Article 12	Procedures for assessing the conformity of machinery
Article 13	Procedure for partly completed machinery
Article 14	Notified bodies
Article 15	Installation and use of machinery
Article 16	CE (China Export) marking
Article 17	Non-conformity of marking
Article 18	Confidentiality
Article 19	Cooperation between member states
Article 20	Legal remedies
Article 21	Dissemination of information
Article 22	Committee
Article 23	Penalties
Article 24	Amendment of Directive 95/16/EC
Article 25	Repeal
Article 26	Transposition
Article 27	Derogation
Article 28	Entry into force
Article 29	Addressees

And annexes:

Annex I	Essential health and safety requirements for the design and construction of machinery
Annex II	Declarations
Annex III	CE-marking
Annex IV	Categories of machinery to which one of the procedures referred to in Articles 12(3) and (4) must be applied
Annex V	Indicative list of the safety components referred to in Article 2(c)
Annex VI	Assembly instructions for partly completed machinery
Annex VII	1. Technical file for machinery 2. Relevant technical documentation for partly completed machinery
Annex VIII	Assessment of conformity with internal checks on the manufacture of machinery
Annex IX	EC type-examination
Annex X	Full quality assurance
Annex XI	Minimum criteria to be taken into account for the notification of bodies
Annex XII	Correlation table

12.3.1.1 Scope of the Machinery Directive

What is a machine?

To gain an insight in the application of the Machinery Directive, it is important to take a closer look at the scope of the Machinery Directive.

Art. 2. The directive's definition of

Machine: In contrast to the definition presented in the introduction, a machine is defined as follows according to the directive:

- an assembly, fitted with or intended to be fitted with a drive system other than directly applied human or animal effort, consisting of linked parts or components, at least one of which moves, and which are joined together for a specific application,
- an assembly referred to in the first indent, missing only the components to connect it on site or to sources of energy and motion,
- an assembly referred to in the first and second indents, ready to be installed and able to function as it stands only if mounted on a means of transport, or installed in a building or a structure,
- assemblies of machinery referred to in the first, second and third indents or partly completed machinery referred to in point (g) which, in order to achieve the same end, are arranged and controlled so that they function as an integral whole,
- an assembly of linked parts or components, at least one of which moves and which are joined together, intended for lifting loads and whose only power source is directly applied human effort.

There is also 'interchangeable equipment' which is defined as a device which, after the putting into service of machinery or of a tractor, is assembled with that machinery or tractor by the operator himself in order to change its function or attribute a new function, in so far as this equipment is not a tool.

Safety component: Safety component is defined as a

- a component: which serves to fulfil a safety function,
- which is independently placed on the market,
- the failure and/or malfunction of which endangers the safety of persons, and
- which is not necessary in order for the machinery to function, or for which normal components may be substituted in order for the machinery to function.

Other elements in the scope: Furthermore, the Machinery Directive also applies to lifting accessories, chains, ropes and webbing for lifting purposes, removable mechanical transmission devices and partly completed machinery but this is not relevant for purposes of this book.

12.3.1.2 Application of the directive machines on exoskeletons

When we look at the above definitions of machinery, the question is to decide whether an exoskeleton is in the scope of the directive or not.

Based on the basic definition: an exoskeleton is

- an assembly: yes
- fitted with or intended to be fitted with a drive system: yes
- other than directly applied human or animal effort: yes/no?
- consisting of linked parts or components: yes

- at least one of which moves: yes
- and which are joined together for a specific application: yes

The third bullet item can give rise to discussion along the lines: is an exoskeleton driven by other than directly applied human effort? Arguments against the application could be that an exoskeleton will not move without human effort and thus the exoskeletons would be excluded based on the fact that the basic definition of a machine states 'fitted with a drive system other than directly applied human or animal effort'. This case would be similar to electric bikes where the European Commission decided in 2010, after long discussions, to put electric bikes in the scope of the directive, although one needs to apply human effort before the bike will move or assist in the movement. The fact that an electric bike contains a source of energy that could give rise to serious risks has led to the decision to include them in the scope of the Machinery Directive. A similar reasoning for exoskeletons would also lead to the same mandatory conclusion the Machinery Directive is applicable.

12.3.1.3 Excluded or not?

The application of the Machinery Directive does not depend only on the definition, but also on the exclusions. The following are excluded from the scope:

- certain safety components
- specific equipment for use in fairgrounds and/or amusement parks
- machinery specially designed or put into service for nuclear purposes which, in the event of failure, may result in an emission of radioactivity
- weapons, including firearms
- several means of transport covered by other directives
- seagoing vessels and mobile offshore units and machinery installed on board such vessels and/or units
- machinery specially designed and constructed for military or police purposes;
- machinery specially designed and constructed for research purposes for temporary use in laboratories;
- mine winding gear;
- machinery intended to move performers during artistic performances;
- electrical and electronic products falling within the following areas, in so far as they are covered by the low voltage directive (see 2014/30/EU for more information):
  - household appliances intended for domestic use,
  - audio and video equipment,
  - information technology equipment,
  - ordinary office machinery,
  - low-voltage switchgear and control gear,
  - electric motors.

Note: When no risks are present, in other words, if none of the essential requirements are applicable, the directive will de facto also not be applicable.

Exoskeletons do not naturally fall under any of the above exclusions. One could argue that some exoskeletons could be electrical/electronical products in the area of household appliances, intended for domestic use. The fact that free-walking exoskeletons are normally not covered by the low voltage directive (voltage range between 50 and 1,000 V AC or between 75 and 1,500 V DC) will always lead to the conclusion that this exclusion is not applicable to exoskeletons. Most of these free-walking exoskeletons have a battery and it should be noted that there is also a Battery Directive (2006/66/EC) which should be fulfilled.

12.3.1.4 What if the exoskeleton is considered as a medical device?

In this case, Article 3 of the Machinery Directive reads as follows:

> *Where, for machinery, the hazards referred to in Annex I are wholly or partly covered more specifically by other Community Directives, this Directive shall not apply, or shall cease to apply, to that machinery in respect of such hazards from the date of implementation of those other Directives.*

In practice, this means that if the medical device directive applies, the Machinery Directive ceases to apply. See also Section 12.4 which covers more specific details about medical device regulations.

12.3.1.5 Division in risk-categories – Annex IV

The Machinery Directive makes a distinction between machines that are included in Annex 4 and machines that are not included in Annex 4. Annex 4 covers 23 types of machines and safety components, for which a specific procedure applies. These machines are mostly manually fed machines [with manual intervention of the operator (loading and/or unloading)] and a number of very specific cases involving special risks (e.g. hydraulic lifts for vehicles, lifting equipment for persons, household refuse collecting vehicles, etc.). For exoskeletons, Annex 4 will never apply.

12.3.1.6 Conformity procedures

The Machinery Directive 2006/42/EC contains two conformity procedures for machines, namely the procedure for machines in Annex 4, and the procedure for the other machines. For the machines of Annex 4, an intervention of a notified body is mandatory in some cases (EC-type examination, etc.). Because exoskeletons are not referred to in Annex 4, we limit ourselves here to the procedure for non-Annex 4 machines. The procedure for non-Annex 4 machines is the following:

- Step 1: The machine must be built in conformity with the essential requirements of Annex 1
- Step 2: The proof of this conformity must be written down in a Technical file by the manufacturer (as defined in Annex 7)
- Step 3: The conformity must be declared by the manufacturer in the declaration of conformity (Annex 2.1.A)
- Step 4: The conformity must be shown through the CE marking (Annex 3).

Step 1: Conformity with the essential requirements
Every machine must be in compliance with the essential requirements of Annex 1 of the directive. This is a mandatory rule and the essential requirements cover the following items:

Preliminary observations

1. Essential health and safety requirements
   (i) General remarks
   (ii) Controls
   (iii) Protection against mechanical hazards
   (iv) Required characteristics of guards and protection devices
   (v) Protection against other hazards
   (vi) Maintenance
   (vii) Indicators
2. Additional requirements for certain categories of machinery
   (i) Agri-foodstuffs machinery
   (ii) Portable hand-held and/or hand-guided machinery
   (iii) Machinery for working wood and analogous materials
3. Particular hazards due to the mobility of machinery
4. Particular hazards due to a lifting operation
5. Machinery intended solely for underground work
6. Hazards due to the lifting or moving of persons.

It is clear that Parts 2 to 6 will not apply to exoskeletons.

Part 1 will be applicable and the exoskeleton must comply with it. To establish which essential requirements apply, the manufacturer must conduct a risk analysis on his machine. The purpose of this risk assessment is to determine which essential requirements apply to the machine. The European standard EN ISO 12100 contains the basic principles to conduct a correct risk assessment.

As mentioned before, the European harmonized standards are very important to establish the conformity with the essential requirements. In the case of the Machinery Directive, the European harmonized standards for machinery are divided into three types, which make their use more transparent and reduce overlaps and repeats:

- Type A is generally applicable standards (the only type A standard that is relevant here is EN ISO 12100 [7] concerning general design principles, risk assessment, etc.)
- Type B is standards covering common aspects (B1) (e.g. EN 60204: electrical equipment of machines) and safety components (B2) (e.g. EN ISO 13850 emergency stop: principles for design)
- Type C is standards covering specific machines or families of machines. For exoskeletons, the EN ISO 13482:2014 [13] exists and applies.

Only the C-type standards can provide a full presumption of conformity with the directive, as the B-type standards only cover some very specific but general safety aspects, such as safety distances (EN ISO 13857), electrical equipment (EN 60204-1), etc. This is clearly defined in Annex Z which is attached to each harmonized standard. The idea behind the structure is that C-type standards refer to B-type standards for common aspects. B-type standards will typically provide all possible situations for the treated common aspect, without specific selection of those situations. The C-type standard will provide information on which parts or solutions that are mentioned in the B-type standards need to be applied.

For example, EN ISO 13857 on safety distances provides a choice between the safety distances for reaching through regular openings for adults (Table 4) and for children (Table 5). The C-type standard should clearly mention which table is to be applied for the specific machine. The full list of European harmonized standards under the Machinery Directive is available on the website of the European Commission [14].

EN ISO 13482 on safety of personal care robots: The standard EN ISO 13482:2014 [13] is an ISO standard which was adopted as an EN standard under the Vienna Agreement, without modifications. The following types of personal care robots are covered in the scope:

- mobile servant robot;
- physical assistant robot (this includes wearable exoskeletons, see Figure 12.1);
- person carrier robot.

It is important to note that the standard draws the attention to the fact that for hazards related to impact (e.g. due to a collision) no exhaustive and internationally recognized data (e.g. pain or injury limits) exist at the time of publication (2014) of the Standard. A B-type standard covering these aspects is currently under development under ISO TC199/WG12 (Human–machine interactions).

All other hazards are covered by the standard, and it will therefore provide a presumption of conformity with the directive.

Step 2: The technical file

Before a machine can be placed on the market, or put into service in Europe, the manufacturer must compile a Technical file according to Annex 7. This file must contain the following elements:

1. A construction file including
   - (i) a general description of the machinery,
   - (ii) the overall drawing of the machinery and drawings of the control circuits, as well as the pertinent descriptions and explanations necessary for understanding the operation of the machinery,
   - (iii) full detailed drawings, accompanied by any calculation notes, test results, certificates, etc., required to check the conformity of the machinery with the essential health and safety requirements,
   - (iv) the documentation on risk assessment demonstrating the procedure followed, including
     - (a) a list of the essential health and safety requirements which apply to the machinery,

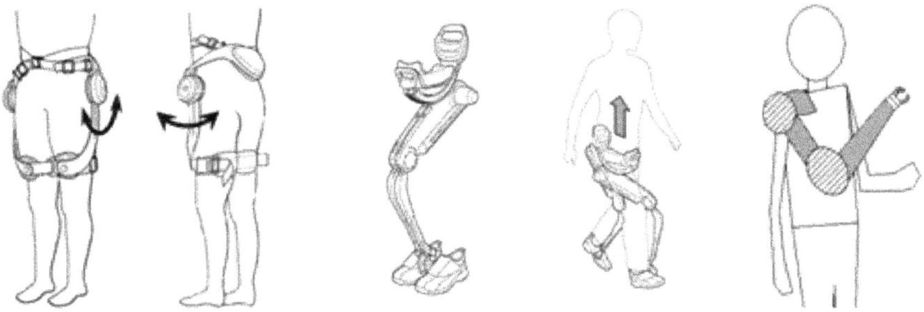

Figure 12.1 Exoskeletons covered by the standard EN ISO 13482

(b) the description of the protective measures implemented to eliminate identified hazards or to reduce risks and, when appropriate, the indication of the residual risks associated with the machinery,

(v) the standards and other technical specifications used, indicating the essential health and safety requirements covered by these standards,
(vi) any technical report (TR) giving the results of the tests carried out either by the manufacturer or by a body chosen by the manufacturer or his authorized representative,
(vii) a copy of the instructions for the machinery,
(viii) where appropriate, the declaration of incorporation for included partly completed machinery and the relevant assembly instructions for such machinery,
(ix) where appropriate, copies of the EC declaration of conformity of machinery or other products incorporated into the machinery,
(x) a copy of the EC declaration of conformity;

2. for series manufacture, the internal measures that will be implemented to ensure that the machinery remains in conformity with the provisions of this directive.

The technical file can be considered as the proof of compliance of the machine according to the manufacturer (and under his responsibility). Besides the legal requirement for the existence of the file for each machine with CE marking and CE declaration of conformity, this idea is important for the way of establishing the file.

The **defence of the compliance must be credible**. This means that use must be made of serious reasoning or motivation, where necessary documented, for example with:

- reference to relevant standards,
- test reports,
- calculation notes,

- experience elements,
- logical reasoning.

If a European (C-type) standard is used, the production of the technical file will become easier, because the proof of compliance with the standard will be sufficient as proof of compliance with the essential requirements of the directive. Care shall be taken to cover all requirements (including references to other standards). The manufacturer must be able to present the relevant parts of the file to the competent authorities, in case of a motivated request by these authorities. The file must be kept available for the authorities by the manufacturer for a period of 10 years (after the last produced machine of a certain type). Thus, the file is not to be delivered with the machine.

Step 3: The declaration of conformity

The Machine Directive distinguishes two types of declarations

- Annex 2.1.a: Declaration of conformity
- Annex 2.1.b: Declaration of the manufacturer for partly completed machinery (intended to be built in or to be assembled with other machines). This type of declaration will normally not be relevant for exoskeletons, since these will not be considered as partly completed machinery.

The declaration of conformity (Annex 2.1.a): For all machinery that is placed on the market or put into service, the manufacturer must deliver a declaration of conformity. This declaration is a legal document that engages the responsibility of the manufacturer. The declaration of conformity contains the following elements:

1. The business name and full address of the manufacturer and, where appropriate, his authorized representative;
2. Name and address of the person authorized to compile the Technical file, who must be established in the Community;
3. Description and identification of the machinery, including generic denomination, function, model, type, serial number and commercial name;
4. A sentence expressly declaring that the machinery fulfils all the relevant provisions of this directive and where appropriate, a similar sentence declaring the conformity with other directives and/or relevant provisions with which the machinery complies. These references must be those of the texts published in the Official Journal of the European Union;
5. Where appropriate, the name, address and identification number of the notified body which carried out the EC type-examination referred to in Annex IX and the number of the EC type-examination certificate;
6. Where appropriate, the name, address and identification number of the notified body which approved the full quality assurance system referred to in Annex X;
7. Where appropriate, a reference to the harmonized standards used, as referred to in Article 7(2);

8. Where appropriate, the reference to other technical standards and specifications used;
9. The place and date of the declaration;
10. The identity and signature of the person empowered to draw up the declaration on behalf of the manufacturer or his authorized representative.

Notes:

- Item 4 states that conformity with other directives must also be declared. In the case of exoskeletons, this will be the electromagnetic compatibility (EMC) directive. The low voltage directive applies, but should not be mentioned (see Section 12.3.3.2)
- Items 5 and 6 of the declaration will not be applicable to exoskeletons, since the latter is not mentioned in Annex 4 of the Machinery Directive.

Step 4: The CE marking

The CE marking is a symbol of free marketability in the European Economic Area (Internal Market). It symbolizes the compliance of the product with the applicable directives (in most cases, there will be more than one directive applicable). The CE marking is a confirmation of the manufacturer towards the authorities that the product complies with the applicable directives, and that he has followed the conformity procedures as described in the relevant directives. It is the finishing touch of the procedure. The CE marking has the format shown in Figure 12.2, which must not be modified.

All markings that can lead to confusion are forbidden (e.g. CE as China Export). The following rules apply:

- If the CE marking is reduced or enlarged, the proportions shown in the above drawing must be respected (not reduced to less than 5 mm)
- The CE marking shall be affixed to the machinery visibly, legibly and indelibly
- The CE marking must be affixed in the immediate vicinity of the name of the manufacturer or his authorized representative, using the same technique.

12.3.2 The Low Voltage Directive (2014/35/EU)

DIRECTIVE 2014/35/EU OF THE EUROPEAN PARLIAMENT AND OF THE COUNCIL of 26 February 2014 on the harmonization of the laws of the Member States relating to the making available on the market of electrical equipment designed for use within certain voltage limits.

As already mentioned, several directives can apply to a machine. This is certainly the case for the low voltage directive.

The new legislative framework: The European commission introduced the new legislative framework in 2010 mainly to improve the market surveillance in the EU. The Low Voltage Directive is one of the first set of directives that have been adapted to meet the new rules of this framework. This modification introduces obligations for importers and distributors.

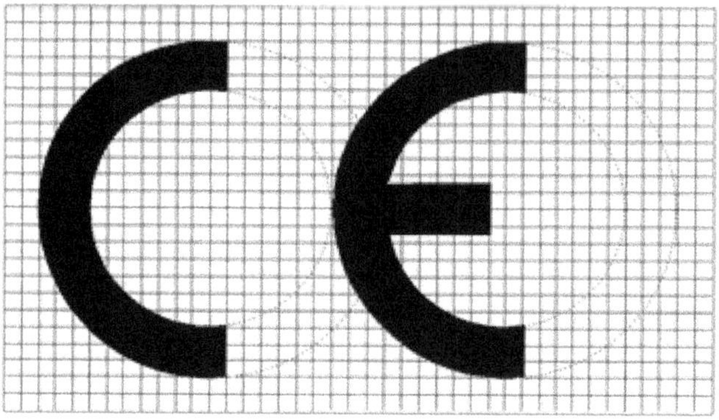

Figure 12.2 CE-marking symbol

Importers shall

- only place compliant electrical equipment on the market
- ensure that the appropriate conformity assessment procedure has been carried out
- ensure that technical documentation has been drawn up, the CE marking is affixed and required documents are provided
- not place electrical equipment in the market which they consider or have reason to believe is not in conformity. They must inform the manufacturer and the authorities if equipment presents a risk
- ensure that the electrical equipment is accompanied by instructions and safety information in a language which can be easily understood by consumers and other end-users
- when deemed appropriate, carry out sample testing of electrical equipment made available on the market
- take corrective measures when they consider or have reason to believe that equipment which they have placed on the market is not in conformity and inform authorities if equipment presents a risk
- keep a copy of EU declaration of conformity and ensure that the technical documents can be made available for a period of 10 years
- to a reasoned request provide the national authorities with all the information and documentation necessary to demonstrate the conformity of electrical equipment

Distributors shall

- act with due care when making electrical equipment available
- verify that equipment bears CE marking and is accompanied by the required documents and by instructions and safety information in a language which can be easily understood by consumers and other end-users

- verify that manufacturer and importer have complied with their requirements
- not make electrical equipment available on the market which they consider or have reason to believe is not in conformity and inform manufacturer/importer and authorities if equipment presents a risk
- make sure that corrective measures are taken when they consider or have reason to believe that equipment which they have placed on the market is not in conformity and inform authorities if equipment presents a risk
- to a reasoned request provide the national authorities with all the information and documentation necessary to demonstrate the conformity of electrical equipment.

12.3.2.1 Scope of Low Voltage Directive

The Low Voltage Directive applies to electrical equipment designed for use with a voltage rating of between 50 and 1,000 V for alternating current and between 75 and 1,500 V for direct current. This means that exoskeletons that are battery driven will normally not fall in the scope of the low voltage directive. Battery loaders will fall in the scope.

12.3.2.2 Conformity procedures

The basic idea is that machinery must comply with the Machinery Directive, including EHSR 1.5.1 (Machinery Directive Annex 1) on the risks related to the electrical supply. Essential requirement 1.5.1 also states that machinery must fulfil the technical requirements of the low voltage directive, but not the procedure. This is mainly because the two procedures are not compatible and it becomes therefore impossible to comply with both procedures. The consequence is of course that the declaration of conformity will only mention the Machinery Directive. The low voltage directive should therefore not be mentioned in the declaration.

These technical requirements in the low-voltage directive are quite limited: they cover protection against hazards arising from the electrical equipment itself and protection against hazards which may be caused by external influences on the electrical equipment.

The steps to be followed in the procedure are like those valid for the Machinery Directive as discussed above. The system of the European harmonized standards is also valid for the Low Voltage Directive (presumption of conformity, see Section 12.3.1). Some standards are harmonized as well under the Machinery Directive and the Low Voltage Directive (e.g. standard EN 60204-1 on electrical equipment for machinery). The full list of harmonized standards under the low voltage can be found under: https://ec.europa.eu/growth/single-market/european-standards/harmonised-standards/low-voltage_en

12.3.3 The EMC Directive (2014/30/EU)

DIRECTIVE 2014/30/EU OF THE EUROPEAN PARLIAMENT AND OF THE COUNCIL of 26 February 2014 on the harmonization of the laws of the Member States relating to EMC.

The EMC Directive has also been adapted to meet the new rules of this new legislative framework. It was published together with the low voltage directive.

12.3.3.1 Scope of EMC Directive

The EMC Directive is applicable to

- 'apparatus', which means any finished appliance or combination thereof made available on the market as a single functional unit, intended for the end-user and liable to generate electromagnetic disturbance, or the performance of which is liable to be affected by such disturbance and
- 'fixed installation' which means a particular combination of several types of apparatus and, where applicable, other devices, which are assembled, installed and intended to be used permanently at a predefined location

It is clear that exoskeletons are to be considered as 'apparatus', and that the EMC directive applies, next to the Machinery Directive.

EMC means the ability of equipment to function satisfactorily in its electromagnetic environment without introducing intolerable electromagnetic disturbances to other equipment in that environment.

So, the equipment must not disturb the environment, where 'electromagnetic disturbance' means any electromagnetic phenomenon which may degrade the performance of equipment; an electromagnetic disturbance may be electromagnetic noise, an unwanted signal or a change in the propagation medium itself; and the equipment must be immune to the environment where 'immunity' means the ability of equipment to perform as intended without degradation in the presence of an electromagnetic disturbance.

Here, we have an overlap with the Machinery Directive: the EMC directive covers the aspects of the immunity of the machine (immunity) whereby the functioning may be disturbed, whereas the Machinery Directive focuses on the immunity aspects that are important for the safety of the machine. The purposes and the criteria for acceptance are not necessarily identical in both directives.

12.3.3.2 Conformity procedures

The steps to be followed in the procedure are like those that are valid for the Machinery Directive (see Section 12.3.1.5). The manufacturer must perform an EMC assessment of his product. The system of the European harmonized standards is also valid for the EMC directive (presumption of conformity, see Section 12.3.1). The full list of harmonized standards under the EMC directive can be found on the website.[1]

[1]https://ec.europa.eu/growth/single-market/european-standards/harmonised-standards/low-voltage_en.

12.4 Regulation for medical exoskeletons

When it comes to the conception and realization of wearable medical devices, different international, national, or regional standards are brought to bear. To some extent, compliance with these is directly or indirectly demanded by national/regional legislation, as, for example, is valid in the European Union using appropriate directives. Annex I of Medical Device Directive (MDD) 93/42/EEC [9] uses the term 'harmonized standards', and this list of harmonized standards is regularly published in the 'Official Journal of the European Union' [15]. These harmonized standards could be used to show the evidence of MDD Annex on Essential Requirements. Besides the Medical Device Directive, manufacturers should be aware that they should have to fulfil all applicable European directives. It should be mentioned that medical devices very often fulfil the definition of a 'machine' discussed in Section 12.3. For wearable medical devices, the EU-Battery-Directive may be important as well. According to this directive, there is also a list of harmonized standards regularly published. It is the obligation of the manufacturer of medical devices to provide the evidence of applicable standards.

In the USA, the FDA [16] is responsible for regulation framework of medical devices. Besides these official directives and regulations, it should be mentioned that there are a lot of other documents available to help and guide manufactures of wearable medical devices but also the national regulators in the field of wearable medical device technology. The links to available guidance documents area are as follows: IMDRF and GHTF [17] and MEDDEV [18].

Additional standards may become relevant in order to provide evidence for fulfilment of certain requirements of the regulations, e.g. Annex I of MDD 93/42/EEC [9]. In particular, the product standards contain details for safety aspects, which can either be of a general or high specific nature. These coherences will be outlined roughly subsequently.

12.4.1 Standards for medical devices

International standards are published from different organizations. In the area of wearable medical devices, the most important international organizations are IEC and ISO. For all users of their standards, it is important that there are no conflicts nor contrary requirements for the same topic and both organizations try to ensure that does not happen. Nowadays, both organizations work together for projects in so-called joint working groups (JWG). Types and scopes of standards for medical devices are described on high-level documents in ISO/IEC Guide 63 [6] and ISO/IEC Guide 51 [4]. According to these documents, international standards for medical devices can be classified into the following listed groups.

12.4.1.1 Product standards

These standards are related to a specific product or group of products. They include standards that state safety or performance parameters and include reference test methods that can be used to demonstrate conformance to those parameters (e.g. IEC

60601-1 [10]) and the associated supplementary standards and special specifications for medical electrical devices. IEC 60601-1 is the key standard for all medical devices which are medical electrical equipment or systems. In fact, IEC 60601-1 has other specific collateral standards included which have a specific focus like usability engineering for all medical electrical equipment's or systems. Details are shown in following sections.

In some cases, international standard organizations write additional TRs for specific standards to give the user of standards more background information and guidance how to use a specific standard. Nowadays, IEC publishes their TR regarding the IEC 60601 series in the IEC 60601-4-'x' format and IEC 60601-4-1 has recently been published to provide guidance for including autonomy into medical electrical equipment [19]. Disclosure and test method standards, where, adherence to declare pass/fail criteria is necessary for safety and performance of medical devices.

12.4.1.2 Process standards

A series of types of standard falls in this category, including:

- Quality management system standards that establish a framework within which the manufacturer is able to design, develop, and produce medical devices that consistently meet specifications [e.g. standards for good manufacturing practice (GMP)]. Quality management standards like ISO13485 [20] cover the whole life cycle for a medical device and ISO 9001 [21] is normally not accepted as an adequate standard for medical device manufacturers. The design and development phase, production, and purchasing of parts; storage; transporting; servicing; document management and other aspects are covered and should be followed. Where a registration procedure for wearable medical device is required, normally the conformity to (national) quality management system standards (ISO 13485 [20]) is a part of the process.
- The ISO 13485 standards for medical device quality management have been updated in 2016, whereas ISO 9001 changed its whole structure. Hence, ISO 13485 is not anymore following the structure from ISO 9001.
- Standards for processes used for the design, development or production of safe and effective medical devices (e.g. sterilization, biological evaluation, clinical investigation; sterility, biocompatibility or risk management and usability engineering).

12.4.1.3 Installation and environmental standards

The standards for installation are generally applicable for medical devices which must be installed (putting into operation is not equal to installation). These can be

- Construction and installation standards
- System standards [addresses the proper precautions and procedures for interconnection of multiple devices into a single system (medical electrical system)]
- Commissioning standards (addresses the proper testing and inspection procedures applying to permanent installed equipment and systems prior to initial use)

- Environmental standards [addresses precautions and testing to ensure that a medical device does not negatively affect its environment and the environment does not degrade or otherwise impair the performance of a medical device (e.g. EMC standards)].

12.4.1.4 In-process standards

These can be

- Routine in-service testing standards to ensure that the safety of medical devices is maintained over the useful life of the medical device
- Quality assurance and calibration standards to ensure the continued proper function and accuracy of medical devices, where relevant to safety.

12.4.1.5 Safety standards

Scopes of safety standards, like ISO 16142 [22], will ensure that each standard is restricted to specific aspects and makes reference to standards of wider application for all other relevant aspects. Such a hierarchy is built on

- Basic safety standards include fundamental concepts, principles and requirements with regard to general safety aspects, applicable to all kinds or a wide range of products, processes and services (basic safety standards are sometimes referred as horizontal standards too). (Note: ISO uses the term 'horizontal' in the same way IEC uses the term 'collateral'.)
- Group safety standards include safety aspects, applicable to several or a family of similar products, processes or services dealt with by two or more technical committees or subcommittees, making reference, as far as possible, to basic safety standards.
- Product safety standards include all necessary safety aspects of a specific or a family of product(s), process(es) or service(s) within the scope of a single technical committee or subcommittee, making reference, as far as possible, to basic safety standards and group safety standards (product safety standards are sometimes referred to as vertical standards).

12.4.2 IEC 60601 standards series

The IEC 60601 standards series essentially defines safety requirements and essential performances for medical electrical devices and medical electrical systems. IEC 60601-1 covers the basic safety and the essential performance of medical electrical devices and medical electrical systems. For the user of this standard, it is important to know which version is accepted and in which country. The last published version is IEC 60601-1 Edition 3 with amendment 1 from 2012 to 2008 (IEC 60601-1:2005+A1:2012) [7].[2] A few countries still require and accept only the previous edition; other countries have a longer transitory period for the new one. In addition to this situation, the user of standards should be aware of national

[2] IEC SC62A decided to start working on the second amendment for IEC 60601-1 Ed 3 and in parallel to work on the 4th Edition.

deviations which are quite normal. This IEC 60601-1 standard is accompanied by a series of further requirements of a general nature (coded as IEC 60601-1-x and named collateral standards) as well as by requirements for certain types of specific medical devices (coded as IEC 60601-2-x and ISO 80601-2-x) and named particular standards.

Standards from the IEC 60601-2-x (series), which could be related directly and particularly to the type of medical devices being considered, should be carefully evaluated. Standards that could be used for medical exoskeletons (in part) are, for example medical electrical equipment – Part 2–10: particular requirements for the safety of nerve and muscle stimulators.

An overview of the currently published IEC 60601-x-x standards among other standards from IEC SC 62 A and SC 62 D which are the relevant subgroups for this topic, contain basic safety and essential performance requirements for all medical devices, provided they are electrically operated and used for all cases described, can be found on the webpage of IEC62 and their subcommittees.[3] All standards from the IEC 60601 family are dealing with basic safety and essential performance. Therefore, it is important to understand the meaning of the following defined terms in IEC 60601-1 [10]:

- Basic safety: freedom from unacceptable risk directly caused by physical hazards when medical electrical equipment is used under normal conditions and single fault condition.
- Essential performance: performance of a clinical function, other than that related to basic safety, where loss or degradation beyond the limits specified by the manufacturer results in an unacceptable risk. Note that essential performance is most easily understood by considering whether its absence or degradation would result in an unacceptable risk.

Medical electrical equipment or a medical electrical system that does not perform properly could result in unacceptable risk for patients, operators or others. In order to achieve its intended use, medical electrical equipment or the medical electrical system needs to perform within certain limits. These limits are usually specified by the manufacturer, collateral or particular standards of the IEC 60601 series.

Examples of essential performance from medical devices are

- Fall protections for patients wearing a medical wearable exoskeleton
- Correct administration of a drug by a syringe pump where inaccuracy/incorrect administration would cause an unacceptable risk to the patient
- The ability of an electrocardiograph/monitor to recover from the effects of the discharge of a defibrillator where the failure to recover could lead to an incorrect response by the medical staff that would present an unacceptable risk to the patient

[3]IEC TC 62 webpage: http://www.iec.ch/dyn/www/f?p=103:29:0::::FSP_ORG_ID,FSP_LANG_ID:1245,25#1.

- Correct operation of an alarm system in an intensive care or operating room monitoring system where an incorrect/missing alarm signal could lead to an incorrect response by the medical staff that would present an unacceptable risk to the patient
- Correct output of diagnostic information from medical electrical equipment that is likely to be relied upon to determine treatment, where incorrect information could lead to an inappropriate treatment that would present an unacceptable risk to the patient.

For purposes of IEC 60601 standards, performance related to basic safety aspects of the medical electrical equipment, such as the performance of basic insulation, is not considered to be essential performances. Based on IEC TR 60601-4-1 [19], essential performance could be interpreted as follows:

- All medical electrical equipment must perform as per their intended use. It must be noted essential performances are not related to all the performance of the medical electrical equipment. Not all these performances can be called essential! Essential performances are related to the clinical functions that must be preserved in normal condition and in single fault condition. If the risk of degradation of performances is found to be unacceptable, then they will be considered as essential performances. For a more complete understanding of what essential performances mean, the reader should read Clause 4.3 and the subclause (informative rational section) of the IEC 60601-1 [10].
- During risk analysis, the manufacturer shall identify the performance of the clinical function(s) of the medical electrical equipment or medical electrical system, other than that related to basic safety, that is necessary to achieve its intended use or that could affect the safety of the medical electrical equipment or medical electrical system.
- The manufacturer shall then specify performance limits between fully functional and total loss of the identified performance in both normal conditions and single fault conditions.
- The manufacturer shall then evaluate the risk from the loss or degradation of the identified performance beyond the limits specified by the manufacturer. If the resulting risk is unacceptable, then the identified performance constitutes an essential performance of the medical electrical equipment or medical electrical system.
- The manufacturer shall implement risk-control measures to reduce the risk from the loss or degradation of the identified performance to an acceptable level.
- If a manufacturer of a medical device claimed to have no essential performances, the result of risk management process according to ISO 14971 [11] should show evidence.

Unfortunately, the term 'clinical function' is not defined in the standards at the moment but it has been defined in IEC 60601-4-1 [19] as medical operation that the medical electrical equipment or medical electrical system is intended to perform.

As this definition has just been published, each manufacturer has had to define the clinical functions for its specific medical devices individually, during risk management process to identify the essential performance of the specific medical devices. Essential performances generally relate to medical devices operating as intended without creating an unacceptable risk. A failure of essential performance can be either a lack of performance (such as life-supporting performance) or incorrect performance (such as delivering an incorrect dose to the patient). Hence, essential performance is based on a risk management process related to performance of clinical functions.

Applicable product standards for wearable medical devices that can be partially or totally used are not found exclusively in the sector of medical devices. They are found, for example, in the area of machines. International standards about treadmills, stationary training equipment or body weight support systems, for example, can be used additionally if wearable medical device standards are not available in a specified area.

In addition, it must be mentioned that, apart from the ISO 60601-x/ISO 80601-x standards series, further individual standards or standard series exist that refer to medical devices. Regular standards research should be carried out in order to ensure the respective current status of information. In this respect, service providers offering an appropriate service can also be called upon.

12.4.2.1 Electromagnetic disturbances

IEC 60601-1-2 [23] concerning EMC, in particular, is to be included, which are to be taken into account for the case under consideration. IEC 61000 series gives much more details about the whole topic of EMC.

12.4.3 Quality management system standards

Quality management system standards are ranked among the process standards. ISO 13485 [20] for international applications should normally be fulfilled. For the American market, the FDA regulations according to 21 CFR § 820 aspects from GMP are mandatory. Here, it should be noted that the requirements from FDA are regulations having a legislative character.

12.4.4 Programmable electrical medical systems

For medical devices that are categorized as PEMS (programmable electrical medical systems), apart from the requirements which IEC 60601-1 sets, those of IEC 62304 [24] concerning the life-cycle requirements for medical device software must also be complied with. The standard defines the life-cycle requirements for medical device software. The set of processes, activities and tasks described in this standard establishes a common framework for medical device software life-cycle processes.

This standard applies to the development and maintenance of processes, to the development and maintenance of processes when software is itself a medical device or when software is an embedded or integral part of the final medical device.

This standard does not cover validation and final release of the medical device, even when the medical device consists entirely of software.

12.4.5 Biocompatibility

For medical devices intended to have direct or indirect contact with biological tissues, cells or body fluids, manufactures should proceed according to the instructions and principles of the ISO 10993 standard family, in order to verify the biocompatibility of the materials utilized, where this is necessary. ISO 10993 series uses a risk management-based approach. ISO 10993-1 [25] gives an overview and explanation which additional standards of the whole series are applicable, based on the intended use of the medical device and nature and duration of body contacting material, either direct or indirect. The ISO 10993 series does not cover testing of materials and devices which do not come into direct or indirect contact with the patient's body nor does it cover biological hazards arising from any mechanical failure.

12.4.6 Usability engineering

Usability engineering requirements which are given in IEC 60601-1-6 [26] standard and the IEC 62366 [27] standards 'Application of Usability Engineering to Medical devices' should be consulted. The reason why usability is important is explained as follows in the standard:

> *Medical practice is increasingly using medical devices for observation and treatment of patients. Use errors caused by inadequate medical device usability have become an increasing cause for concern. Many of the medical devices developed without applying a usability engineering process are non-intuitive, difficult to learn and to use. As healthcare evolves, less skilled users including patients themselves are now using medical devices and medical devices are becoming more complicated. In simpler times, the user of medical device might be able to cope with an ambiguous, difficult-to-use user interface. The design of a usable medical device is a challenging endeavor, yet many organizations treat it as if it were just 'common sense'. The design of the user interface to achieve adequate (safe) usability requires a very different skill set than that of the technical implementation of the interface.*
>
> *The usability engineering process is intended to achieve reasonable usability, which in turn is intended to minimize use errors and to minimize use-associated risks. Some, but not all, forms of incorrect use are amenable to control by the manufacturer.*

Figure 12.3 shows the relationship between the risk management process according ISO 14971 [11] and the usability engineering process according IEC 62366 [27].

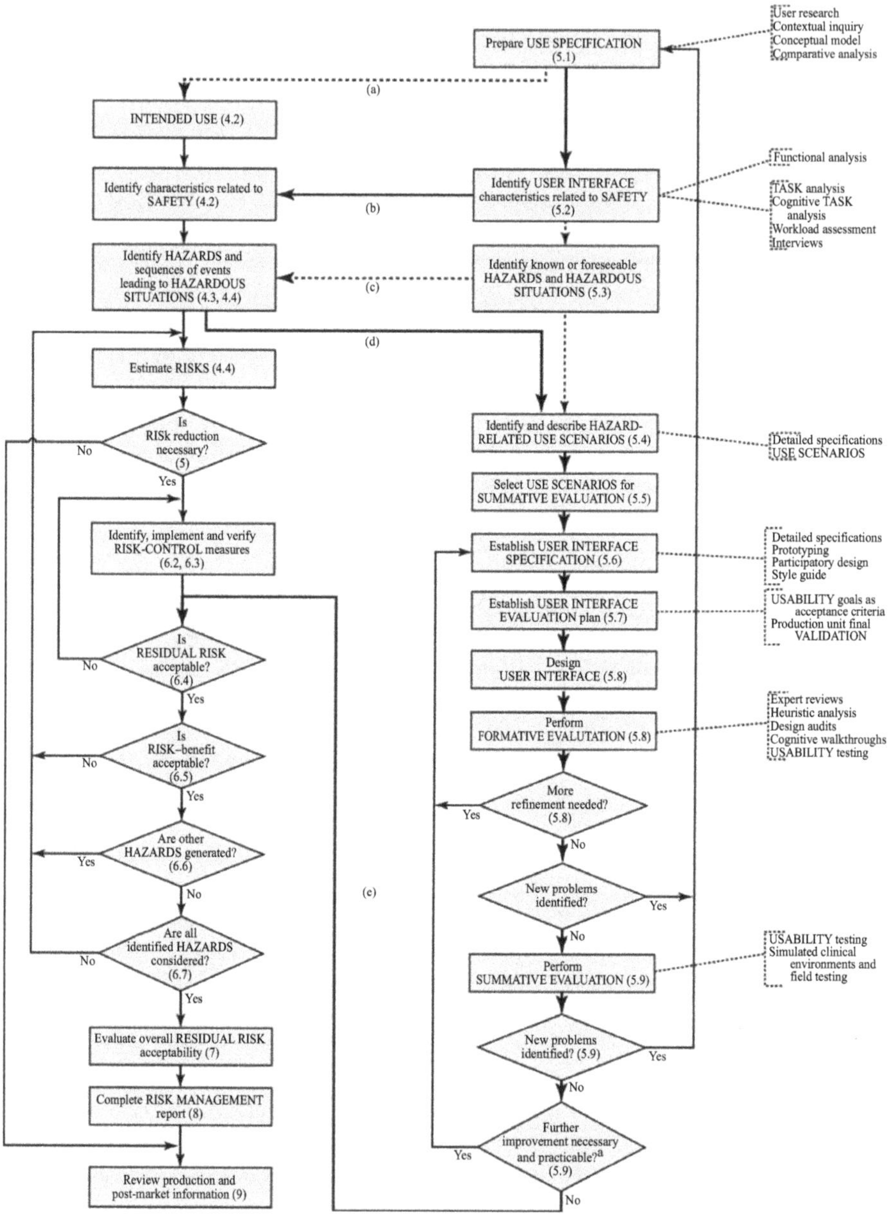

Figure 12.3 The relationship between the risk management process (ISO 14971) and the usability engineering process (IEC 62366-1). (a–e) Represent information flow between the two processes. The heavy solid lines (b, d, e) represent information flow required by usability engineering process. If new problems are identified these should be interpreted to mean new hazards, hazardous situations or

hazard-related use scenarios discovered or implemented risk control is ineffective[4]

[Key: (a) Use Specification is an input to ISO 14971:4.2; (b) identified user interface characteristics related to safety (see 5.2); (c) identified foreseeable HAZARD *and* HAZARDOUS SITUATIONS *(see 5.3); (d) identified sequences of events leading to* HAZARDOUS SITUATIONS *from ISO 14971:2007, 4.4, are an input to determining* HAZARD-RELATED USE SCENARIOS *(see 5.4); (e) evaluate* RESIDUAL RISK. *IEC 62366-1 ed. 1.0 (© 2015 IEC Geneva, Switzerland. www.iec.ch; used with permission). The authors thank the International Electrotechnical Commission (IEC) for permission to reproduce Information from its International Standards. All such extracts are copyright of IEC, Geneva, Switzerland. All rights reserved. Further information on the IEC is available from www.iec.ch. IEC has no responsibility for the placement and context in which the extracts and contents are reproduced by the author, nor is IEC in any way responsible for the other content or accuracy therein]*

12.4.7 Recurrent test and test after repair

Concerning construction and environmental standards, the standards IEC 62353 [28] for periodic tests should be considered. This standard is not applicable to the assembly medical electrical system. For assembling medical electrical system, see Clause 16 of IEC 60601-1 [10]. IEC 62353 [28] does not define requirements for repair, exchange of components and modification of medical electrical equipment or medical electrical system.

12.4.8 Home healthcare environment

Wearable medical devices which are designed for usage at patients' homes should also fulfil the IEC 60601-1-11 [29]. The definition of home healthcare environment is given as: dwelling place in which a patient lives or other places where patients are present, excluding professional healthcare facility environments, where operators with medical training are continually available, when patients are present. Note professional healthcare facilities include hospitals, physician offices, freestanding surgical centres, dental offices, freestanding birthing centres, limited care facilities, multiple treatment facilities and emergency medical services. According to that standard, the manufacturer decides if a wearable medical device is intended to be used in the home healthcare environment.

12.5 New and future standards for medical electrical devices

It should be mentioned that ISO and IEC (ISO/TC 299 and IEC SC 62A) JWG9 has started to address the subject of medical robot safety and published a

[4]The authors thank the International Electrotechnical Commission (IEC) for permission to reproduce Information from its International Standard IEC 62366-1 ed. 1.0. All such extracts are copyright of IEC, Geneva, Switzerland. All rights reserved. Further information on the IEC is available from www.iec.ch. IEC has no responsibility for the placement and context in which the extracts and contents are reproduced by the author, nor is IEC in any way responsible for.

TR IEC/TR 60601-4-1 [19] where a degree of autonomy is introduced into medical devices. Such medical devices with a degree of autonomy are already used and so such devices will be affected by the TR. This TR should be read carefully from researchers who are developing medical wearable devices which include a degree of autonomy. Also under the lead of IEC, new relevant standardization activities are in the pipeline. A first standardization project about rehabilitation robots for medical wearable devices is currently underway within IEC SC62 D & ISO 299 JWG 36. The draft title is 'IEC 80601-2-78: Particular requirements for basic safety and essential performance of medical robots for rehabilitation, assessment, compensation or alleviation'. The industrial robots utilized today fundamentally differ most of the time with respect to the safety concept of wearable medical devices. Such robots are today predominantly shielded from their surroundings and do not come into contact with humans, apart from for maintenance and repair purposes.

On the other side, robots in the medical wearable devices sector are in direct contact directly with patients and indirectly with a user. Consequently, safety concepts from industrial applications cannot be directly transferred to wearable medical devices. Something between medical application and industrial use are robots, which are so-called service robots; they are intended to fulfil their purpose in the environment of humans, but not as wearable medical devices. Nevertheless, standards from the responsible ISO TC 299 WG2 (personal care robot safety subcommittee) could be helpful to understand safety principles.

12.6 Safety aspects for medical electrical devices

If safety concepts are being considered for wearable medical devices, it is not possible to avoid getting involved with certain definitions. What is basic or first failure safety, what is a hazard, and what is accepted for a specific wearable medical device? How does one attain safety, with what measures, and under what acceptable residual risks? Colloquially, safety is probably mostly equated with expressions such as 'freedom from danger' or 'freedom from risk'. ISO/IEC documents explain some terms which are essential for further understanding of safety aspects.

Basic safety (IEC 60601-1; [10]): Freedom from unacceptable risk directly caused by physical hazards, when medical electrical equipment is used under normal condition and single fault condition.

Additional explanations given in Annex A of IEC 60601-1 [10]: basic safety relates to a device not resulting in harm incidental to its operation. Basic safety is often a passive form of protection (such as radiation shielding or electrical grounding).

Essential performances generally relate to medical electrical equipment or medical electrical system operating as intended without creating hazards. A failure of essential performances can be either a lack of performance (such as life-supporting performance) or incorrect performance (such as delivering an incorrect dose to the patient).

In general, basic safety relates to product properties that are not device specific, and essential performances relate to a class of products (such as defibrillators being able to deliver the correct electrical shock). While the terms basic safety and essential performances are generally considered to be mutually exclusive, there are some hazards defined in ISO/IEC Guide 51 [4] directly caused by physical hazards, when medical electrical equipment is used under normal condition and single fault conditions. These include the following:

- Harm: injury or damage to the health of people or damage to property or the environment
- Hazard: potential source of harm
- Hazardous situation: circumstance in which people, property or the environment is/are exposed to one or more hazards
- Inherently safe design: measures taken to eliminate hazards and/or to reduce risks by changing the design or operating characteristics of the product or system
- Intended use: use in accordance with information provided with a product or system, or, in the absence of such information, by generally understood patterns of usage
- Patient: living being (person or animal) undergoing a medical, surgical or dental procedure. Note that a patient can be an operator and an elderly person is not a patient as age is not a disease, injury or disability.
- Residual risk: risk remaining after risk reduction measures have been implemented
- Risk: combination of the probability of occurrence of harm and the severity of that harm
- Safety: freedom from risk which is not tolerable.

The term 'safe' is often understood by the general public as the state of being protected from all hazards. However, this is a misunderstanding; 'safe' is rather the state of being protected from recognized hazards that are likely to cause harm. Some level of risk is inherent in products or systems. Note that these defined terms of ISO/IEC Guide 51 [4] and IEC 60601-1 [10] are not used in all standards for medical devises in exactly the same way. The risk associated with a particular hazardous situation depends on the elements shown in Figure 12.4.

Besides the (fundamental) requirements for safety of a medical device, nearly identical requirements are made for the performance and effectiveness, or so to speak, on the efficacy of the wearable medical device. From the medical-therapeutic viewpoint, this is understood to mean the medical efficacy so that the wearable medical device should deliver the results expected for treatment or diagnosis. These should be considered during the risk management process to identify the clinical functions of a specific wearable medical device. Similarly, the wearable medical device should render the specified services in the form of defined physical properties, for example, speeds or forces. In most countries, clinical trials or clinical evaluations are required to show evidence regarding efficiency (see ISO 14155 [30] for more information).

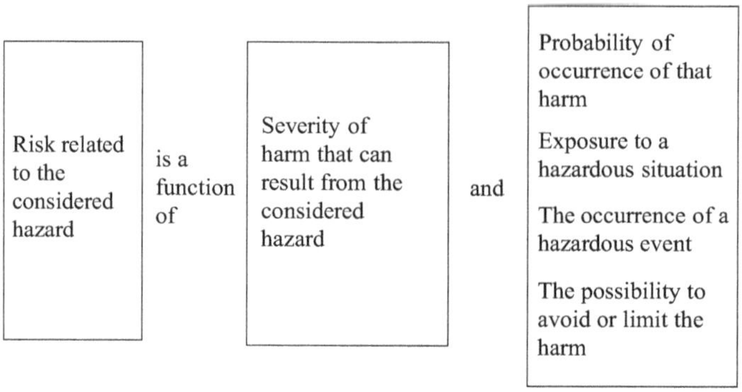

Figure 12.4 Relationship between risk and a particular hazardous situation

The requirements for safety and medical effectiveness and technical efficiency cannot be considered apart from each other. The success of a treatment or even life and health of the patient or user could be endangered by a wearable medical device, if it possesses hazardous capabilities or if it does not function or is not used as intended by the manufacturer.

An example often quoted in the standardization literature about a medical device makes this impressively clear. A defibrillator can save a patient's life, if used correctly and can counteract a ventricular fibrillation. At the same time, if such a defibrillator is improperly used, there is a certain risk for the patient, user and third parties that can lead to a life-threatening situation or even death in case of the wrong indication. It becomes clear that there must be a middle course between 'freedom from risk' and other requirements for a wearable medical device, and thus an acceptable degree of risk or the freedom from unjustifiable hazards must be aimed for. In order to successfully progress along this middle course, it is necessary to draw up a product-specific risk analysis, from which measures to control risk can be implemented. Apart from ascertaining the hazards associated with a certain type of wearable medical device, risk analysis also includes the specification of the essential performance characteristics.

Acceptable risk is partly based on the realization partly, that complete absence of risk is unattainable and, that the degree of risk must be sufficiently low. Risk is a combination of frequency of occurrence and the resulting hazard for patients, user, third party, and, if need be, objects. All wearable medical devices must perform as per their intended use. It must be noted that essential performances are not related to all the performance of the wearable medical device. It has long been recognized that medical electrical equipment or a medical electrical system that does not perform properly could result in unacceptable risk for patients, operators (users) and others. Hence essential performance is based on a risk management process related to performance of clinical functions.

Tests that could applied to decide whether a risk is acceptable or not also determine whether:

1. The risk is so high or the consequences are so unacceptable that it must be rejected as a whole.
2. The risk is so low or made so low that it is negligible.
3. The risk lies between (1) and (2), after it has been reduced to the lowest practical rank, being conscious of the benefits that results, taking the costs of any further reduction of risk into account.

Risks must be reduced to a level as low as reasonably practicable (using the ALARP principle: As Low As reasonably Practicable). If, for example, a risk falls between the two extremes 'not acceptable' and 'insignificant' and the ALARP principle is applied, the resulting risk is an acceptable risk for the application being considered. Although the main considerations for determining the acceptable degree of risk are the extent of damage and the probability, other factors also have to be taken into consideration, e.g.

- How often the prerequisites for the hazard occurrence can be expected (e.g. frequency of the device usage or number of patients treated)?
- The feasibility of further improvements.
- The costs of further improvements.
- Clinical constraints and boundary conditions.
- The benefits that arise by the application of the wearable medical device.
- Public acceptance/customer acceptance.

Since complex medical electrical systems or equipment cannot be exhaustively assessed by tests, their correctness (functionality) and reliability must be assessed in other ways. IEC 62304 requires that the manufacturer shall develop and document an architecture for the interfaces between the software items and the components external to the software items (both software and hardware), and between the software items. Different risk classes (A–C) give a much more clear view which software items are essential to risk control and the standard give advice how these could be handled. Certainty about this is attained by applying suitable procedures during the design process, which have to be transparent and universally consequently applied. The growing realization that unlimited safety cannot be reached has led to the development of risk management concepts. More detailed information on the subject of risk management for wearable medical devices can be found in IS0 14971 [11].

IEC 60601-1 [10] and IEC 60601-2-x already specify most of the general hazards for a wide variety of medical devices. A large number of hazards have already been listed:

- Acceptable configurations of safety-relevant systems (e.g. systems that contribute to safety, such as basic insulation plus a protective earth connection as a reliable configuration for avoiding electric shock)
- The exclusion of certain events in the normal state or in case of a single fault.

A requirement formulated according to one of these two types states that a risk is acceptable. Fault conditions that have to be taken into consideration can be categorized as follows:

- Some faults can be recognized by the user (e.g. external physical damage can be noticed by the careful operator; a broken wire will cause an obvious malfunction in several types of medical electrical devices).
- Some faults cannot be observed, not even by the careful user, but they can be detected by regular maintenance (e.g. partial breakdown of the insulation between the main connection and the protective earth connection in medical electrical devices).
- Some fault conditions can be neither detected by the user nor discovered by regular maintenance (e.g. breakdown of an insulation layer in double insulation).

Only in the fewest cases are there investigations about the actual probabilities of different hazards; instead trust is widely placed on the 'philosophy of the first fault', which can be set out as follows:

- No hazards may result in any of the listed 'conditions of the first fault'.
- All instrument parts that are there to provide safety must be 'appropriately reliable' so that the probability of an 'initial fault' is low.
- Then the probability of two 'initial faults' is very low, and thus the hazard risk caused by a multi-fault condition is acceptable.
- If an initial fault immediately causes others, the probability of these faults is the same as those of the initial fault, and the medical device must remain safe (direct aftereffect on another component caused by the breakdown of an initial component).
- If under certain circumstances two faults arise from a common cause (e.g. bridging of both insulation layers in a double insulation by a conducting liquid or metal objects), the probability of these two faults is the same as the common cause.
- If a fault cannot be discovered at reasonable cost, with workable maintenance procedures and it is not likely that it will be noticed by the user, because it does not influence the device function, the high probability that the fault will remain unnoticed for a long period of time must be taken into account, when developing the safety requirements.

Indeed the probability of simultaneous occurrence of two 'initial faults' is not zero. For wearable medical devices, it is presently considered to be sufficient to guarantee that hazards cannot occur with a 'single fault'. In the case of a double fault, a hazard can occur, but the risk is considered to be slight. The first fault philosophy implies that in general, it is expected that a wearable medical device will have two measures as a protection measurement against each and every hazard. Then it will be assumed that the risk is negligible, provided that the probabilities of faults in the individual systems are low. This implied demand for 'two measures of defence' cannot be covered by redundancy of the same safety systems in every case.

The specific circumstances should be taken into consideration, along with the components, their life cycles and their typical signs of ageing. The use of differently designed safety systems, that utilize different technologies, has proven useful.

The safety of medical electrical devices often demands an integral approach, in which the manufacturer and operator implement a combination of measures, including the following:

- Prerequisites fulfilled by the design of the wearable medical device
- Additional measures such as installation requirements, formal commissioning, regular maintenance and safety checks
- Measures that make the operator aware of the necessity of special precautions, when using certain types of medical electrical devices or with certain applications.

The safe use of medical electrical devices depends on a series of influences, including:

- The construction of the wearable medical device, which must allow and contain the facilities for avoiding hazards
- Appropriate validation of design of hardware and software prior to the start of production
- Application of 'GMP' during the production of the wearable medical device
- Selection of the correct wearable medical device for the respective medical application
- User's familiarity of the wearable medical device and its application, which can be dependent on training or labels on the wearable medical device and manufacturer's instructions
- Use of accessories that are suitable for the wearable medical device
- Connection of the wearable medical device to suitable supply network (e.g. electrical power supply, central gas supply)
- Preventive maintenance of the wearable medical device
- Utilization of specified replacement parts when repairing wearable medical devices.

12.6.1 Safety aspects of wearable medical electrical devices

As explained in the introduction, the severity of the patient's lesions and their cognitive and functional limitations must be taken into consideration so that sufficient attention has already been paid to appropriate reflections in the design and risk management process. It may be assumed that for the majority of the patients, there is a limited reaction and perception capacity. Risk assessments combined with the usability engineering in this respect and options for reducing risk are therefore to be designed appropriately. In addition, many of the patients in therapy often suffer from a series of secondary impairments or aftereffects, which are direct or indirect results of the illnesses or injuries. Risk analyses must therefore be accompanied by clinically experienced persons, familiar with handling the patient population in question. The usability of wearable medical devices should also be considered from this perspective. Patients who will be possibly rehabilitated with the wearable medical devices being considered here are often cognitively and functionally severely impaired, which again demands a high level of care and concentration from the user of the wearable medical devices and distracts from the actual operation of the corresponding wearable medical devices. This must also be taken into account in the design and risk management process and verified by adequate usability engineering.

As a result of the above-given patient population, the benefits and drawbacks that a patient could experience in a therapy using a wearable medical device must be very carefully assessed. For this, the physician or therapist, who is not familiar in any case with the application of modern medical technology, should be given enough information in the form of indication, contraindication, and possible side effects, residual risks and advice in handling of the wearable medical device which enable him to provide the correct and optimal form of therapy. The physician or therapist must be able to make the correct risk–benefit assessment for the well-being of his patient, taking into account, on the one hand, the desired therapy progress for his patient and on the other hand a possible risk of deterioration of the patient's state of health. For this, there must be sufficient information and descriptions of the existing risks, which should be available to him in the user's instructions. There is sometimes a considerable fear of contact on the part of potential users with modern wearable medical devices, along with inexperience with the utilization of technological processes compared to the conventional manual therapies, and this, too, should be taken into consideration in the design of wearable medical devices. Furthermore, the correct measures should be provided in order to introduce the physician or therapist to the new technology and adequately bring him closer to the application of the wearable medical device. This should already be taken into consideration at the conception and risk and usability management and must be systematically implemented. The physician or therapist of a wearable medical device is an important factor not to be neglected when it comes to ensuring safety and effectiveness of a wearable medical device. This can be taken into account by adequate training of the future physician or therapist of the wearable medical device. It is recommended to adapt the duration of training to the prior medical and technology knowledge of the user and to the complexity of the wearable medical device. Regular further training courses and exchange-of-experience workshops reinforce a deeper understanding about the effective application of this type of wearable medical device. For wearable medical devices with a higher degree of autonym, the role of the physician or therapist changes more in the direction of supervision of the treatment. He should always be aware of unforeseeable situation where his spontaneously intervention could be needed.

This autonomy is implemented by the use of detectors, sensors, control loops, software controls and algorithms, just to mention a few aspects of this complex interplay, and mostly without the influence of human interactions. The latter is the prerequisite for the given autonomy. Of course, there are certain pre-settings to be effected, which are essential for the patient, his particular neurological and general medical situation, and to establish his capability. In addition, a corrective intervention by the users should be possible at any time. This inherent autonomy of wearable medical device however requires additional consideration in the design and risk management process.

Special importance is, therefore, attached to software in this type of wearable medical device. Here the software architecture, in particular, must be mentioned. If this can be transparently and exactly built up and displayed coherently in itself and within the whole wearable medical device, it will facilitate the verification and validation effort to concentrate on the right, safety-relevant modules. Clause 14 of IEC 60601-1 [10] describes the requirements of such a PEMS. More guidelines for a corresponding development life cycle are given in IEC 62304 [24] in which, among other things, again draws upon ISO 14971 [11].

Software architecture is mandatorily prescribed standard and must cover the following:

- Components with characteristics of high reliability
- Fail-safe functions
- Redundancy
- Diversity
- Separation of functionality
- Defensive design, e.g. limitation of possible hazardous effect by limiting the available output capacity or by installation of resources that limit the movement of actuators.

The architecture specifications must take the following into consideration:

- Allocation of measures and risk control to PEMS subsystems and components.
- Subsystems and components include sensors, actuators, programmable electronics subsystem (PESS), and interfaces.
- Modes of failure of components and their repercussions.
- Malfunctions with a common cause.
- Systematic malfunctions.
- Length of inspection intervals and the degree of coverage of the function diagnosis.
- Maintainability.
- Protection against reasonably predictable misuse.
- Specification of the network/data sharing if applicable.

IEC 62304 [24] describes processes that have to be included in the software development cycle for the development of safe software for wearable medical devices. In order to determine which functions create or control risk, it is necessary to completely identify the PEMS (PESS) requirements. It is not possible to carry out an appropriate risk assessment without a complete specification of the requirements and an architecture design that satisfies this specification. The requirements should include the following, if applicable to the PEMS software:

- Functional performance requirements including essential performance characteristics in compliance with IEC 60601-1 [10]

- Physical characteristics and the conditions of their surroundings under which the software should run
- External interfaces with the software
- Safety requirements, including risk-control measures for hardware breakdowns and possible software defects and specifications regarding the method of operation and maintenance, environmental influences and risk control
- Software-controlled alarm signals, warnings and operator messages
- Safety requirements, where any gaps in safety could affect overall safety
- Ergonomic requirements regarding the use of PEMS, including those that refer to the following elements: support when operating manually, human–machine interactions, limitations of personnel and areas where intense human attention is required and which are susceptible to human error and training
- Data definitions and requirements for the database
- Installation and acceptance requirements for the PEMS software
- The documentation that has to be drawn up
- Operation and design requirements
- Maintenance requirements.

More information about PEMS structure, PEMS development life-cycle management, and documentation is given in IEC 60601-1; Annex H [10]. Also, Appendix II listed additional Guidelines about Software and Software Development Cycle.

12.7 Conclusions

The chapter has covered the regulatory standardization issues for wearable exoskeletons from machinery and medical device perspectives. Wearable medical device technology is a new, innovative field of activity and is in a rapid and continuous state of development currently. Thus, an overview of standards and safety aspects for machines and medical electrical devices based exoskeletons can only represent a snapshot in time.

As stated above, new standardization efforts are in progress for physical assistant exoskeletons and for medical exoskeletons. Attempts to generally portray the 'state of the art' or the standards are subject to constant change. National regulations tend to drift apart, instead of moving together toward a uniform global procedure. Additionally, the standard organizations and their member organizations push more and more to actualize existing standards in shorter periods. These organizations are also under the pressure from national regulators for wearable medical devices to create additional standards which sometimes have a more political than safety background. Only a permanent systematic observation and adaptation to the various constraints can ensure that products comply with the constraints from the regulated wearable medical device sector.

Society in general and the health sector in particular expect highly efficient products that have a greater and better functionality. As stated above, significant efforts have to be placed on extending the provision of proof for safety and effectiveness here and on more detailed documentation, safeguarding in all directions.

The more complex wearable medical devices become, the costlier the whole process will be, and pressure from the healthcare system regarding cost is worldwide increasing. The world of wearable medical device technology will have to grapple with this development, and wearable medical device manufacturers, in particular, will have to adjust to the increasing constraints from various sources. Environmental, hazardous materials and recycling requirements, which also have to be met by the entire electronics sector, have been intentionally omitted from this chapter.

In the future, the art will consist of achieving a balance between increasing national and international regulation demands, expressed in increasing constraints, with the ability to promptly introduce innovative treatment alternatives in an acceptable cost frame.

References

[1] International Organization for Standardization, www.iso.org, accessed 31 July 2017.

[2] International Electrotechnical Commission, www.iec.ch, accessed 31 July 2017.

[3] ISO Guides, www.iso.org/iso-guides.html, accessed 31 July 2017.

[4] ISO/IEC Guide 51, Safety aspects – guidelines for their inclusion in standards, 2014.

[5] ISO/IEC Guide 78, Safety of machinery – rules for drafting and presentation of safety standards, 2012.

[6] ISO/IEC Guide 63, Guide to the development and inclusion of safety aspects in International Standards for medical devices, 2012.

[7] ISO 12100, Safety of machinery – general principles for design – risk assessment and risk reduction, 2010.

[8] EC Machinery Directive, Directive 2006/42/EC of the European Parliament and of the Council of 17 May 2006 on machinery, and amending Directive 95/16/EC.

[9] EC Medical Device Directive, Commission communication in the framework of the implementation of the Council Directive 93/42/EEC of 14 June 1993 concerning medical devices; Publication of titles and references of harmonized standards under the directive 2010/C 183/03. Amended by Directive 2007/47/EC of the European Parliament and of the Council of 5 September 2007.

[10] IEC 60601-1:2005+AMD1:2012 CSV, Medical electrical equipment – Part 1: General requirement for basic safety and essential performance, 2012.

[11] ISO 14971:2007, Medical devices – application of risk management to medical devices, 2007.

[12] FDA document, https://www.fda.gov/RegulatoryInformation/Guidances/ucm070627.htm on guidance for medical device manufacturers.

[13] ISO 13482:2014, Robots and robotic devices – safety requirements for personal care robots, 2014.

[14] European Harmonised Standards under the Machinery Directive, https://ec.europa.eu/growth/single-market/european-standards/harmonised-standards/machinery_en, accessed 31 July 2017.

[15] Official Journal of the European Union, http://ec.europa.eu/growth/single-market/European-standards/harmonised-standards/medical-devices/index_en.htm, accessed on 31 July 2017.

[16] Food and Drug Administration, www.fda.gov/MedicalDevices, accessed on 31 July 2017.

[17] International Medical Device Regulators Forum, www.imdrf.org, accessed on 31 July 2017.

[18] Medical Device (MEDDEV) Documents, http://ec.europa.eu/geninfo/query/resultaction.jsp?QueryText=MEDDEVþdocuments&sbtSearch=Search&swlang=en, accessed on 31 July 2017.

[19] IEC/TR 60601-4-1, Medical electrical equipment – Part 4–1: Guidance and interpretation – medical electrical equipment and medical electrical systems employing a degree of autonomy, 2017.

[20] ISO 13485:2016, Medical devices – quality management systems – requirements for regulatory purposes, 2016.

[21] ISO 9001:2015, Quality management systems – requirements, 2015.

[22] ISO 16142-1:2016, Medical devices – recognized essential principles of safety and performance of medical devices – Part 1: General essential principles and additional specific essential principles for all non-IVD medical devices and guidance on the selection of standards, 2016.

[23] IEC 60601-1-2:2014, Medical electrical equipment – Part 1–2: General requirements for basic safety and essential performance – collateral standard: electromagnetic compatibility – requirements and tests, 2014.

[24] IEC 62304:2006, Medical device software – software life cycle processes, 2006.

[25] ISO 10993-1:2009/Cor 1:2010, Biological evaluation of medical devices – Part 1: Evaluation and testing within a risk management system and additional parts, 2010.

[26] IEC 60601-1-6:2013, Medical electrical equipment – Part 1-6: General requirements for basic safety and essential performance – collateral standard: usability, 2013.

[27] IEC 62366: with Amendment 1: 2014, Medical devices – application of usability engineering to medical devices, 2014.

[28] IEC 62353:2014, Medical electrical equipment – recurrent test and test after repair of medical electrical equipment, 2014.

[29] IEC 60601-1-11:2015 RLV, Medical electrical equipment; general requirements for basic safety and essential performance; collateral standard: requirements for medical electrical equipment and medical electrical systems used in the home healthcare environment, 2015.

[30] ISO 14155:2011, Clinical investigation of medical devices for human subjects – good clinical practice, 2011.

Chapter 13

Test methods for exoskeletons—lessons learned from industrial and response robotics

Roger Bostelman,** and Tsai Hong**

Abstract

Exoskeletons are devices that can assist the human wearer's limbs to provide functional, normal, or amplified human capabilities. Research on exoskeletons has dramatically increased recently. However, measurements of these devices have yet to show long-term safety and other effects on humans. Safety standards now allow, through risk assessment, both manufacturing and wearable robots to be used, although performance standards for both systems are still lacking. Much can be learned from industrial and response robot safety and performance research and standards activities that can cross over into the exoskeleton arena. For example, ongoing research to develop standard test methods to assess performance of manufacturing robots and emergency response robots can inspire similar test methods for exoskeletons. This chapter first lists exoskeleton performance metrics and standards for collaborative industrial robots, response robots, and also physical assistance robots (i.e., exoskeletons). Then, it describes measurements of joint axis rotation location using an industrial robot simulating a human arm, as well as mobile manipulator and response robot test method developments that could also apply to exoskeletons. These methods and others are then integrated into recommendations for exoskeleton test methods.

Keywords: Exoskeleton, crossindustry, industrial robot, response robot, artifact standards

13.1 Introduction

Exoskeletons[1] (e.g., wearable robots, passive, counterweighted) are devices that can assist the human wearer's limbs to provide functional (i.e., perhaps below

*National Institute of Standards and Technology, Engineering Laboratory, Intelligent Systems Division, USA

**Le2i, Université de Bourgogne, France

[1]Disclaimer: Commercial equipment, software, and materials are identified in order to adequately specify certain procedures. In no case does such identification imply recommendation or endorsement by the National Institute of Standards and Technology nor does it imply that the materials, equipment, or software are necessarily the best available for the purpose.

normal), normal, or amplified human capability. Both passive (spring-supported and/or counterweighted) and active (electrically powered) human–worn devices are considered as exoskeletons. Capabilities provided by these devices include normal human lifting and/or other movements for extended time periods, for example, typical and beyond typical work days, as well as increased human strength, speed, agility, etc. for amplified exoskeleton versions. Recent research on exoskeletons has dramatically increased as seen in, for example, [1,2] and has driven the current global exoskeleton market to include over 50 manufacturers. However, performance, ergonomic, healthcare impacts, long-term safety, and other effects on humans have yet to be studied and understood.

International Organization for Standardization (ISO) 13482 [3] was developed over several years and published in 2014 to address the safety concerns for robots, including exoskeletons, used as personal care devices. Medical exoskeletons, such as those for rehabilitation, are not considered here. However, no normative requirements on data collection and analysis are included in the safety standard to be used as a basis for understanding long-term effects. Test methods that formalize means of collecting and analyzing performance data are therefore needed that allow manufacturers and users to replicate standard safety procedures in a uniform manner. The data from these test methods will enable the understanding of the ramifications from use of exoskeletons. Exoskeleton performance is interlaced with some safety considerations for these devices. For example, what are the issues when an exoskeleton is applied to the lower extremity (e.g., legs and hips) to assist or augment a human in carrying out tasks? As an exoskeleton user stands up from a crouched position or walks for an entire day carrying heavy loads, does the device provide full lift, overdrive human joints, cause the wearer strain or chafed skin, or other harm and if so, how should the device instead be designed to be safe? Safety test methods can help provide this measured data for most any exoskeleton manufacturer for comparison of capability to perform the task. However, performance test methods may additionally provide task-specific measurements of how well the device can provide, for example, improved movement, increased or longer lift capability, or combined lift with precision positioning of a heavy load. Unfortunately, there are currently no performance standards or test method procedures to provide such exoskeleton measurement data to be included in either safety or performance standards when they are developed. In addition, there are no commonly agreed upon physiological measurements of the human user to provide baseline measurements of before and after long-term use of exoskeletons.

The National Institute of Standards and Technology (NIST), Robotic Systems for Smart Manufacturing (RSSM) Program [4] develops test methods and measurement science for stationary and mobile robot arms as well as vehicles to support calibration, measurement, and advance understanding of current and target performance of emerging new capabilities for manufacturing applications. The program develops and deploys advances in measurement science that enhance US innovation and industrial competitiveness by improving robotic system performance to achieve dynamic production for assembly-centric manufacturing. NIST also has the objective to advance the capabilities of remotely operated emergency response robots by developing the measurement and standards infrastructure necessary to quantitatively evaluate system capabilities under the Emergency Response Robots Project [5].

One example of the work being done under the RSSM program is the development of performance metrics and test methods for mobile manipulators that can be used for assembly. Assembly operations performed by a mobile manipulator require accurate position and orientation (pose) relative to a work piece, e.g., to insert pegs in holes or mesh gears with other gears. Advanced measurement tools and artifacts are being used at NIST to develop affordable and repeatable performance measurement and test methods. For example, test methods are being developed that can measure how well a mobile manipulator can align with a work piece so that the peg-in-hole operation can succeed. Another test method being developed can determine the exact location of industrial robot joints from outside the robot so that devices can properly adapt to robots.

Lessons can be learned from these industrial robot measurement methods to apply to exoskeletons where, as exemplified above, load positioning (versus pose of a peg held by a mobile manipulator) or human knee or elbow joint (versus external robot joint) measurement should be performed in a repeatable way. This and many other tests are required to fully understand how exoskeletons can maintain and/or improve human performance throughout an entire day and for many days.

This chapter initially provides a list of potential exoskeleton performance metrics which could be used as a basis for the development of test methods. The chapter then discusses current standards for collaborative industrial robots, response robots, and physical assistance robots (i.e., exoskeletons). The chapter then provides background, a literature survey, and description of experiments and results of joint axis location measurement using a robot arm simulating a human limb and an optical tracking system (OTS). The test method can also apply to measurement of the human knee or elbow rotational axis locations which are important to how well an exoskeleton fits to the human body. From these industrial/response robot test methods and experiments, recommendations for exoskeleton test methods are then considered, followed by a summary and conclusions.

13.2 Exoskeleton performance metrics

It is important to categorize performance metrics of systems-under-test to benchmark and compare results across systems, as well as to compare system applications to tasks. Industrial robot metrics currently typically do not include human(s)-in-the-loop since most robot functions are automatic and disconnected from human control. This is changing as collaborative robots that function side-by-side with humans are becoming available. Response robot metrics do include humans since the robots are mainly remotely controlled by humans and tasks may include finding and interacting with victims. From [6], we extract common metrics for task-oriented mobile robot human–robot interaction that may also apply to exoskeletons.

Below, we provide their list and ask associated questions for each metric when applied to exoskeletons:

- *Navigation*: can the exoskeleton assist the wearer in remaining stable with a short or long stepping gait pattern, when walking, running, or even crawling if specified by the manufacturer, and also when normal gait activity is interrupted or various motion tasks initiated?
- *Perception*: can the exoskeleton access body measurements, perceive these inputs, and make use of the perceived input(s) to control the next intended motion(s)?
- *Management of tasks*: can the exoskeleton react to sudden obstacles, e.g., input sensory information from the body and then react appropriately to such emergency situations for avoiding collision or mitigate their effects?
- *Manipulation*: when using upper-body exoskeletons, can the robot move the arm appropriately for the task, both with and without loading and how does it change the user for when they are not wearing the robot?

Other effects, or what [6] termed as "bias effects," that perhaps adjust the above metrics are: communication (delay, jitter, and bandwidth), response from the robot (reaction to sensory input from the wearer), and the user (i.e., size, weight, profile, signal strength, expected signals). Further complicating matters, exoskeleton metrics that include the human-in-the-loop must be compared to the baseline of when the human is not wearing, controlling, or affected by the robot. Therefore, additional exoskeleton performance metrics suggested by the authors include the following:

- *Duration*: maximum time that a task can be performed with the use of an exoskeleton as compared to the task being performed without the exoskeleton
- *Speed*: velocities that can be achieved and sustained with the use of an exoskeleton as compared to the task being performed without the exoskeleton
- *Acceleration/Deceleration*: accelerations/decelerations that can be achieved with the use of an exoskeleton, as well as expected rapid movement impeded by the exoskeleton, as compared to the task being performed without the exoskeleton
- *Pose uncertainty*: accuracy/resolution (e.g., precision to move to a commanded location and orientation) and repeatability (e.g., move to the same commanded location more than once) for the exoskeleton to position and orient the operator's arm or leg as commanded. Positioning error of a tool or device when held by the controlled arm or leg is the measured component
- *Back-drivability or control force*: force required to resist component reaction or move any or all components of the exoskeleton when they are both driven or not driven
- *Vertical maneuvering (see navigation above)*: capability, speed (for lower body exoskeletons) to traverse inclines, steps, undulating terrain and (for upper body exoskeletons) to lift loads
- *Horizontal maneuvering (see navigation above)*: capability, speed (for lower body exoskeletons) to move the body or torso forward, back, side-to-side

and (for upper body exoskeletons) to move the arm(s) forward, back, side-to-side

- *Ergonomics*: measure of comfort (pain, fatigue) and posture of the body when wearing an exoskeleton
- *Ingress/Egress complexity*: difficulty in putting on or taking off the exoskeleton
- *Ease of use*: simplicity of initial training and ease of control of the exoskeleton as it allows or improves task completion performance
- *Other*: cost, portability, battery life, range of use, environmental, cyber security.

Complications arise when ergonomics are included since comfort and posture are not typical metrics considered in industrial or response robotics. Combining ergonomics with, for example, duration, speed, etc. provides limitations to these metrics. One such example may be that task duration and speed are increased when using an exoskeleton until a period of time or a maximum speed is reached where the body starts to feel pain or other discomfort due to muscle fatigue. In [7], tests using exoskeletons for welding and painting were "continued until the quality reached an unacceptable level" during operator fatigue measurements. Tests were stopped when: the subject felt too much pain, the quality-of-work score dropped below a threshold, the subject used other parts of his/her body to continue the task, or the examiner felt that the subject was at a safety risk. In the last "other" category, a list of miscellaneous metrics is provided that may affect the potential user's decision whether to procure exoskeletons and/or how well a subject wearing an exoskeleton can perform tasks.

Additional measurements that are not explicitly discussed here, but that can be conducted in conjunction with the proposed tests include clinical measures of walking performance [8] and other measures of human performance, such as might be used in the rehabilitation community. Furthermore, there are measures of vitals, including heart rate, blood pressure, oxygen demand, etc. [9] that should also be considered.

13.3 Standards

Safety standards for exoskeletons are well underway and are briefly described in the chapter on regulations for medical and nonmedical exoskeletons. However, performance standards for exoskeletons have not started to be formulated in a coordinated manner. Beyond the current standards section is a section on cross-industry performance standards with subsections listing industrial robot and response robot performance standards efforts and publications. These efforts may also provide guidance for future developments of exoskeleton standards.

13.3.1 Safety standards

ISO 13482:2014 [3] safety requirements for personal care robots cover safety protocols for three robot types: mobile servant, person carrier, and physical assistant. Exoskeletons are described as physical assistant robots and their safety risk assessment and mitigation procedures are described. Figure 13.1 shows the basic assistive device designs from 13482 Appendix D. The classes of physical assistant robots show only example designs of what is an exoskeleton and not how well, how long, or how safely they can perform. The functional tasks for each assistive exoskeleton designs are

- *Leg motion assistive exoskeleton*: Applying cooperative control to a user's thighs in order to control the stride and to achieve comfortable walking.
- *Body weight supportive exoskeleton*: Reducing the load on legs, hips, knees, and ankles while standing or walking by supporting part of or fully the user's bodyweight.
- *Exoskeleton wearable robot*: Physically supporting a human and manipulating body parts through direct interaction and fixtures to the person, e.g., via straps or clamps. Enabling the user to carry loads similar to or above average human strength.

Exoskeletons provide assistance to the user as perhaps described in these three designs. The body weight supportive device is, however, similar to a motor bike where the user "rides" on the exoskeleton system and this is therefore not considered wearable. The leg motion assistive exoskeleton design allows it to rest on the user, potentially without straps or other means of attachment. Therefore, only the exoskeleton wearable robot is attached to the body and is the physical assistant robot design addressed in this chapter.

Experts from China, France, Germany, Japan, Korea, United Kingdom, USA, and other countries are developing a standard for exoskeleton robots to provide test methods for ISO 13482 within ISO Technical Committee 299, Working Group two [10]. The current ISO/Committee Draft Technical Report (TR) 23482-1 Robotics—Application of ISO 13482—Part 1: Safety-related test methods [11] draft covers

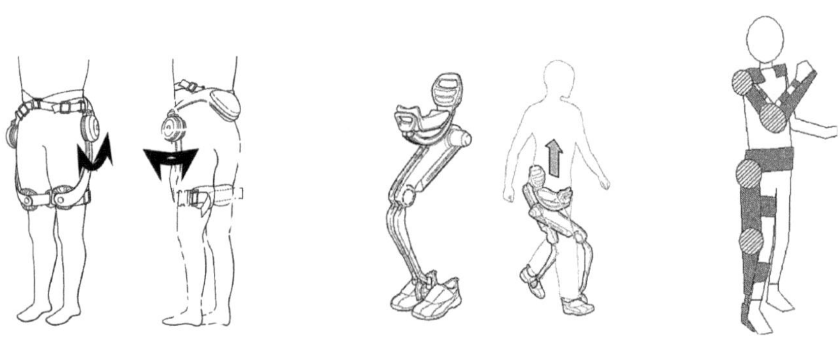

(a) (b) (c)

Figure 13.1 Basic exoskeleton designs shown in ISO 13482 Appendix D: (a) leg motion assistive device, (b) body weight supportive device, and (c) exoskeleton wearable robot physical hazard characteristics for "wearable robots," stress for skin, principle, apparatus, and procedures sections. The principle "stress for skin" test method describes three steps that utilize simulation and a test dummy to define, test, and record contact states and maximum load on the human body at the location where the wearable robot is worn. A "wearable robot cuff" is described in the apparatus and the procedure involves attaching the cuff to a dummy having artificial skin. A three-dimensional (3D) load device moves the cuff for the length of load time specified. The load device generates load patterns specified where "the pattern of maximum load on the human body skin simulation device is measured during the operation." The load measurements are then recorded with the data logging process also described in the standard.

Additionally, the draft standard provides a test method that "inspects visible damage (e.g., fracture, deformation, or disengagement of parts) and functional damage (e.g., abnormality of control system) of a robot from continuous locomotion in order to estimate durability throughout its design life." The test covers robots worn by a person to provide walking assistance where the device may harm the wearer. The test includes set-up, test motion, and inspection as well as test support surface (simulation, e.g., treadmill, of the intended robot-worn or robot-use environment), test dummy, and supporting device apparatus as needed for the test. The test method also includes accelerated testing (e.g., increased speed of the treadmill).

As mentioned in Section 13.1, there are currently no exoskeleton performance standards. As a potential crossover from other standards, we provide in this section a list of current and developing industrial and response robot performance standards.

13.3.2 Crossindustry performance standards

13.3.2.1 Industrial robots

Performance standards are currently being developed for driverless automatic guided industrial vehicles (i.e., ASTM F45 [12]) and are already published for industrial robots (i.e., ISO 9283 [13]). The industrial robot performance standard ISO 9283 provides methods for measuring performance of robot arms. ASTM F45 includes five subcommittees developing the following documents for any type of industrial vehicle, generically termed "automatic/automated/autonomous-unmanned ground vehicles (A-UGVs)":

- F45.01 WK54576—Standard Practice for Recording Environmental Effects
- F45.02 WK48955—Standard Test Method for Navigation: Defined Spaces for A-UGVs
- F45.03 WK54662—Standard Test Method for Grid-Video Test Method for A-UGVs
- F45.04 WK54431—Standard Practice for Communication and Integration Interruptions for A-UGVs
- F45.91 ASTM F3200-16—Standard Terminology for Automatic Guided Industrial Vehicles

Documents designated by "WK" are working drafts, whereas "ASTM" denotes an approved standard. WK48955 discusses test methods with physical and virtual barriers defining test spaces for A-UGVs to navigate within. Test spaces are intersection or "L-shaped" with curved or chamfered corner designs to test industrial vehicles. WK54576 can be applied to any test method, e.g., WK48955, providing a consistent and practical recording method to describe the environment during the test. A similar navigation test method with varying environmental conditions can be developed and used to test humans wearing exoskeletons. The number of repetitions is outlined in WK48955 to provide a statistical confidence level for the user. As the exoskeleton may affect the wearer during a variety of human motions such as walking, running, crawling, sit-to-stand, or vice versa transfers, the test should be completed with the subject initially not wearing the exoskeleton as a baseline, then wearing it while completing the test method, and repeated where timing or other metric is used to measure and record performance.

WK54576 provides example recording of lighting, external sensor emission, temperature, ground surface, air quality, humidity, and electrical interference. Similarly, WK54431 provides a practice for recording types and locations of interruptions in communication when A-UGVs are deployed, either indoors or outdoors depending on their applications. Exoskeleton communication with external devices or with body measurement sensors could be similarly tested and recorded.

F3200-16 defines terminology that is useful for F45 test method development and implementation. A similar type of document would be useful for exoskeleton manufacturers, users, and test method developers to ensure consistent and unambiguous definitions of terms.

As outlined in ISO 9283, a series of industrial robot methods are described for specifying and testing performance characteristics, including: pose, distance and path accuracy, and repeatability; position stabilization, overshoot, and drift; deviations; pose timing; compliance; and weaving deviations. All of these characteristics are potentially useful when measuring performance of the human–worn exoskeleton.

13.3.2.2 Response robots

Several performance standards have been created through the ASTM International E54 Committee for Homeland Security Applications [14]. Specifically, the E54.09 Subcommittee-developed Standard Test Method Suite for Evaluating Emergency Response Robot Capabilities focuses on measuring capabilities of robots with respect to mobility, energy/power, radio communication, durability, logistics, safety, human–system interaction (HSI), sensors, and autonomy, although most response robots are teleoperated [15]. This suite of standards can provide cross-industry test methods that may apply to wearable robots and passive systems. Below are the potentially relevant standards (noted by "ASTM"), working documents under development (i.e., indicated by "WK" prior to a number), and planned standards for future development that may also apply to exoskeletons:

Mobility, Confined Area Terrains and Obstacles:

- Gaps (ASTM E2801),

- Hurdles (ASTM E2802),
- Inclined Planes (ASTM E2803),
- Stair/Landings (ASTM E2804),
- Gravel (WK35213),
- Sand (WK35214)
- Continuous Pitch/Roll Ramps (ASTM E2826)
- Crossing Pitch/Roll Ramps (ASTM E2827)
- Symmetric Stepfields (ASTM E2828)
- Mud (WK54403)

Human–systems interaction:

- Maneuvering, Sustained Speed (ASTM E2829)[2]
- Maneuvering Tasks, Towing Grasped/Hitched Sleds (ASTM E2830)
- Maneuvering Tasks, Post/Hole Slaloms
- Search Tasks, Random Mazes with Complex Terrain (ASTM E2853),
- Navigation Tasks: Hallway Labyrinths with Complex Terrain (WK33260)
- Confined Space Voids with Complex Terrain (WK34434)

Sensors:

- Localization and Mapping: Hallway Labyrinths with Complex Terrain, (planned)
- Localization and Mapping: Wall Mazes with Complex Terrain, Sparse Feature Environments (planned)

Manipulation:

- Door Opening and Traversal Through Door (WK27852)
- Heavy Lifting: Grasp, Lift, and Place (WK41323)
- Extract (WK54274)
- Touch or Aim (WK54272)
- Place Object (WK54283)
- Inspect (WK54271)
- Dexterous Extract (planned)

13.4 Crossindustry measurements applicable to exoskeletons

13.4.1 Joint rotation axis location

13.4.1.1 Background

Exoskeletons normally provide torque assistance at human joints through passive (e.g., springs, counterweights) and/or active (e.g., electrically powered) mechanisms. As humans vary broadly in size and shape, adapting exoskeletons to them is complex, creating a challenge to manufacturers and users to properly fit these

[2]Maneuvering Tasks are under the HSI category because they are performed at a standoff distance by the operator, requiring high levels of situational awareness to perform successfully.

machines to the wearer. Without proper fit, exoskeletons can produce potentially uncomfortable and/or worse, unsafe use systems for the wearers. For example, [16] describes the sizing restrictions and the sizing process and a check-list for the exoskeleton fit to the user. If the user feels discomfort, the user is to make adjustments, although there are no specifics provided on exactly how much to adjust for safe use. The goal, for example, for good fitting at the knee is to have the axis of rotation for the exoskeleton's knee remain co-located with the axis of rotation of the wearer's knee which is known to move with knee rotation (see Figure 13.2(d)).

A goniometer device has been used to measure joint rotation axis location since the early 1900s. The location of the axes highly depends on the alignment of the device to the limb. In [17], the researchers suggest that accuracy error can be up to 10°. They also suggest that through the use of a camera positioned approximately perpendicular to the leg, errors can be approximately 3° or less. Therefore, new technology can provide improvements to joint measurement.

Additionally, once properly fitted, the exoskeleton must not change position, for example by sliding down the limb due to human motion or gravity. Assuming the machine is truly fixed to the wearer, ideally, it is known exactly where human joint axes (i.e., shoulder, elbow, hip, knee, ankle, etc.) are located, where even on a single human there may be variability between right and left joint location, rotation, etc. Figure 13.2(a) shows an image of a human body reference frame [18], useful

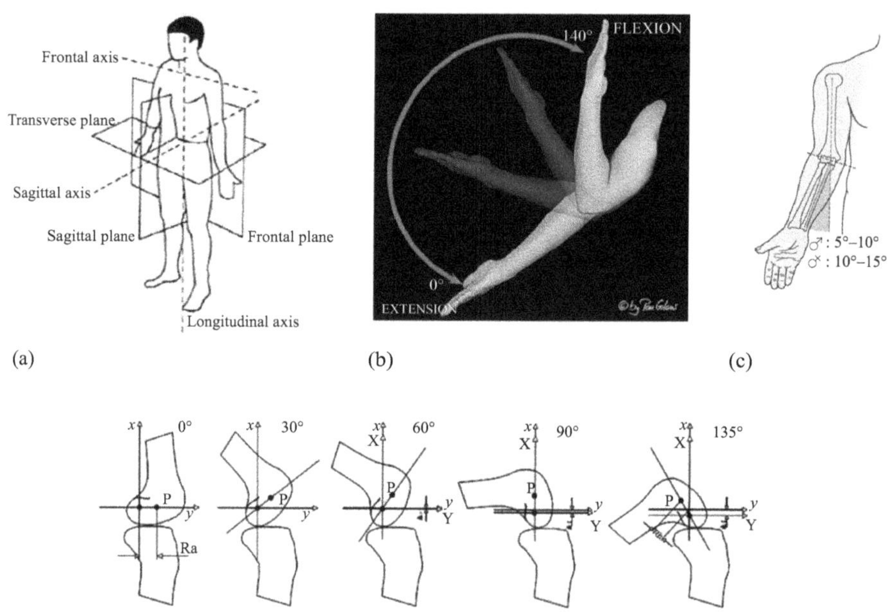

Figure 13.2 (a) Human body reference planes [18]. (b) Elbow extension through flexion angle [19]. (c) Side-to-side joint rotation between the humerus and the ulna during normal carrying angle [21]. (d) Compounded knee rotation and lift at various joint angles [22]

for further joint angle discussion. An example elbow flexion/extension shown in Figure 13.2(b) [19] shows a single axis rotation of 140°. Normal maximum elbow, hip, knee, and shoulder rotations are described in [20] as

- Elbow—Flexion: 150°, Extension: 180°, Supination: 90°, and Pronation: 90°
- Hip—Flexion: 110°–130°, Extension: 30°, Abduction: 45°–50°, Adduction: 20°–30°, Internal rotation: 40°, External rotation: 45°
- Knee—Flexion: 130°, Extension: 15°, Internal rotation: 10°
- Shoulder—Abduction: 180°, Adduction: 45°, Horizontal Extension: 45°, Horizontal Flexion: 130°, Vertical extension: 60°, Vertical flexion: 180°.

A slight complication to the sagittal plane rotation is that the elbow or synovial hinge joint angle moves along the frontal plane between the humerus (upper arm) and the ulna/radius (lower arm) bones as shown in Figure 13.2(c) [21]. This occurs when extended during what is referred to as the normal carrying angle. Further, in many joints, such as the knee shown in Figure 13.2(d) [22], multiple joint axes occur as the knee rolls about a noncircular surface which also provides lift and also causes leg-length increase, both in the sagittal plane. The shoulder and ankle are capable of more spherical rotation motions which must also be considered for exoskeleton adaptability to humans.

13.4.1.2 Literature survey of human body measurement

Researchers have attempted various techniques to identify the joint rotation axis location considering the simple measurement case of a human hinge joint motion. Deland *et al.* [23] used flashing lights captured on photographic film to measure joint rotation axis location (see Figure 13.3(a)) on five cadaver arms. Bottlang *et al.* [24] used electromagnetic motion-tracking with radiography of an inserted screw at the estimated joint of seven cadaver arms. More recent measurement techniques use optical tracking to mainly measure human body motion. Examples of marker locations are the Plug-in-Gait, Helen Hayes (Davis), and Body Segment CM marker

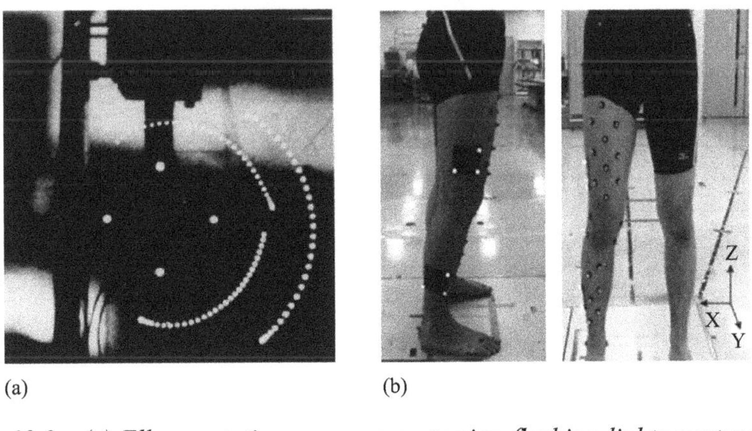

(a) (b)

Figure 13.3 (a) Elbow rotation measurement using flashing lights captured by a camera [23] and (b) rigid (left) versus point cluster (right) marker sets being used to measure gait [28] placement techniques [25]. Plug-in-Gait places markers only in the frontal plane. Helen Hayes (Davis) adds a marker on, for example, the upper foot and heel. Body Segment CM places a marker on each body segment. Various tests have used these and other marker placements measured by OTSs to measure motion [26], bone bending [27], or to compare rigid marker sets to point clusters (see Figure 13.3(b)) [28,29] to estimate anatomical landmarks based on markers attached to a segment. The rigid marker sets are meant to limit measurement errors due to muscle and skin motion [30], although when using many markers as shown in Figure 13.3(b), errors were minimized. However, positioning the markers and keeping the markers attached to the human provide challenges. Rosenhahn *et al.* [31] compared using markers to markerless motion capture resulting in root mean square errors of less than 3° between methods. Zhang *et al.* [32] used a hybrid of both marker and markerless video capture to measure body motion. Additionally, Bakhshi [33] used a marker tracking system as ground truth for comparing to an inertial measurement unit (IMU) to measure human joint angles where IMU average errors ranged from 0.08° to 3.06° for a variety of body motions. And even further, long-term joint motion was considered by [34] using electrodes sown within spandex fabric and worn by a human in motion.

Of key importance to all of the previous references is the notion of how certain and reliable are the measurements and the techniques used to make the measurements. For example, in the above list of normal maximum rotations from [20], it is unknown how the measurements were made and therefore, how much uncertainty there is in each measurement noting that the angles are nearly all listed in 10° increments. "Uncertainty is a measure of the 'goodness' of a result. Without such a measure, it is impossible to judge the fitness of the value as a basis for making decisions relating to health, safety, commerce or scientific excellence" [35]. The references also measure joint angles and do not describe measurement of the rotation axis location that could perhaps be derived from the angle, assuming the exact centroid axes of the links connected to the joint are known. The following subsections discuss similar techniques to the cited research although the subsections include a measure of uncertainty.

13.4.1.3 Robot joint measurement

Mori and Malik [36] considered that one might take "a single two-dimensional (2D) image containing a human figure, locate the joint positions, and use these to estimate the body configuration and pose in 3D space." Similarly, we used 2D artifacts in two experiments to measure joint rotation axis location using an industrial serial-link manipulator or robot arm and an OTS. The measurements were used to consider OTS capability for measuring human–worn artifacts and to consider two test methods using an OTS and artifacts that could enable proper fitting of exoskeletons to human joints to support exoskeleton design and testing for safe use of these systems.

Industrial serial-link manipulators can have simple motion similar to the motion of human legs or arms and much can be learned using this robot system as a simulation tool. An example drawing of these simple motions is shown in

Figure 13.4. The drawing shows labels of the human body links and joints and the unknown joint axis rotation locations for the shoulder/hip and elbow/knee joints. For simplicity, we will continue discussion of only the arm, specifically the elbow joint. An area typically not considered in industrial or other robot applications is to externally measure robot joints rotation axes locations, as was previously shown for human limbs. Robot kinematics provide the mathematical basis to properly control robots to ensure their end-effectors or attached tools are located at expected positions and orientations in joint or world space coordinates. However, it may also be necessary to physically measure the joint rotation axis location to adapt external

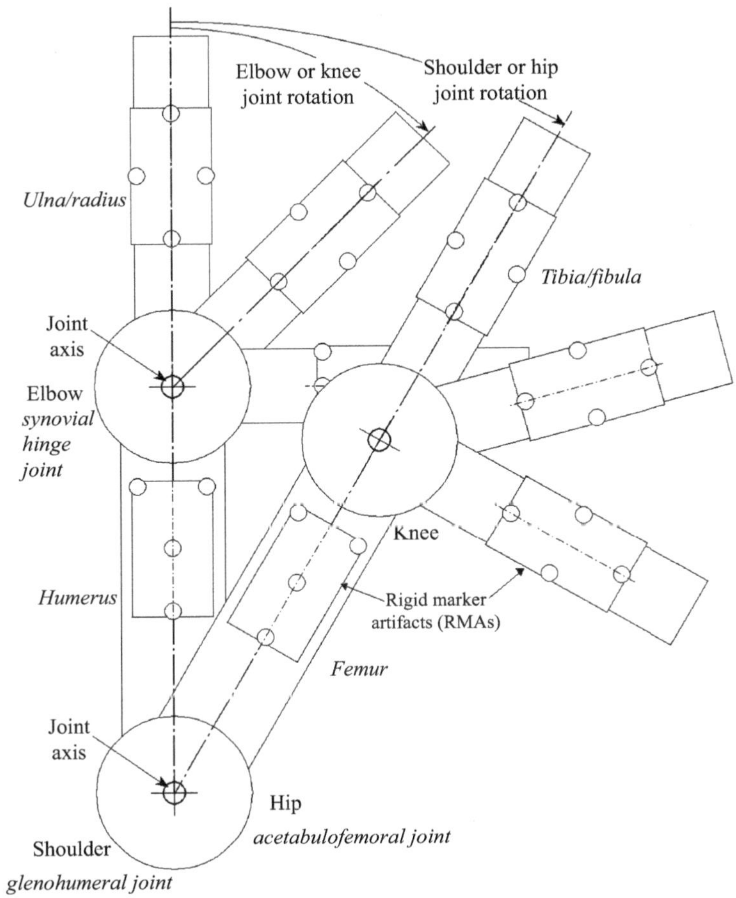

Figure 13.4 Joint axis locations and rotations for a robot with human body labels on links and joints. Left side labels correspond to human arms and right side labels correspond to human legs. The center of the shoulder/hip and elbow/knee joints are marked. Note that only the rigid marker artifacts (RMAs) attached to the upper link were used in experiments described in this chapter

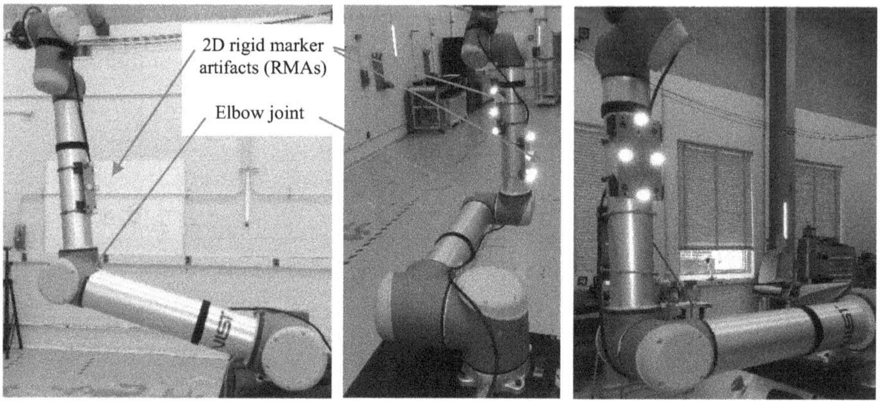

(a) (b) (c)

Figure 13.5 *(a) RMA posed on a robot arm for experiment I circle measurement method for determining elbow joint rotation axis location. (b) Shoulder-to-elbow view and (c) side view (top RMA visible, bottom RMA not visible) of illuminated (from camera flash) photos of two RMAs posed on the robot wrist-to-elbow link for experiment II for measuring uncertainty of joint rotation axis location*

apparatus to the robot, such as hard-mounted devices/tools or sensor sleeves/skins, and/or to accurately know the joint rotation in the case of tight-tolerance robot accessibility to openings. Using new OTS technologies and applying a concept for an alternative test method for measuring joint rotation axis locations of both robots and humans, an experiment was performed at NIST on a 1,300-mm (51.2 in.) long robot arm.

Similar to [23] who used light markers and film to measure joint rotation axis location, NIST researchers used reflective, passive markers and an OTS of 12 cameras [37] to track the joint rotation. And similar to both [28,29], rigid marker artifacts (RMA) were used. Two 89 × 152 mm (3.5 × 6 in.) flat aluminum RMAs, each with four markers in unique patterns, were symmetrically mounted onto two 3D printed arcs that matched the robot links radius-of-curvatures. The RMA was designed to also be considered as a potential fitting on to a human arm or leg.

Experiment I included attaching one RMA to the wrist-to-elbow link, parallel to the joint rotation, and rotating the link, similar to the arm motion shown in Figure 13.2(a). Figure 13.5(a) shows the experiment I setup. The RMA was mounted with its center at approximately 300 mm from the elbow joint center. The wrist-to-elbow link was rotated and measured at 5° and 10° increments from −120° to +90° at 22 different angles.

13.4.1.4 Results

A circle was fitted to the data set and a histogram of the circle's fit error was calculated (see Figure 13.6). Results showed that the data provided a mean fitting error of 0.34 mm with an uncertainty of 0.27 mm, i.e., a relatively high confidence

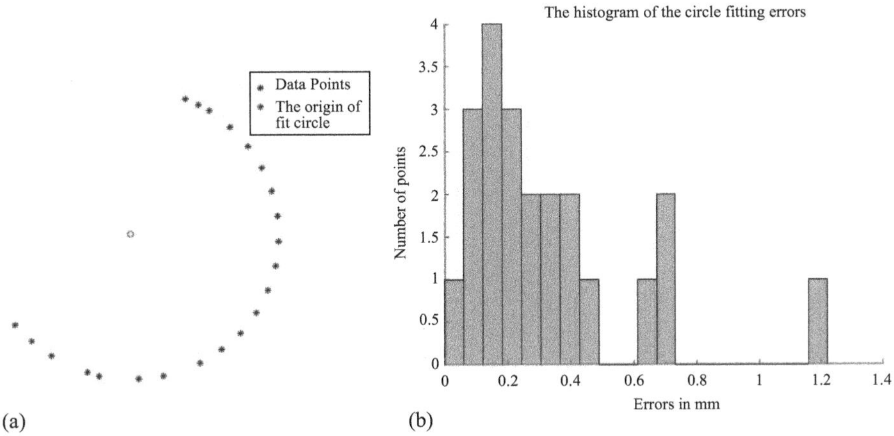

Figure 13.6 *(a) Plot of data points and calculated center joint rotation location after circle fit. (b) Histogram of experiment I data fit to a circle to find the uncertainty for the joint rotation axis location*

in the OTS measurement method. No specification was provided for the backlash in the robot elbow, although the amount was considered negligible as compared to measured error. Hence, based on measurement with the OTS, the actual joint rotation axis location is within this uncertainty in the 2D plane along the X and Y axes.

Assuming that a human arm was measured in the same manner, the shoulder-to-elbow link would require clamping (e.g., [23] clamped a cadaver arm) with no motion while rotating the wrist-to-elbow joint at multiple angles to get results similar to experiment I. This is difficult with a live human with soft tissue in the arm and of course, the use of a clamp to allow no shoulder-to-elbow link motion.

Ideally, a single joint angle is used to measure joint rotation axis location, especially as described in Section 13.4.1.3, since the location changes with flexion or extension of the limbs. Therefore, in experiment II, the elbow was set to 90° using the robot controller, the two RMAs were attached parallel to one another and perpendicular to the joint rotation on the wrist-to-elbow link (see Figure 13.5(b) and (c)), and a random 11 measurements were performed using the OTS. A computer program was designed and used to analyze the data to determine the uncertainty of the joint rotation axis location measurement. Also, electronic calipers were used to measure marker center attached to the RMA and RMA (top) to RMA (bottom) distances. As per [34], the propagation of standard deviation uncertainty is calculated using

$$\sqrt{U(\mathrm{EWtop})^2 + U(\mathrm{EWbottom})^2 + U(\mathrm{caliper})^2} \tag{13.1}$$

where U(EWtop) = the uncertainty of fitting a plane to 11 OTS measurements of the upper RMA relative to the link center, U(EWbottom) = the uncertainty of fitting a plane to 11 OTS measurements of the bottom RMA relative to the link center, and U(caliper) = the uncertainty of five electronic caliper measurements of the {[marker center to RMA] × 2 + [RMA(top) to RMA(bottom)]}/2 = center location of the robot link. The [RMA(EWtop) to RMA(EWbottom)] measurement include two measurements on either side of the link to mathematically cancel out nonparallel RMA pose.

The resulting uncertainty for experiment II was therefore: $\sqrt{1.04^2 + 1.5^2 + 1.9^2} = 1.93$ mm or nearly seven times larger than the results of experiment I. Also, by comparison, several RMA pose concepts were tested to determine the lowest uncertainty for locating the joint rotation axis location, including

- Perpendicular RMAs,
- One RMA on the shoulder-to-elbow link and one RMA on the wrist-to-elbow link parallel to one another, perpendicular to the joint rotation, and offset by the robot links,
- One RMA on the shoulder-to-elbow link and one RMA on the wrist-to-elbow link parallel to one another, perpendicular to the joint rotation, and aligned in the same plane using an offset mount equal to the robot links offset.

Upon analysis, none of these methods provided lower uncertainty than the parallel RMA method.

13.4.2 Industrial mobile manipulator

As discussed in Section 13.4.1, measurement of exoskeleton performance can also utilize measurement concepts developed for industrial robots. Specifically, in this section, the use of artifacts to measure mobile manipulator performance is discussed where ground truth measurement from an OTS was compared to making use of an artifact [37]. Crossindustry measurements used on industrial robots are envisioned to also be useful for exoskeleton performance measurement where exoskeleton wearers would walk up to an artifact, demonstrate pose performance for performing peg-in-hole and other tests.

Measurement science for smart manufacturing robotics is being researched at NIST [4]. As part of this research, simple, cost-effective, repeatable performance measurement methods are being developed and tested and applied toward developing potentially new performance standards for mobile manipulators (i.e., robot arms onboard mobile robot bases) [38–40].

Figure 13.7 shows examples of the mobile manipulator moving toward an artifact, called the reconfigurable mobile manipulator artifact (RMMA-1). In Figure 13.7(a), the RMMA-1 is horizontal while in Figure 13.7(b), it is angled at 45°. The reconfigurability of the artifact allows it to provide the means to measure manipulator alignment of a laser carried by the manipulator with 1 mm diameter or larger reflectors positioned in simple-through-complex geometric patterns on the RMMA-1. Without making contact, the mobile manipulator can be measured to a known artifact when posed at an infinite number of vehicle orientations. Future mobile manipulator measurements will utilize the RMMA-2 shown in

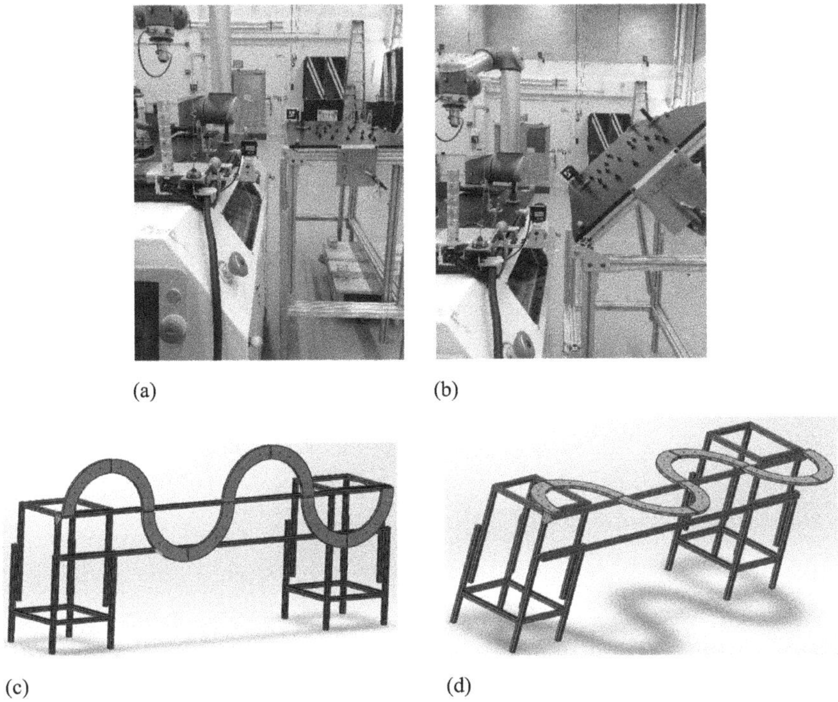

Figure 13.7 (a) Automatic guided vehicle with onboard robot arm moving to dock with the NIST reconfigurable mobile manipulator artifact (RMMA-1) angled (a) horizontal and (b) at 45°. RMMA-2 computer-aided design models angle in the (c) vertical and (d) horizontal configurations

Figure 13.7(c) and (d). This artifact will allow measurement of a dynamic mobile base as well as manipulator performance. Both RMMAs are also potentially useful models of how human–worn exoskeleton performance can be measured.

13.4.3 Response robots

As with industrial robot performance measurement, response robot performance is also being measured and standard test methods are being developed for these robots [41]. These test methods are expected to make it simple to measure, for example, how well a robot navigates around an obstacle on a level floor. Incrementally more challenging conditions can also be tested, for example, to measure how well a robot navigates on inclined planes, steps, undulating floors or more complex or unstructured terrains, and around obstacles as illustrated in Figure 13.8. Additionally, the navigation and obstacle avoidance tests can be combined with vision tests since most response robots are teleoperated. This combination also provides a human-in-the-loop test where a robot's pitch and roll can skew the operator's reference frame for the images provided by the onboard camera(s), thus can hinder

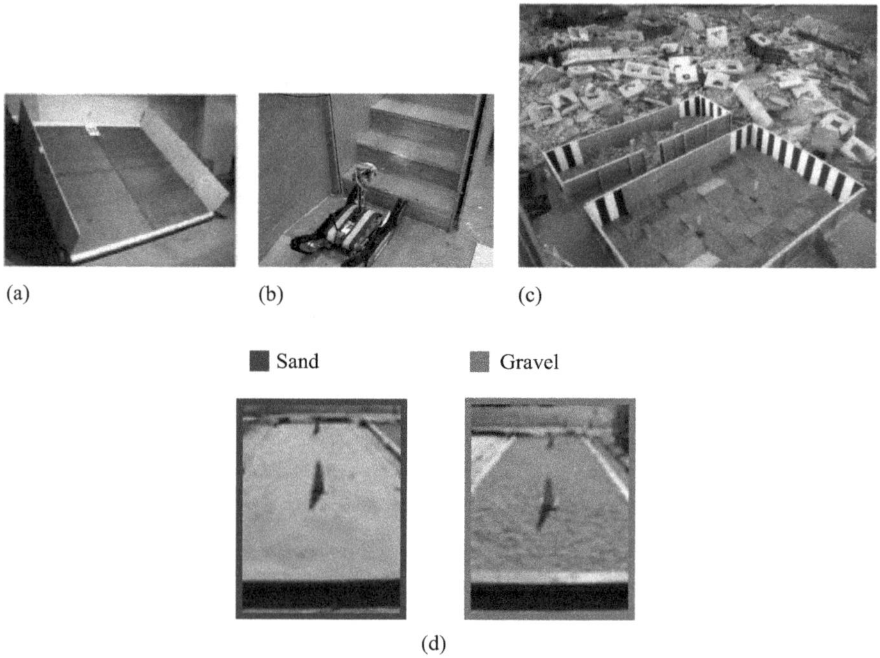

Figure 13.8 Examples of (a) inclined planes, (b) stairs, and (c) varying terrain test apparatus and actual varying terrain (above the apparatus). (d) Example artifacts of sand and gravel terrains

robot control. Each test generically simulates a particular capability which response robots must possess to be useful in critical situations. For example, undulating floors or complex terrains may appear in collapsed buildings where search and rescue robot missions are normally required.

13.5 Recommended test methods for exoskeletons

Safety and performance measurements of exoskeletons can be intertwined; for example, how safe is the system to the wearer after walking, changing direction, stepping on soft versus hard surfaces. As described in Section 13.2, there are many metrics to consider when describing the effects of exoskeletons, both on the human body and on how well the exoskeleton can help humans perform the wide variety of motion tasks. Repeatable test methods can help exoskeleton manufacturers and users highlight capabilities of their systems, compare these exoskeletons to their motion tasks, show design flaws or enhancements, and help with procurement requirements. Ideally, these test methods are not only repeatable, but also standardized, as with the ISO TR 23482-1 being developed, such that both manufacturers and users can simply select a document that describes how to perform the test method no matter which exoskeleton they make or use. Sections 13.3 and 13.4

(a) (b)

Figure 13.9 APA shown with (a) docking pose tray and (b) docking bolt pattern

described test methods for other industries that can cross the industry boundary to also apply to exoskeletons. As a direct application to exoskeletons, we recommend in this section the performance measurement concepts that should be considered for development of performance test methods for exoskeletons. Variability in generic loads, positioning heights, defined motion spaces, etc. can also be detailed in follow-on research and standards from these recommended test methods to help exoskeleton designers fit the tasks to the potentially wide variety of exoskeleton wearers.

13.5.1 Load handling

A docking test apparatus can be used to measure a variety of exoskeleton-worn tasks, such as: load carry, position, and orient; peg-in-hole insertion; and tooling forces. Figures 13.9 and 13.10 show computer-aided design (CAD) models, developed by the authors, of the load-handling device and docking test apparatus concepts pictorially, respectively, and the tasks are explained in the following subsections.

13.5.1.1 Load carry, position, and orient

Exoskeletons are being developed to not only support the human but to also allow load handling equal to and above normal human load carrying capacity. Loads can dramatically vary dependent upon the application. However, load handling includes not only picking up but also placing loads, sometimes with relative precision (e.g., installing a wheel on a vehicle axle bolt circle). Ideally, a single, replicable artifact is used for docking tests. Figure 13.9 shows a concept for a generic artifact with variable width measuring from approximately 500 to 1,200 mm wide called the Adjustable Payload Artifact (APA) that can be used to measure, using pass/fail scoring criteria, how well the exoskeleton wearer can place, in both position and orientation (pose) the APA.

Different types of load grasping can be achieved with the APA, including handles (e.g., equipment), knurled ends (e.g., vehicle tires), and underneath with both arms (e.g., boxes). The APA can be placed in a rectangular tray(s) of variable sizes (see Figure 13.9(a)), with chosen tolerance where simultaneous alignment of APA edges and corners to the tray are required. Similarly, as with the task of installing a wheel on an axle bolt circle, the APA includes a hole pattern (see Figure 13.9(b)) to mate to a bolt pattern mounted, for example, on a wall or frame.

13.5.1.2 Peg-in-hole

The docking test apparatus in Figure 13.10 measures ≈3,600 mm wide × 3,000 mm high × 1,200 mm deep and shows tubes mounted to the back wall for peg-in-hole testing. Similar to the industrial mobile manipulator testing with the RMMA (see Section 13.4.3), the subject wearing an exoskeleton could maneuver simple pegs or a tool, such as a drill, with drill bit to insert into the tubes as an additional position and orientation test. Ideally, the load handling and peg-in-hole tests require that the exoskeleton wearer stands in one location or must move to other locations to position and orient the APA, pegs, or tools to match tray, hook, or tube locations as part of the test method.

13.5.1.3 Tool force

The docking test apparatus shows force plates mounted on the floor, wall, and an adjustable height (and perhaps angle) ceiling mount (see Figure 13.10). As tools, such as grinders, saws, drills, are potentially used by exoskeleton users, test methods are needed that measure the user without wearing the exoskeleton as baseline and wearing the exoskeleton to compare performance. The user can walk up to the force plates with tools, apply the tool to the force plates for a test period of

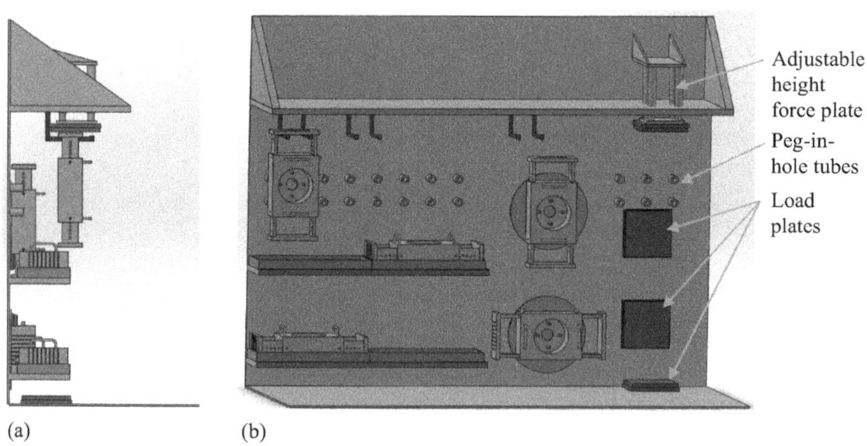

Figure 13.10 The figure shows (a) the side view and (b) the front view of a test apparatus with the APA trays and bolt patterns, as well as holes for peg-in-hole insertion, hooks for hanging the load, and load plates for applying tool control forces. Pass/fail position and orient and elapsed time could be measured for multiple repetitions

time, and measure force applied to user-fatigue time periods. Applied force versus time at a number of heights can be measured with this concept.

13.5.1.4 Navigation

The exoskeleton user can navigate along many different paths and subject orientations to test the safety and performance of the exoskeleton. Again, ideally, a small set of artifacts or test apparatus are used in test methods to measure safety and performance for relatively simple reproducibility. ASTM F45 is developing a navigation test method and an environmental effects practice and ASTM E54.09 combines navigation and environmental effects into methods that can also apply to exoskeletons. Figure 13.11 shows an example of several F45.02 navigation-straight aisle sections that also include various (undulating, sand, and stone) terrains within the sections. An exoskeleton user can test balance and timing to traverse this type of course. Touching the walls could be considered a failed navigation test and passing considered when no contact is made throughout the entire test course. Tape switches mounted to the walls or cameras can monitor wall contact. Alternatively, virtual barriers such as laser lines as described in ASTM F45.02 WK48955 could be used instead of physical barriers to detect when the exoskeleton user crosses the barrier.

Similarly, F45.02 WK48955 navigation-perpendicular aisle and constant radius curve sections, defined space, navigation test methods can be assembled to produce a more complex navigation performance test apparatus. The apparatus may or may not include various terrains as shown in Figure 13.11. The same wall contact or virtual barrier measurement and test pass/fail due to no contact/contact or barrier-cross as described above can be used. Additionally, in Figure 13.12, the docking test apparatus is shown at each end of the constant radius curve sections as a docking area for the subject wearing an exoskeleton. Combined navigation and docking test methods can be useful for testing load carrying through a complex maze and terrain, followed by accurate positioning of the load. Further, defined-space sections can be thought of as building blocks to assemble even longer or more complex navigation defined-space apparatus.

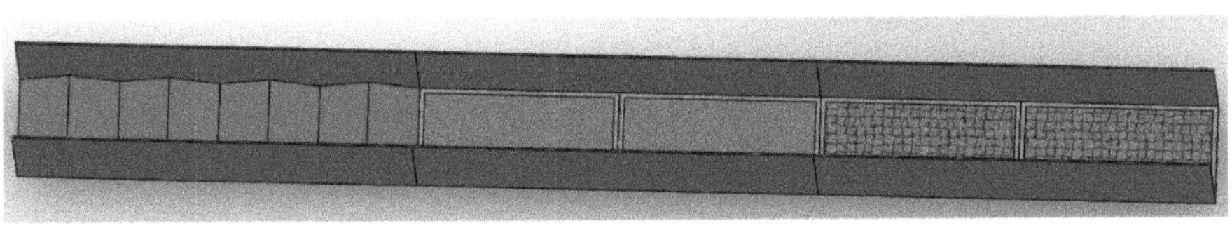

(a) (b) (c)

Figure 13.11 Example exoskeleton test method using F45.02 WK48955 navigation-straight aisle sections defined space navigation test methods with example response robot terrain types, i.e., (a) undulating, (b) sand, and (c) stone, in pans within sections

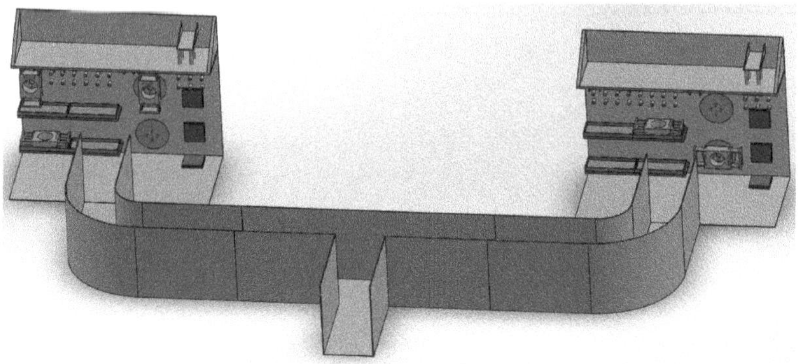

Figure 13.12 Example exoskeleton test method using F45.02 WK48955 navigation-perpendicular aisle section (center) and constant radius curve sections (left and right) defined space navigation test methods leading to docking tests

13.5.1.5 Test dummy

Test dummies can be useful for testing mean time-between-failure of exoskeletons, skin effects, effects of off-axis joint rotation, and other areas where fatigue or unsafe conditions may be evident for human-in-the-loop tests. Hip with torso and/ or single or dual: ankle, knee, shoulder, or elbow joint test dummies, along with their links, can be useful for these measurements. As described in Section 13.4.1, Robot Joint Axis Location Measurement, the concept for measuring joint axis rotation location combined with a test dummy could provide useful information about exoskeleton fit with off rotation-axis and on rotation-axis force effects on the user prior to exoskeletons worn by humans. The development of robots with inline joints, such as humanoid robots, could be useful, so long as they directly mimic sizes and motions of the variety of humans who intend to use exoskeletons.

Figure 13.13 shows a CAD model, developed by the authors, of a modular upper body test dummy for testing exoskeletons. The arms and shoulders are actuated similar to human biomechanical motions, as shown by the opening in the arm at the elbow joints. A similar modular lower body can also be added or independently used. The joints are in-line with the actuators inside the hollow shell. The dummy is modular, potentially even 3D printed components, so that components with smaller or larger shoulder span, arm length and diameter, torso diameter and more human-like shapes, etc. can be modified and refastened to the overall frame to test a certain size/shaped person. The torso can be divided into further segments to allow for torso twist between the shoulders and hips. Padding, that can mimic human flesh and able to change shape, can be added to the modular components as desired. Similarly, sensors to measure exoskeleton effects and electrical and magnetic activity to output to exoskeleton components that use electrocardiography or electromyography [42] for exoskeleton control can also be added as needed. Since the test dummy control can be simple or complex, a system's engineering approach can also be applied through concepts such as systems modeling language (SysML) [43].

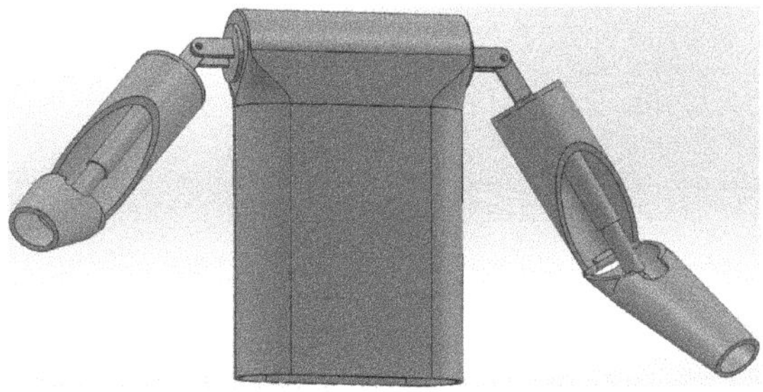

Figure 13.13 Modular human upper body test dummy for testing exoskeletons

Joint(s) actuation, sensing, and combined system motions, including feedback from the exoskeleton can all be integrated into a model and potentially controlled using SysML.

Ideally, a passive measurement-exoskeleton strapped to a range of humans representing various sizes, shapes, fitness-levels, and both genders can measure actual human motion with data for replication in test dummies to mimic human biomechanical motions when testing exoskeletons. Sensors integrated within the test dummy, like those used on the passive measurement-exoskeletons, can also measure chafing, device movement, torques and other forces, etc. to provide additional information. Eventually, two or more test dummy joints, such as ankle and knee, ankle and knee and hip, joints with adjustable links can be combined into more complex test dummies for repeatable and improved measurements of exoskeleton effects on humans. These concepts may also provide useful short- and long-term effects measurements on humans and on exoskeletons to improve wearable exoskeleton designs to be safe and increase performance.

13.6 Summary and conclusions

Much can be learned from the industrial robot and response robot sectors that can be used to support safety and performance measurements of exoskeletons prior to or when they are used by humans. This chapter builds on referenced human–robot interaction metrics that may apply to exoskeletons to include typical industrial robot metrics of speed, pose uncertainty, control force, etc. Additionally, more human-related metrics of ergonomics, ease-of-use, ingress/egress complexity, etc. are added and defined by the authors that more relevantly apply to exoskeletons. The ISO 13482 safety standard for physical assistant robots directly addresses exoskeletons. However, only recently the ISO TR 23482-1 TR is under development to support the safety requirements standard with how appropriate test methods can be developed. There are currently no test methods for safety and performance measurements of exoskeletons and lessons can be learned from recent industrial performance standards developments under ASTM F45 and response robot performance standards developments under E54.09. Many of these standards and working documents were listed in this chapter.

To aid in exoskeleton design, development, and use was a discussion on cross-sector measurements that also include experiences from industrial robot and response robot domains. Joint rotation axis location can dramatically vary from joint-to-joint and from person-to-person. Additionally, limitations, joint motion, and over-extension caused by exoskeletons must be addressed before humans can safely use them. For example, an exoskeleton moving an upper and lower leg about the knee must move as the human knee moves. Additionally, the location of the limb-mover must be aligned properly with the human joint and known angles of motion must not be exceeded. Relatively old goniometer technology and referenced measurement concepts may suffice for exoskeleton joint positioning should the uncertainty be known, although minimal uncertainty information for joint location has been published. NIST performed experiments with rigid marker apparatus using an industrial robot arm to determine uncertainty when using a state-of-the-art OTS. The results for two concepts are 0.27 and 1.93 mm. Hence, potentially minimal unsafe torque would be applied to the human joint from an exoskeleton that was positioned with these offsets from the actual joint axis should today's OTSs be used in conjunction with the exoskeleton being fit to the human.

Further, tests using mobile manipulator and response robot artifacts can provide relatively low-cost environments and methods for measuring performances of these systems. A direct crossover can be applied to exoskeletons as provided in the recommended concepts in this chapter. Load handling, navigation, and perhaps other test concepts can be made into generic methods for testing system performance. Test dummies are another pre- or parallel test method for performance measurement of exoskeletons to test mean-time-between-failure, how failures can occur, etc. without having people in the wearable robot in potentially unsafe conditions. Test dummies and computer models can be designed to match human physique, speed, joint motion characteristics, sensing, control, and many other biomechanical parameters to fully understand the exoskeleton effects on humans. Future efforts should include developing some or all of the recommended concepts suggested to develop the safest, highest performance, and most cost-effective exoskeletons to support or enhance human activities.

Acknowledgment

The authors thank Sebti Foufou, Qatar University, Doha, Qatar for his guidance on industrial robot arm and mobile manipulator measurement research discussed in this paper.

References

[1] Berkeley Robotics & Human Engineering Laboratory, http://bleex.me.berkeley.edu/research/exoskeleton/, accessed August 21, 2017.

[2] Research Exoskeleton Report, http://exoskeletonreport.com/category/research-and-academia/, accessed October 28, 2016.

[3] ISO 13482:2014 Robots and robotic devices—Safety requirements for personal care robots, http://www.iso.org, 2014.
[4] NIST Robotic Systems for Smart Manufacturing Program, https://www.nist.gov/programs-projects/robotic-systems-smart-manufacturing-program, accessed August 21, 2017.
[5] NIST Emergency Response Robots Project, https://www.nist.gov/programs-projects/emergency-response-robots, accessed August 21, 2017.
[6] Steinfeld A, Fong T, Kaber D, Lewis M, Scholtz J, Schultz A, Goodrich M, "Common metrics for human–robot interaction", Proceedings of the 1st ACM SIGCHI/SIGART conference on human–robot interaction, pp. 33–40, ACM, March 2, 2006.
[7] Butler T, "Exoskeleton Technology – Making Workers Safer and More Productive", Professional Safety, Journal of the American Society of Safety Engineers, www.asse.org, pp. 32–36, September 2016.
[8] Lonini L, Shawen N, Scanlan K, Rymer WZ, Kording KP, Jayaraman A, Accelerometry-enabled measurement of walking performance with a robotic exoskeleton: a pilot study. Journal of NeuroEngineering and Rehabilitation. 2016;13:35. doi:10.1186/s12984-016-0142-9.
[9] Asselin P, Knezevic S, Kornfeld S, Heart rate and oxygen demand of powered exoskeleton-assisted walking in persons with paraplegia. Journal of Rehabilitation Research and Development. 2015;52.2:147.
[10] ISO/TC 299 Robotics, https://www.iso.org/committee/5915511.html, accessed October 26, 2017.
[11] ISO/CD TR 23482-1 Robotics—Application of ISO 13482—Part 1: Safety-related test methods, https://www.iso.org/standard/71564.html, accessed October 26, 2017.
[12] ASTM International Committee F45 Driverless Automatic Guided Industrial Vehicles, http://www.astm.org/COMMITTEE/F45, 2016, accessed August 21, 2017.
[13] ISO 9283:1998 Manipulating industrial robots—Performance criteria and related test methods, https://www.iso.org/standard/22244.html, accessed October 26, 2017.
[14] ASTM International Committee E54, http://www.astm.org/COMMITTEE/E54.htm, 2015, accessed August 21, 2017.
[15] Bostelman R, Messina E, Foufou S, "Cross-Industry Standard Test Method Developments – from Manufacturing to Wearable Robots", Journal of Zhejiang University—Science C (Computers&Electronics), February 21, 2016.
[16] Lockheed Martin, Fortis Exoskeleton User's Manual, Version 1.0, 2016.
[17] Bennett D, Hanratty B, Thompson N, Beverland D, Measurement of knee joint motion using digital imaging. International Orthopaedics (SICOT). 2009;33:1627. doi:10.1007/s00264-008-0694-9.
[18] Krishnan RH, Devanandh V, Leyland A, Brahma AK, Pugazhenthi S, Estimation of mass moment of inertia of human body, when bending forward, for the design of a self-transfer robotic facility. Journal of Engineering Science and Technology. February 2016;11(2):166–176.

[19] Malagelada F, Dalmau-Pastor M, Vega J, Golano P, "Elbow Anatomy", In Sports Injuries, doi:10.1007/978-3-642-36801-1_38-1, Springer-Verlag, Berlin Heidelberg, 2014.
[20] Massachusetts Institute of Technology, http://web.mit.edu/tkd/stretch/stretching_8.html#SEC84, accessed October 28, 2016.
[21] ClinicalGate, http://clinicalgate.com/elbow-3/, Chapter 6, March 16, 2015, accessed August 21, 2017.
[22] Pellis G., Di Cosmo F., "The ROTOTRANSLATORY motion: experimental studies, mathematical analysis, orthopedic device", Calzetti Editore, Perugia, atti del XVIII edizione del Convegno di Traumatologia e Riabilitazione Sportiva, Bologna, 2009.
[23] Deland J, Walker P, Sledge C, Farberov A, Treatment of posttraumatic elbows with a new hinge-distractor, orthopedics. Nursing & Allied Health Database. June 1983;6(6):732.
[24] Bottlang M, Madey SM, Steyers CM, Marsh JL, Brown TD, Assessment of elbow joint kinematics in passive motion by electromagnetic motion tracking. Journal of Orthopaedic Research. 2000;18:195–2002.
[25] LifeModeler, "Marker Placement Protocols", http://www.lifemodeler.com, 2010, accessed August 21, 2017.
[26] SedkyAdly A, Abdelhalim MB, AmrBadr, Analyzing and measuring human joints movements using a computer vision system. International Journal of Computer Applications (0975–8887). May 2012;45(20):21–29.
[27] Nesbitt RJ, Bates NA, Karkhanis TD, Schaffner G, Shearn JT, Impacts of robotic compliance and bone bending on simulated in vivo knee kinematics. American Journal of Biomedical Engineering. 2016;6(1):12–18. doi:10.5923/j.ajbe.20160601.02.
[28] Fukaya T, Mutsuzaki H, Ida H, Wadano Y, Two different protocols for knee joint motion analyses in the stance phase of gait: correlation of the rigid marker set and the point cluster technique. Rehabilitation Research and Practice. 2012;2012. Article ID 586348, doi:10.1155/2012/586348.
[29] Robertson GE, "Motion Capture Essentials", University of Ottawa, Ottawa, Ontario, Canada, 2016.
[30] Deschamps JE, Dudum KC, Dandekar EM, Hazelwood SJ, Klisch SM, "Pseudo-rigid body method for reducing soft tissue artifact: validation and application to gait", Summer biomechanics, bioengineering and biotransport, conference, Snowbird Resort, Utah, June 17–20, 2015.
[31] Rosenhahn B, Kersting UG, Smith AW, Gurney JK, Brox T, Klette R, "A system for marker-less human motion estimation", In Pattern Recognition, Springer LNCS 3663, W. Kropatsch, R. Sablatnig, and A. Hanbury (Eds.), pp. 230–237, Vienna, Austria, August 2005, Springer-Verlag Berlin Heidelberg 2005.
[32] Zhang L, Brunnett G, Rusdorf S, Real-time human motion capture with simple marker sets and monocular video. Journal of Virtual Reality and Broadcasting. 2011;8(1).

[33] Bakhshi S, Mahoor MH, Davidson BS, "Development of a body joint angle measurement system using IMU sensors", 33rd Annual international conference of the IEEE EMBS, Boston, Massachusetts, USA, August 30–September 3, 2011.

[34] Gibbs PT, Asada HH, Wearable conductive fiber sensors for multi-axis human joint angle measurements. Journal of NeuroEngineering and Rehabilitation 2005;2:7. doi: 10.1186/1743-0003-2-7, 02 March 2005.

[35] NIST/SEMATECH e-Handbook of Statistical Methods, http://www.itl.nist.gov/div898/handbook/, October 30, 2013.

[36] Mori G, Malik J, Recovering 3D human body configurations using shape contexts. IEEE Transactions on Pattern Analysis and Machine Intelligence. July 2006;28(7):1052–1062.

[37] Bostelman R, Falco J, Shah M, Hong Hong T, "Dynamic metrology performance measurement of a six degree-of-freedom tracking system used in smart manufacturing", In Autonomous Industrial Vehicles: From the Laboratory to the Factory Floor, ASTM Book, Chapter 7, 2016.

[38] Bostelman R, Hong T, Marvel J, "Performance measurement of mobile manipulators", SPIE 2015, Baltimore, MD, April 2015.

[39] Bostelman R, Foufou S, Legowik S, Hong Hong T, "Mobile manipulator performance measurement towards manufacturing assembly tasks", 13th IFIP international conference on product lifecycle management (PLM16), Columbia, SC, July 11–13, 2016.

[40] Bostelman R, Hong T, Legowik S, "Mobile robot and mobile manipulator research towards ASTM standards development", SPIE 2016, Baltimore, MD, USA, April 2016.

[41] Robotics Test Facility, http://www.nist.gov/el/isd/ms/ roboticsbldg.cfm, 2014.

[42] Wehner M, "Man to machine, applications in electromyography", In EMG Methods for Evaluating Muscle and Nerve Function, Schwartz M (Ed.), InTech, DOI: 10.5772/26495, 2012.

[43] Friedenthal S, Moore A, Steiner R, "A Practical Guide to SysML, Third Edition: The Systems Modeling Language", Elsevier, ISBN: 9780128008003, October 22, 2014.

Chapter 14
Ekso Bionics

Russ Angold[1]

14.1 Business overview

Ekso Bionics designs and develops commercial exoskeletons for the healthcare, industrial, military, and consumer markets. Our exoskeleton systems work in conjunction with the operator to enhance strength, endurance, and mobility. These systems serve multiple markets and can be used both by persons with physical disabilities and by able-bodied users as well. We have sold, rented, or leased devices that (a) enable individuals with neurological conditions affecting gait [e.g., spinal cord injury (SCI) or stroke] to rehabilitate and to walk again; (b) allow industrial and construction workers to perform heavy duty work with increased efficiency and reduced strain; and (c) permit soldiers to carry heavy loads for long distances while mitigating lower back, knee, and ankle injuries.

Thanks to recent advancements in material technologies, electronic and electrical engineering, control technologies, and sensor and software development, the commercial opportunity for exoskeleton systems is accelerating. Taken individually, many of these advancements have become ubiquitous in peoples' everyday lives. At Ekso Bionics, we have learned how to integrate these existing technologies and efficiently, elegantly and safely, wrap the result around a human being. Supported by an industry leading intellectual property portfolio, we believe we can continue to develop solutions across a broad spectrum of applications, from able-bodied users to persons with paralysis.

While advancements in technology will continue to drive commercial interest in and further development of exoskeleton systems, we also recognize that we are in the early stages of exoskeleton commercialization. To increase adoption of our exoskeleton technology, we intend to focus our efforts on the following key initiatives:

- Drive robotic exoskeleton rehabilitation to become the standard of care for both in-patient and out-patient rehabilitation for patients with some form of weakness or lower limb paralysis, with an enhanced focus on sales and marketing commercial execution in North America and Europe through the implementation of new processes, strategies, and results orientation.

[1]Ekso Bionics, USA

- Introduce new features in rehabilitation for our Ekso GT, such as SmartAssist Software, which could expand access to care to more patients, and EksoPulse Analytics, which aids in providing more personalized care in rehabilitation sessions.
- Continue patient enrollment in our company-sponsored clinical trial Walking Improvement for SCI with Exoskeletons Study.
- Continue leveraging our exoskeleton research and development work and commercial experience with the Ekso GT to develop a next generation medical device for use outside of a rehabilitation setting. We are striving to produce a product that will have greater levels of independence and functionality than any exoskeleton currently on the market.
- Build upon our momentum in industrial markets with an enhanced focus on commercial rollout of EksoZeroG for aerial work platforms (AWP) and scaffolding to reduce work-related injuries, as well as introduce our next generation innovation for improving overhead work.

14.2 Rehabilitation robotics

Our current focus at Ekso Bionics is on rehabilitation robotics. We are leveraging our patented exoskeleton technology to develop and market products intended to enable patients with some form of lower limb paralysis to rehabilitate earlier and with better outcomes than the current standard of care.

14.2.1 Ekso GT

Our current product, the Ekso GT, is a wearable bionic suit that allows our hospital and rehabilitation customers to provide in-patients and out-patients with "SCI" and hemiplegia due to stroke the ability to stand and walk over ground with a full weight-bearing, reciprocal gait using a cane, crutches or a walker under the supervision of a physical therapist. Walking is achieved by a user shifting their body to activate sensors in the device which in turn initiate steps. Battery-powered motors drive the legs, detecting the deficient neuromuscular function and providing that level of assist for a user to complete their step. Users can expect to walk with aid from the device the first time they put on the Ekso exoskeleton (after passing an assessment). Physical therapists can transfer patients to or from their wheelchair and don or remove the Ekso in less than 5 min.

The Ekso GT incorporates Variable Assist™, our proprietary adaptive software that detects a user's level of motor loss and dynamically adjusts and provides 0%–100% power to either side of the body, depending on the users' unique needs. Variable Assist can promote a larger number of high-quality steps in a short time period and support the early relearning of correct step patterns and weight shifts, potentially mitigating compensatory behaviors. Variable Assist has significantly expanded the spectrum of our customer's patients that can potentially benefit from robotic rehabilitation.

Another important feature of our Ekso GT is a real-time data capture program called Ekso Pulse. Ekso Pulse gathers and transmits device information and statistics during Ekso rehabilitation sessions. This information can be used to monitor patient progression and device utilization. Data such as steps, step size, speed, and other settings along with all error logs and operating parameters are recorded on the Ekso. Data is then wirelessly sent securely to our servers where it is available for customers to view, filter, and export through a secure web portal. This feature allows visibility into each rehabilitation session enabling more thorough patient care while reducing manual data entry. It also enables a higher level of customer service through early identification and thorough reporting of device errors, saving the time and expense of unnecessary on-site visits.

The Ekso GT is used by customers in both in-patient and out-patient settings. Our customers believe that for patients with some motor ability intact (e.g., an incomplete SCI, or after a stroke), the Ekso GT offers unique benefits to help therapists teach proper weight shifts and step patterns, allowing patients potentially to mobilize earlier and ultimately to walk again. By allowing individuals to stand and walk in a full weight-bearing setting, early clinical evidence is also beginning to show that the Ekso exoskeleton may offer potential healthcare benefits (including for patients with complete SCI) such as reducing postinjury medical costs through reduction in secondary complications such as urinary tract infections, pressure sores, pneumonia, bowel problems, and other respiratory issues, bone loss/osteoporosis, psychological disorders, and cardiovascular disease.

As of March 1, 2017, over 71 million steps are taken in Ekso Bionics exoskeletons. The company has now shipped to over 160 rehabilitation facilities or customers worldwide. The number of units utilized at a center varies from one to six and is driven by the number of beds and rehabilitation sessions a hospital can offer and that hospital's adoption of robotics within its rehabilitation protocols.

14.2.2 Market overview

The primary market for the Ekso GT is SCI rehabilitation and stroke clinics with significant patient populations. In the United States there are about 5.9 million SCI and stroke rehabilitation sessions conducted annually on about 680,000 SCI and stroke patients at approximately 16,900 facilities. Global estimates for SCI and stroke populations are more than double those in the United States.

Due to the chronic nature of the conditions resulting in lower limb impairment, we believe these diagnoses have an enormous clinical and economic impact on both people with the conditions and the healthcare system. The American Heart Association states that there are approximately 795,000 strokes per year with approximately 7 million people living in the United States who have suffered from a stroke. Direct and indirect costs associated with those who have suffered a stroke total approximately $60 billion annually. Similarly, according to the National Spinal Cord Injury Statistical Center, there are approximately 12,500 incidences of SCI per year in the United States with approximately 275,000 people living with SCI. Direct and indirect costs associated with those who have suffered SCI total approximately $18.5 billion annually.

While the market opportunity for robotic exoskeleton rehabilitation is large, we also recognize that the process for medical devices to become standard of care is long and challenging. We believe our ability to accelerate adoption will also be based, in part, on our ability to build on our and our partners' early efforts to expand clinical evidence and to drive toward standard of care. We are already seeing customers appreciate that one way for stroke patients at in-patient facilities to receive the recommended amount of rehabilitation per guidelines is by using an Ekso GT, the only device currently in the market that has the versatility to provide an over-ground gait training intervention that is task-specific, high intensity and allows for a margin of error, across the continuum of care.

14.2.3 Clinical evidence and reimbursement

Many of our early clinical customers have undertaken research to evaluate the use in rehabilitation of exoskeletons in general and specifically our Ekso robotic exoskeleton. Although these studies have primarily focused on feasibility and safety and have relied on small sample sizes, initial study findings have been favorable. Ekso Bionics is aware of eight completed case studies for SCI, four for stroke and one for multiple injury states, with a total patient count for all such studies of approximately 110 patients. Additionally, we are aware of 14 investigator-initiated studies currently underway, covering stroke, SCI (complete and incomplete), acquired brain injury, Multiple Sclerosis and Cerebral Palsy with a total patient enrollment goal of over 500. Two studies recently announced in 2016 include:

- The MOST (mobility improved after stroke when a robotic device was used in comparison to physical therapy) study—Moritz Klinik, Germany; Professor Dr. med F. Hamzei. This study follows early observations from clinical use of the Ekso GT and is investigating the impact of gait training with the Ekso GT on functional independence of 80 patients with stroke related impaired gait.
- Robotic Exoskeleton Gait Training during Acute Stroke Rehabilitation—Kessler Institute of Rehabilitation; Karen J. Nolan, PhD. This study will seek to enroll 96 inpatients that are within 2 weeks of stroke onset to investigate the potential value of the Ekso GT in post stroke rehabilitation.

We intend to continue our work with rehabilitation centers and clinicians studying the benefits of robotic exoskeleton rehabilitation using the Ekso. We believe that additional clinical evidence will help treating physicians to better understand the benefits of rehabilitation with the Ekso GT and will support our efforts to achieve reimbursement for the Ekso GT. To this end, we intend to make additional investments in clinical data generation in 2016. Specifically, we plan to initiate a registry study and one or more company-sponsored clinical trials. We expect to begin enrollment in a company-led, prospective, multi-center trial with chronic, incomplete SCI patients in the third quarter of 2016.

We believe that reimbursement by the Centers for Medicare/Medicaid Services and third party insurers will play an important role in the long-term success of the commercial adoption of our Ekso GT and to make the Ekso GT a standard of care for rehabilitation for patients with some form of lower limb weakness or paralysis.

To gain coverage and payment by payers, Ekso Bionics and its competitors must generate both clinical and economic evidence demonstrating the benefits of robotic exoskeletons. We believe that the investments we are making in clinical trials will assist in generating this evidence. Generally, reimbursement for professional services performed at the hospital by physicians is reported under separate billing codes issued by the American Medical Association known as Current Procedural Terminology ("CPT") codes. Currently, generic codes exist that provide some modest reimbursement for the use of our technology in the rehabilitation setting; however, we are not aware of a CPT code that is specifically applicable to the use of the Ekso GT. We may determine to pursue an application for a new CPT code. We have engaged the services of expert consultants with extensive experience in the CPT, coverage and payment decision processes to assist us in our reimbursement strategy.

The European Union also requires a two-track approach to market penetration and subsequent coverage, requiring separate claims for purchasing the device and for requests for training. Our competition has had initial success in Germany with four of the top private payer insurance companies purchasing a personal device.

14.2.4 Current sales and marketing efforts

Our key marketing goal today is to achieve broad-based commercial adoption of our Ekso GT in rehabilitation settings. We are focusing our go-to-market protocols and collateral on our three target audiences: medical directors/therapists, medical administrators, and patients. Working closely with thought leaders, we will continue to build upon our early user-group exchanges, develop clinical education programs, and grow our medical advisory council. We plan to create centers of excellence in the United States and Europe, the Middle East and Africa ("EMEA") that are committed to exoskeleton education and to developing the quantifiable metrics and results by which the effectiveness of exoskeletons may be measured. We are also implementing a customer experience program to increase adoption and utilization in new and existing accounts and to generate more multiple device customers.

Our sales efforts continue to focus on in-patient and out-patient centers that provide stroke and SCI rehabilitation. Geographically, the priorities remain North America (Canada, the United States and Mexico) and EMEA. Currently, we utilize a direct sales force for the United States, Canada, the United Kingdom, Spain and the German-speaking countries of Europe. We also have a distributor network that currently covers 19 countries (an increase from seven countries at year end 2014). Our three largest distributors based on Ekso sales are based in Italy, Poland, and Mexico.

The sales and marketing team is principally based in the United States and Germany and is structured as follows:

- One national account manager for the United States and one EMEA-based manager for our distributors;
- US and EMEA sales professionals that pursue new prospects and organizes demonstrations;

- Clinical professionals/physical therapists that provide peer-to-peer demonstrations and trainings;
- Marketing professionals, graphic designers, and consultants to build awareness and generate demand;
- Ambassadors with SCI that provide demonstrations and personal experiences.

The sales cycle for the Ekso GT averages approximately 8 to 12 months for a first device and 2 to 4 months for subsequent devices. Our typical sale is the Ekso GT complete package, which includes the device and all relevant components, two sets of batteries for continuous run-time, training through two levels of certification, and our Variable Assist software. Customers also typically purchase Ekso Care, which is our 1- to 4-year after-sales service package.

14.2.5 After sales service

We provide service for the Ekso GT at our facility in Richmond, California or by having one of our Ekso field technicians visit customers at their places of business. When maintenance or service is required, a customer schedules service by contacting us and we then arrange for the appropriate service, depending on the level of Ekso Care for which a customer has contracted. The Ekso GT is designed with Ekso Pulse, which allows us to diagnose a variety of customer service issues remotely.

14.2.6 Manufacturing and supply chain

We assemble the Ekso GT and manufacture certain components that are critical to our know-how at our facilities in Richmond, California. We currently run one line for one shift per day and believe we have the capacity to eventually run up to four lines for two shifts per day should we deem it appropriate.

The Ekso GT uses over 700 purchased parts, which we source globally from over 70 suppliers. Whenever possible, we seek to secure dual source suppliers for our components.

14.3 Home mobility

The dynamics and product requirements of the home mobility market are different from those of the rehabilitation clinic. While we believe the home mobility market opportunity is sizable, it will only be served once new technology is brought to market that is cost effective for individuals because reimbursement is available and has a level of functionality that enables independent mobilization. Home mobility exoskeletons should fit a specific patient and be designed for all-day use. In addition, we believe they must be easily transportable, have improved dynamic stability, user interfaces, and terrain navigation to allow the home mobility users to confidently walk through their daily life with little or no assistance. Given our commercial experience with a medical exoskeleton that has recorded over 75 million steps, coupled with recent research and development advancements in exoskeleton and related technologies, we are now proactively investing resources to design such a commercial product and developing our go-to-market approach for mass adoption of home mobility devices. We are also collaborating with world-class commercial and academic institutions to refine our technology and to apply the latest technological breakthroughs to the advancement of human ambulation.

In addition to implementing the technological changes necessary in an exoskeleton designed for the home mobility market, we are working with payers and ensuring our (and where possible, our partners') trials are and will be generating clinical and economic evidence on the benefits of exoskeletons for home mobility use. Lastly, our go-to-market strategy will likely be very different than our current rehabilitation markets sales and marketing approach. Critical to our success will be implementing such a strategy, possibly with partner(s), which is sustainable to address the potential size of the market.

14.4 Able-bodied industrial applications

In December 2014, we introduced our first prototype of a passive exoskeleton intended for industrial applications. During 2015, we began investing resources to support requests for demonstrations and in-depth field-testing in real world conditions with advanced prototypes.

Our feedback indicates a growing imperative among manufacturing and construction companies to drive adoption of improved safety and health practices. Furthermore, based on initial market research and customer field-testing, we believe industrial exoskeletons have the potential to help prevent workforce injuries, improve productivity and over time reduce workmen's compensation and related costs. According to a Bureau of Labor Statistics Report (2012), direct costs related to injuries associated with overexertion in the workplace total over $21.1 billion per year.

In addition, human augmentation technology is being viewed by senior managers of companies that have participated in field-testing as an opportunity to extend the careers of experienced and skilled workers, increase the demographic of workers that can perform the tasks, and attract new workers to the construction and manufacturing sector.

Last year, we introduced the EksoZeroG, a new product innovation for AWP and scaffolding, which is intended to significantly improve workforce productivity while dramatically reducing workplace-related injuries to keep workers healthy, strong, and safe. EksoZeroG is a mobile arm mount that makes heavy tools feel weightless and enables workers to be safer and more productive. In 2017, we are focusing on increasing sales of the EksoZeroG by pursuing alternative channels such as rental and construction equipment providers as well as focusing more effort on commercial execution, awareness generation, and tradeshow attendance.

We believe there is additional mid-to-long-term potential in the industrial markets, and accordingly, we will continue our development efforts to expand our EksoWorks product offerings.

14.5 Ekso Labs

Ekso Labs, our engineering services division, is focused on new technology development and future applications. It is an exoskeleton laboratory that integrates emerging technologies into new product applications and expands on it for our partners. To date, the majority of our Ekso Labs revenue has been in the form of grants from government organizations including United States Special Operations Command, the Defense Advanced Research Projects Agency, the National Science Foundation, and the National Institute of Health. These projects fund research and development on new exoskeleton systems, providing us with novel intellectual property and exoskeleton designs that have the potential for commercialization.

In addition to furthering exoskeleton technology for our current medical applications, Ekso Labs' research and development work may have potential use in future, able-bodied models of the Ekso human exoskeleton. Many of the research projects funded by grants are focused on researching future medical applications and capabilities not yet ready for commercial development. Other projects, often funded by commercial partners or the US military, focus on able-bodied human exoskeleton applications.

In early 2016, we made the strategic decision to shift our engineering resources away from the billable engineering services of Ekso Labs and to our internal development efforts both for our next generation home/wellness device and for able-bodied industrial offerings. As a consequence, in the near term, we expect Ekso Labs to play a lesser role than historically.

14.6 Intellectual property

Ekso Bionics has established an extensive intellectual property portfolio that includes various US patents and patent applications. The table below provides a summary of US patents by issuing status and ownership status.

License status	**Issuing status**		
	Issued patents	**Pending applications**	**Provisional applications**
Licensed to Ekso Bionics	13	2	–
Exclusively licensed to Ekso Bionics	6	–	–
Coowned with Regents of the University of California, exclusively licensed to Ekso Bionics	4	–	–
Coowned with the Regents of the University of California	2	1	–
Sole ownership by Ekso Bionics	3	30	3
Total: 64	**28**	**33**	**3**

Pending applications mean a complete application has been filed with the applicable patent authority and additional action is pending. Provisional applications mean that we have filed a short form application to establish an early filing date in anticipation of completion and submission of a complete application in the future.

Many of our applications have also been filed internationally as appropriate for their respective subject matter. As of February 8, 2017, 112 applications have issued or have been allowed as patents internationally. All told, our patent portfolio contains 214 cases that have issued or are in prosecution in 24 countries.

Ekso Bionics' patent portfolio includes method and product type claims, since the devices that we produce and the processes performed by those devices are patentable. Our patents encompass technologies relevant to our devices, including medical exoskeletons, commercial exoskeletons, actuators, and strength-enhancing exoskeletons. The earliest priority date of the portfolio reaches back to 2003, and new applications continue to be filed.

14.7 Competition

The medical technology and industrial robotics industries are characterized by intense competition and rapid technological change. We believe several other companies are developing competitive technology and devices for both the able-bodied and medical fields of use and many of these competitors have significantly more financial and other resources than we possess.

In the medical field, we face competition from companies that are focused on technology for rehabilitation of patients suffering from stroke and related neurological disabilities as well as from companies that are focused on SCI. In stroke, Cyberdyne has developed ambulatory exoskeletons with a current commercial focus in Japan and Germany, while Hocoma, AlterG, Aretech, and Reha Technology are selling treadmill-based gait therapies. In SCI, ReWalk Robotics and Parker Hannafin sell ambulatory exoskeletons. Other companies who have announced plans to commercialize robotic exoskeletons include: Bionik Laboratories, US Bionics, and ExoAtlet.

Technologies developed by competitors in the areas of stroke rehabilitation and SCI represent therapeutic interventions with utility at varying points of the continuum of care. Clinically, the Ekso is unique in its broad ability to mobilize pre- or even nonambulatory patients using a full weight bearing, over ground, task-based platform. From a practice management perspective, the Ekso is less expensive than many other systems, has a smaller footprint, the ability to move around the hospital, and uses standard power requirements that make it easy to integrate into existing infrastructure. Other over-ground exoskeletons were initially designed for an individual to achieve ambulation reliant on the device. By contrast, the Ekso's design accommodates patients with complete paraplegia and additionally includes features that are optimized to assist therapists in helping patients with some motor ability learn to walk again in a clinical setting, treating several patients and indications in a single day.

Notwithstanding the foregoing, the most pressing challenges we face are not necessarily competitive technologies, but rather achieving rapid market awareness and adoption of this emerging technology while acclimating prospects to a fundamentally new paradigm in neurorehabilitation and mobility. In addition, it may be difficult for the rehabilitation department of a hospital or clinic to secure the funds for acquisition of an Ekso device in an environment where capital expenditures of this magnitude are not commonly incurred by those rehabilitation departments.

In the able-bodied field, Lockheed Martin, Raytheon, BAE Systems, Panasonic, Honda, Daewoo, Noonee, Revision Military, and Cyberdyne—among others—are each developing some form of exoskeleton for military and/or industrial applications.

The field of robotic exoskeleton technology remains in its infancy. As this field develops, we believe we will face increased competition based on product features, price, services, clinical outcomes, and other factors. Our competitive position will depend on multiple, complex factors, including our ability to achieve market acceptance for our products, develop new products, implement production and marketing plans, secure regulatory approvals for products under development and protect our intellectual property. In some instances, competitors may also offer, or may attempt to develop, alternative therapies for disease states that may be delivered without a medical device.

14.8 Research and development

The company engages in research-and-development activities to enhance the effectiveness, ease of use, safety and reliability of our commercial exoskeletons and to expand their applications. The company's research and development expenditures were \$8.9 million, \$6.5 million, and \$3.9 million in 2016, 2015, and 2014, respectively.

14.9 Governmental regulation and product approval

14.9.1 US regulation

The US government regulates the medical device industry through various agencies, including but not limited to, the US Food and Drug Administration (FDA), which administers the Federal Food, Drug, and Cosmetic Act (FDCA). The design, testing, manufacturing, storage, labeling, distribution, advertising, and marketing of medical devices are subject to extensive regulation by federal, state, and local governmental authorities in the United States, including the FDA, and by similar agencies in other countries. Any medical device product that we develop must receive all requisite regulatory approvals or clearances, as the case may be, before it may be marketed in a particular country.

Device development, marketing clearance, and approval. The FDA classifies medical devices into one of three classes (Classes I, II, or III) based on the degree of risk the FDA determines to be associated with a device and the extent of control deemed necessary to ensure the device's safety and effectiveness. Devices requiring fewer controls because they are deemed to pose lower risk are placed in Class I or II. Class I devices are deemed to pose the least risk and are subject only to general controls applicable to all devices, such as requirements for device labeling, premarket notification, and adherence to the FDA's current good manufacturing practice requirements, as reflected in its Quality System Regulation (QSR). Class II devices are intermediate risk devices that are subject to general controls and may also be subject to special controls such as performance standards, product-specific guidance documents, special labeling requirements, patient registries, or post-market surveillance. Class III devices are those for which insufficient information exists to assure safety and effectiveness solely through general or special controls, and include life-sustaining, life-supporting, or implantable devices, and devices not "substantially equivalent" to a device that is already legally marketed. Most Class I devices, and some Class II devices are exempted by regulation from the 510(k) clearance requirement and can be marketed without prior authorization from FDA. Class I and Class II devices that have not been so exempted are eligible for marketing through the 510(k) clearance pathway. By contrast, devices placed in Class III generally require premarket approval (PMA) prior to commercial marketing.

To obtain 510(k) clearance for a medical device, an applicant must submit a premarket notification application to the FDA demonstrating that the device is "substantially equivalent" to a predicate device, which is typically a Class II device that is legally marketed in the United States. A device is substantially equivalent to a predicate device if it has the same intended use and (a) the same technological characteristics or (b) has different technological characteristics and the information submitted demonstrates that the device is as safe and effective as a legally marketed device and does not raise different questions of safety or effectiveness. A showing of substantial equivalence sometimes, but not always, requires clinical data. Generally, the 510(k) clearance process can exceed 90 days and may extend to a year or more. After a device has received, 510(k) clearance for a specific intended use, any modification that could significantly affect its safety or effectiveness, such as a significant change in the design, materials, method of manufacture or intended use, will require a new 510(k) clearance or [if the device as modified is not substantially equivalent (NSE) to a legally marketed predicate device] PMA approval. While the determination as to whether new authorization is needed is initially left to the manufacturer, the FDA may review this determination and evaluate the regulatory status of the modified product at any time and may request the manufacturer to cease marketing and recall the modified device until 510(k) clearance or PMA approval is obtained. The manufacturer may also be subject to significant regulatory fines or penalties.

The second, more comprehensive, approval process applies to a new device that is NSE to a predicate device or that is to be used in supporting or sustaining life or preventing impairment. These devices are normally Class III devices.

For example, most high-risk implantable devices are subject to the PMA approval process. Two steps of FDA approval are generally required before a company can market a product in the United States that is subject to approval, as opposed to clearance, as a Class III device. First, a company must comply with investigational device exemption (IDE) regulations in connection with any human clinical investigation of the device. These regulations permit a company to undertake a clinical study of a "nonsignificant risk" device without formal FDA approval. Prior express FDA approval is required if the device is a significant risk device. Second, the FDA must review Ekso Bionics' PMA application, which contains, among other things, clinical information acquired under the IDE. Additionally, devices subject to PMA approval may be subject to a panel review to obtain marketing approval and are required to pass a factory inspection in accordance with the current "good manufacturing practices" standards in order to obtain approval. The FDA will approve the PMA application if it finds there is reasonable assurance that the device is safe and effective for its intended use. The PMA process takes substantially longer than the 510(k) process, approximately one to 2 years or more.

In some instances, the FDA may find that a device is new and NSE to a predicate device but is also not a high risk device as is generally the case with Class III PMA devices. In these instances, the FDA may allow a device to be reclassified from Class III to Class I or II. The de novo reclassification option is an alternate pathway to classify novel devices of low to moderate risk that had automatically been placed in Class III after receiving a "NSE" determination in response to a 510(k) notification. The FDCA has also been amended to allow a sponsor to submit a de novo reclassification request to the FDA for novel low to moderate risk devices without first being required to submit a 510(k) application. These types of applications are referred to as "Evaluation of Automatic Class III Designation" or "de novo." In instances where a device is deemed NSE to a Class II predicate device, the candidate device may be filed as a de novo application which may lead to delays in regulatory decisions by the FDA. FDA review of a de novo application may lead the FDA to identify the device as either a Class I or II device and subject to or exempt from 510(k) premarket notification.

Clinical trials are generally required to support a PMA or de novo reclassification application and are sometimes required for 510(k) clearance. Clinical trials generally require an IDE application, or IDE, approved in advance by the FDA for a specified number of patients and study sites, unless the product is deemed a nonsignificant risk device eligible for more abbreviated IDE requirements. Clinical trials are subject to extensive monitoring, recordkeeping and reporting requirements. Clinical trials must be conducted under the oversight of an institutional review board (IRB), for the relevant clinical trial sites and must comply with FDA regulations, including but not limited to those relating to good clinical practices. To conduct a clinical trial, we also are required to obtain the patients' informed consent in form and substance that complies with both FDA requirements and state and federal privacy and human subject protection regulations. We, the FDA or the IRB, could suspend a clinical trial at any time for various reasons, including a belief that the risks to study subjects outweigh the anticipated benefits. Even if a trial is completed, the results of clinical testing may not adequately demonstrate the safety and efficacy of the device or may otherwise not be sufficient to obtain FDA approval to market the product in the United States. Similarly, in Europe the clinical study must be approved by a local ethics committee and in some cases, including studies with high-risk devices, by the ministry of health in the applicable country. To date, the Ekso device has been the subject of several clinical studies, some sponsored by Ekso Bionics, as well as non-Ekso-sponsored independent studies conducted by rehabilitation institutions. In addition, we are currently conducting several studies to investigate additional indications for use for the Ekso device, as well as to evaluate clinical and nonclinical outcomes of using the Ekso device.

While we believe that Ekso Bionics' robotic exoskeleton has been appropriately marketed as a Class I 510(k) exempt Powered Exercise Equipment device since it was first sold in the United States in February 2012, on June 26, 2014, the FDA announced the creation of a new product classification for Powered Exoskeleton devices. On October 21, 2014, the FDA published the summary for the new Powered Exoskeleton classification and designated it a Class II medical device, which requires the clearance of a 510(k) notice.

On October 21, 2014, concurrent with the FDA's publication of the reclassification of Powered Exoskeleton devices, the FDA issued the company an "Untitled Letter" which informed the company in writing of the agency's belief that this new product classification applied to our Ekso GT device. On December 24, 2014, the company filed a 510(k) notice for the Ekso robotic exoskeleton, which was accepted by the FDA for substantive review on July 29, 2015.

On April 4, 2016, we received clearance from the FDA to market our Ekso GT robotic exoskeleton for use in the treatment of individuals with hemiplegia due to stroke, individuals with spinal cord injuries at levels T4 to L5, and individuals with spinal cord injuries at levels of T3 to C7 (ASIA D), in accordance with the device's labeling. On July 19, 2016, we received clearance from the FDA to expand/clarify the indications and labeling to expressly include individuals with hemiplegia due to stroke who have upper extremity function of at least 4/5 in only one arm. Our prior cleared indications for use statement required that individuals with hemiplegia due to stroke have upper extremity function of at least 4/5 in both arms.

Pervasive and continuing regulation. After a device is placed on the market, numerous regulatory requirements apply. These include:

- Product listing and establishment registration, which helps facilitate FDA inspections and other regulatory action;
- Quality System Regulation, or QSR, which requires manufacturers, including third-party manufacturers, to follow stringent design, testing, control, documentation, and other quality assurance procedures during all aspects of the manufacturing process;
- Labeling regulations and FDA prohibitions against the promotion of products for un-cleared, unapproved or off-label use or indication;

- 510(k) clearance of product modifications that could significantly affect safety or efficacy or that would constitute a major change in intended use of one of our cleared devices;
- Medical device reporting regulations, which require that manufacturers comply with FDA requirements to report if their device may have caused or contributed to a death or serious injury, or has malfunctioned in a way that would likely cause or contribute to a death or serious injury if the malfunction of the device or a similar device were to recur;
- Postapproval restrictions or conditions, including postapproval study commitments;
- Postmarket surveillance regulations, which apply when necessary to protect the public health or to provide additional safety and effectiveness data for the device;
- The FDA's recall authority, whereby it can ask, or under certain conditions order, device manufacturers to recall from the market a product that is in violation of governing laws and regulations;
- Regulations pertaining to voluntary recalls; and
- Notices provision regarding corrections or removals.

Advertising and promotion of medical devices, in addition to being regulated by the FDA, are also regulated by the Federal Trade Commission and by state regulatory and enforcement authorities. Recently, promotional activities for FDA-regulated products of other companies have been the subject of enforcement action brought under healthcare reimbursement laws and consumer protection statutes. In addition, under the federal Lanham Act and similar state laws, competitors and others can initiate litigation relating to advertising claims. If the FDA determines that our promotional materials or training constitutes promotion of an uncleared or unapproved use, it could request that we modify our training or promotional materials or subject us to regulatory or enforcement actions. It is also possible that other federal, state or foreign enforcement authorities might take action if they consider our promotional or training materials to constitute promotion of an unapproved use, which could result in significant fines or penalties under other statutory authorities, such as laws prohibiting false claims for reimbursement. In that event, our reputation could be damaged and adoption of the products would be impaired.

The FDA has broad postmarket and regulatory enforcement powers. We are subject to unannounced inspections by the FDA to determine our compliance with the QSR and other regulations.

From September 2, 2015 to September 11, 2015, the Division of Bioresearch Monitoring of the FDA's Office of Compliance conducted an inspection of the company's facility in Richmond, California. At the conclusion of the inspection, the FDA issued a Form FDA 483 with observations pertaining to informed consent requirements, reporting of events to FDA, and records maintenance. These observations are inspectional and do not represent a final FDA determination of noncompliance. On October 2, 2015, we responded to the FDA describing the corrective and preventive actions that we have implemented and continue to implement to address the FDA's observations. Due to the nature of the findings, we do not expect that the Form FDA 483 will result in a warning letter or other action that could interfere with our operations. On March 30, 2016, the FDA accepted our corrective actions for the Form 483 observations that were generated during the FDA inspection.

Since July 1, 2015, we have been informed of seven events with respect to our Ekso GT devices that are reportable pursuant to the FDA's medical device reporting, or MDR, regulations. There were no reported patient injuries related to any of these events, and in each case we have filed the required adverse event reports with the FDA. We have analyzed the root causes of these issues and have improved the design and strengthened our manufacturing processes as a result. In addition, we have proactively adjusted the device maintenance schedules based on field usage to address these issues.

Failure to comply with applicable regulatory requirements can result in enforcement action by the FDA or other regulatory authorities, which may result in sanctions including, but not limited to

- untitled letters, warning letters, fines, injunctions, consent decrees, and civil penalties;
- unanticipated expenditures to address or defend such actions
- customer notifications for repair, replacement, refunds;
- recall, detention, or seizure of our products;
- operating restrictions or partial suspension or total shutdown of production;
- refusing or delaying our requests for 510(k) clearance or PMA of new products or modified products;
- operating restrictions;
- withdrawing 510(k) clearances that have already been granted;
- refusal to grant export approval for our products; or
- criminal prosecution.

14.9.2 Foreign regulation

In addition to regulations in the United States, we will be subject to a variety of foreign regulations governing clinical trials and commercial sales and distribution of our products in foreign countries. Whether or not Ekso Bionics obtains FDA approval for a product, we must obtain approval of a product by the comparable regulatory authorities of foreign countries before Ekso Bionics can commence clinical trials or marketing of the product in those countries. The approval process varies from country to country, and the time may be longer or shorter than that required for FDA approval. The requirements governing the conduct of clinical trials, product licensing, pricing, and reimbursement vary greatly from country to country.

The policies of the FDA and foreign regulatory authorities may change and additional government regulations may be enacted which could prevent or delay regulatory approval of our products and could also increase the cost of regulatory compliance. We cannot predict the likelihood, nature or extent of adverse governmental regulation that might arise from future legislative or administrative action, either in the United States or abroad.

14.10 Corporate information

We were incorporated as PN Med Group Inc. in Nevada on January 30, 2012. Prior to the Merger and Split-Off (each as defined below), our business was to distribute medical supplies and equipment in Chile.

On January 15, 2014, our wholly owned subsidiary, Ekso Acquisition Corp., a corporation formed in the State of Delaware on January 3, 2014 merged with and into Ekso Bionics, Inc., a corporation incorporated in the State of Delaware on January 19, 2005. Ekso Bionics was the surviving corporation in the Merger and became our wholly owned subsidiary. All of the outstanding Ekso Bionics stock was converted into shares of our common stock.

In connection with the Merger and pursuant to a split-off agreement and general release, we transferred our pre-Merger assets and liabilities to our pre-Merger majority stockholders, in exchange for the surrender by them and cancellation of 2,497,586 shares of our common stock (the "Split-Off"), after adjusting to effect to the 1-for-7 reverse stock split, discussed in Note 13 in the notes to our consolidated financial statements under the caption Capitalization and Equity Structure—Reverse Stock Split.

As a result of the Merger and Split-Off, we discontinued our pre-Merger business and acquired the business of Ekso Bionics, and have continued the existing business operations of Ekso Bionics as a publicly traded company under the name Ekso Bionics Holdings, Inc.

Our principal executive office is located at 1414 Harbour Way South, Suite 1201, Richmond, CA, USA, 94804 and our telephone number is +1 510 984 1761.

Index

www.ingramcontent.com/pod-product-compliance
Lightning Source LLC
LaVergne TN
LVHW051046210926
848486LV00001B/39